W0082525

annual reports in organic synthesis – 2002

ANNUAL REPORTS IN ORGANIC SYNTHESIS

ANNUAL REPORTS IN ORGANIC SYNTHESIS-1970
John McMurry and R. Bryan Miller, Eds.
ANNUAL REPORTS IN ORGANIC SYNTHESIS-1972
John McMurry and R. Bryan Miller, Eds.
ANNUAL REPORTS IN ORGANIC SYNTHESIS-1973
R. Bryan Miller and Louis S. Hegedus, Eds.
John McMurry, Series Editor
ANNUAL REPORTS IN ORGANIC SYNTHESIS-1974
Louis S. Hegedus and Stephen R. Wilson, Eds.
R. Bryan Miller, Series Editor
ANNUAL REPORTS IN ORGANIC SYNTHESIS-1975
R. Bryan Miller and L.G. Wade, Jr., Eds.
ANNUAL REPORTS IN ORGANIC SYNTHESIS-1976
R. Bryan Miller and L.G. Wade, Jr., Eds.
ANNUAL REPORTS IN ORGANIC SYNTHESIS-1978
L.G. Wade, Jr., and Martin J. O'Donnell, Eds.
ANNUAL REPORTS IN ORGANIC SYNTHESIS-1980
L.G. Wade, Jr., and Martin J. O'Donnell, Eds.
ANNUAL REPORTS IN ORGANIC SYNTHESIS-1981
L.G. Wade, Jr., and Martin J. O'Donnell, Eds
ANNUAL REPORTS IN ORGANIC SYNTHESIS-1982
L.G. Wade, Jr., and Martin J. O'Donnell, Eds.
ANNUAL REPORTS IN ORGANIC SYNTHESIS-1983
Martin J. O'Donnell and Louis Weiss, Eds.
ANNUAL REPORTS IN ORGANIC SYNTHESIS-1984
Martin J. O'Donnell and Louis Weiss, Eds.
ANNUAL REPORTS IN ORGANIC SYNTHESIS-1985
Martin J. O'Donnell and Eric F.V. Scriven, Eds.
ANNUAL REPORTS IN ORGANIC SYNTHESIS-1986
Eric F.V. Scriven and Kenneth Turnbull, Eds.
ANNUAL REPORTS IN ORGANIC SYNTHESIS-1987
Eric F.V. Scriven and Kenneth Turnbull, Eds.
ANNUAL REPORTS IN ORGANIC SYNTHESIS-1989
Kenneth Turnbull and Daniel M. Ketcha, Eds.
ANNUAL REPORTS IN ORGANIC SYNTHESIS-1990
Kenneth Turnbull, Philip M. Weintraub, Daniel M. Ketcha,
and James Keay, Eds.
ANNUAL REPORTS IN ORGANIC SYNTHESIS-1991
Philip M. Weintraub and Kenneth Turnbull, Eds.
ANNUAL REPORTS IN ORGANIC SYNTHESIS-1992
Philip M. Weintraub, Kenneth Turnbull,
Daniel M. Ketcha, and Raymond S. Gross, Eds.
ANNUAL REPORTS IN ORGANIC SYNTHESIS-1993
Philip M. Weintraub, Kenneth Turnbull,
Daniel M. Ketcha, Raymond S. Gross,and Tony Yantao Zhang, Eds.
ANNUAL REPORTS IN ORGANIC SYNTHESIS-1994
Philip M. Weintraub, Kenneth Turnbull,
Daniel M. Ketcha, Raymond S. Gross,and Tony Yantao Zhang, Eds.
ANNUAL REPORTS IN ORGANIC SYNTHESIS-1995
Philip M. Weintraub, Kenneth Turnbull,
Daniel M. Ketcha, Raymond S. Gross,and Tony Yantao Zhang, Eds.
ANNUAL REPORTS IN ORGANIC SYNTHESIS-1996
Philip M. Weintraub, Kenneth Turnbull,
Daniel M. Ketcha, Raymond S. Gross,and Gary W. Morrow, Eds.
ANNUAL REPORTS IN ORGANIC SYNTHESIS-1997
Philip M. Weintraub, Kenneth Turnbull,
Daniel M. Ketcha, Raymond S .Gross,and Gary W. Morrow, Eds.
ANNUAL REPORTS IN ORGANIC SYNTHESIS-1998
Philip M. Weintraub, Kenneth Turnbull,
Daniel M. Ketcha, Raymond S. Gross,and Gary W. Morrow, Eds
ANNUAL REPORTS IN ORGANIC SYNTHESIS-1999
Philip M. Weintraub, Kenneth Turnbull, Daniel M. Ketcha,
Gary W. Morrow, and Allan Pinhas, Eds.
ANNUAL REPORTS IN ORGANIC SYNTHESIS-2000
Philip M. Weintraub, Kenneth Turnbull, Daniel M. Ketcha,
Mark McMills, and Eric Fossum, Eds.
ANNUAL REPORTS IN ORGANIC SYNTHESIS-2001
Philip M. Weintraub, Kenneth Turnbull,
Daniel M. Ketcha, and Jeffrey Sabol, Eds.

annual reports in organic synthesis – 2002

by

Philip M. Weintraub
Aventis Pharmaceuticals
Bridgewater, New Jersey

Kenneth Turnbull
Wright State University
Dayton, Ohio

Jeffrey Sabol
Aventis Pharmaceuticals
Bridgewater, New Jersey

Peter Norris
Youngstown State University
Youngstown, Ohio

ACADEMIC PRESS
An imprint of Elsevier Science

Amsterdam • Boston • London • New York • Oxford • Paris
San Diego • San Francisco • Singapore • Sydney • Tokyo

This book is printed on acid-free paper.

© 2002 Elsevier Science (USA)

All rights reserved.
No part of this publication may be reproduced or transmitted in any form or by any means, electronic or mechanical, including photocopy, recording, or any information storage and retrieval system, without permission in writing from the Publisher.

The appearance of the code at the bottom of the first page of a chapter in this book indicates the Publisher's consent that copies of the chapter may be made for personal or internal use of specific clients. This consent is given on the condition, however, that the copier pay the stated per copy fee through the Copyright Clearance Center, Inc. (222 Rosewood Drive, Danvers, Massachusetts 01923), for copying beyond that permitted by Sections 107 or 108 of the U.S. Copyright Law. This consent does not extend to other kinds of copying, such as copying for general distribution, for advertising or promotional purposes, for creating new collective works, or for resale. Copy fees for pre-2002 chapters are as shown on the title pages. If no fee code appears on the title page, the copy fee is the same as for current chapters. 0065-7743/02 $35.00

Explicit permission from Academic Press is not required to reproduce a maximum of two figures or tables from an Academic Press chapter in another scientific or research publication provided that the material has not been credited to another source and that full credit to the Academic Press chapter is given.

Academic Press
An Elsevier Science Imprint
525 B Street, Suite 1900, San Diego, California 92101-4495, USA
http://www.academicpress.com

International Standard Book Number: 0-12-040832-5
ISSN: 0066-409X

Add Transferred to Digital Printing 2006

02 03 04 05 06 MP 9 8 7 6 5 4 3 2 1

Contents

PREFACE

One of the more difficult problems facing chemists today is that of "keeping up with the literature." For several reasons, the problem is particularly severe for the synthetic organic chemist. Bits of information of potential use are scattered throughout common chemistry journals and can be found in any paper, not just those dealing strictly with synthesis. Thus, synthetic chemists must read a large number of journals and must organize and index what they read to make the information available for future reference. Fewer chemists are doing this as electronic searching becomes more prevalent. The ability to leaf through a concise compendium of the recent literature still remains important. Thus, the reader of such a collection can quickly see new developments in several areas of synthesis and, additionally, an abstract may catch his/her eye and catalyze a new research idea.

The problem, however, is shared to some extent by all. Most organic chemists are at some time faced with the problem of synthesizing a desired material, and for many the problems are formidable. Nonspecialists faced with the synthetic problem are not likely to have kept pace with the developments in synthetic chemistry that may well solve their problems, and they will not have the necessary information in their files, despite the capabilities of on-line searching

Thus, we feel an organized annual review of synthetically useful information should prove beneficial to nearly all organic chemists, both specialists and nonspecialists in synthesis. It should help relieve some of the information storage burden of the specialist and should enable the non specialist who is seeking help with a specific problem to rapidly become aware of recent synthetic advances. Ideally also, it should appear as promptly as possible after the close of the abstracting period. As in the past years, we have placed particular emphasis on keeping the abstracts as concise as possible, while indicating the generality of the reactions involved. We have tried to combine similar publications into inclusive abstracts. This practice has allowed us to include a larger number of references without a substantial increase in the book's length. It should be noted that where multiple references are included in the abstract, the first mentioned refers to the equation shown. The remaining references are related but not identical. To further aid the readers, we have separated related but less similar references from that represented by the graphic by the phrase "see also:". We have allowed for two such separations per graphic. In a number of cases we have attempted to further elucidate the contents of these multiple references by including a statement below the graphic. If this statement is enclosed in square brackets [e.g. II.A.1-1 and II.A.1-6] then it pertains to data from the references following the lead reference. If round brackets are employed (e.g. I.A.1-1 and I.G.3-6) then further information about the lead reference is being provided.

The year has been omitted from each reference as presumably all are from 2001. Any references from 2000 (journals received after our February 1 cutoff date) are noted appropriately. In an effort to be more space efficient, we have adopted letter abbreviations for the journal references from Katritzky's Handbook of Heterocyclic Chemistry. See the List of Abstracted Journals for definitions of these letter abbreviations; they are alphabetized by the abbreviations rather than the journal name. In producing *Annual Reports in Organic Chemistry–2002* we have abstracted 44 primary chemistry journals, selecting useful synthetic advances. Only the common journals received by our libraries have been abstracted.

We have also exercised selectivity in choosing which papers to abstract. Our general guidelines have been to include reactions and methods that are new, synthetically useful, or reasonably general. We have tried to present the information in an organized manner, emphasizing rapid visual retrieval. The purpose of this emphasis is to aid the reader in scanning the book. The mind is capable of absorbing a whole picture in an instant, but is considerably slowed by having to read sentences. If the pictures presented catch the reader's interest, he or she should then seek details from the original paper.

The author index is based on the name of the senior author(s) or sometimes the first author. No subject index is included because we feel the Table of Contents serves that function. Chapters I–III are organized by reaction type and, hopefully, the organization is self-explanatory; thus, there should be no difficulty in locating a new method of oxidation or a new cyclopropanation procedure. Chapter IV deals with methods of synthesizing heterocyclic systems. Where fused ring systems bearing multiple heterocyclic rings are shown, we have chosen to categorize the heterocyclic system by the ring formed in the reaction. Chapter V covers the use of protecting groups. Chapter VI deals with those synthetically useful transformations that do not fit easily into the first three chapters. In Chapter VII, the reviews have been divided into sections to help the reader to quickly find a review on a specific topic. Other area specific reviews may be found at the end of the section. Heterocyclic reviews may be found at the end of Chapter IV; Fullerene chemistry reviews are in VII.A.5; Dendrimers, etc reviews are in VII.C.6; and Combinatorial chemistry reviews are in VII.F.7. Chapter VIII, Selected Topical Areas, consists of topical areas we felt were "hot" topics, and collected titles of papers in these areas. While not an all inclusive listing, we hope it will prove useful. We organized the section on Total Syntheses of Selected Natural Products (VIII.D) so the synthesized target names (with associated author and reference) have been sorted and grouped alphabetically to help the reader locate a particular compound as quickly as possible. To keep the Annual to a reasonable size, only one author is listed in this section. We used our editorial prerogative to select from the many applicable syntheses.

Because of the increasing number of papers being published and the constraints placed on the number of pages we want *AROS-2002* to be, we made further changes to the formating allowing more abstracts to be presented. We tightened the spaces between abstracts even further, and no longer use "and" when two or more authors are cited. We divided the section on Combinatorial Chemistry (VIII.F) into seven sections: 1. Supports, Linkers & Protecting Groups; 2. Supported Reagents, Catalysts Ligands & Scavengers; 3. Solid-Phase Heterocyclic Synthesis; 4. Solid -Supported Organic Reactions; 5. Targeted Libraries Synthesis; 6. Novel Techniques in Combinatorial Chemistry; and 7. Reviews.

Any undertaking of this type involves a series of compromises. We have chosen to emphasize reasonable cost and rapid visual retrieval of information at the admitted expense of detail and beauty.

Comments (negative or preferably positive) or suggestions from the reader will be well received by the senior editor.

Senior and Contributing Editor
Philip M. Weintraub

Contributing Editors
Kenneth Turnbull
Jeffrey S. Sabol
Peter Norris

LIST OF JOURNALS ABSTRACTED

AA	Aldrichimica Acta
ACR	Accounts of Chemical Research
ACS	Acta Chemica Scandinavia
AG(E)	Angewandte Chemie International Edition in English
AJC	Australian Journal of Chemistry
BCJ	Bulletin of the Chemical Society of Japan
BMCL	Bioorganic and Medicinal Chemistry Letters
CC	Journal of the Chemical Society Chemical Communications
CCC	Collection of Czechoslovakian Chemical Communications
CEJ	European Journal of Chemistry
CI(L)	Chemistry and Industry (London)
CJC	Canadian Journal of Chemistry
CL	Chemistry Letters
CPB	Chemical and Pharmaceutical Bulletin
CRV	Chemical Reviews
CSR	Chemical Society Reviews
EJOC	European Journal of Medicinal Chemistry
EJOC	European Journal of Organic Chemistry
H	Heterocycles
HCA	Helvetica Chimica Acta
JACS	Journal of the American Chemical Society
JCR(S)	Journal of Chemical Research (S)
JCS(P1)	Journal of the Chemical Society (Perkin I)
JCS(P2)	Journal of the Chemical Society (Perkin II)
JHC	Journal of Heterocyclic Chemistry
JMC	Journal of Medicinal Chemistry
JOC	Journal of Organic Chemistry
JOM	Journal of Organometallic Chemistry
M	Monatschefte fur Chemie
OL	Organic Letters
OM	Organometallics
OPP	Organic Preparations and Procedures International
OS	Organic Synthesis
PAC	Pure and Applied Chemistry
RCR	Russian Chemical Reviews
S	Synthesis

GLOSSARY OF ABBREVIATIONS

4-PP   pyrrolidinopyridine
9-BBN   9-borabicyclo[3.3.1]-
         nonane
18-Cr-6 = 8-C-6 18-crown-6
AA   amino acid
Ac   acetyl
acac   acetonylacetone
ad   adamantanyl
ADC   9,10-anthracene
         dicarbonitrile
ADDP   1,1'-(azadicarbonyl)-
         dipiperidine
AIBN   azobisisobutyronitrile
ALB   aluminum lithium
         bis(binaphthoxide)
All   allyl
Alloc = ALOC allyloxycarbonyl
An   *p*-anisyl
aq   aqueous
Ar   aryl
ATD   aluminum tris(2,6-di-*tert*-
         butyl-4-methylphenoxide)
ATPH   aluminum tris(2,6-
         diphenylphenoxide)
AZMB   2-(azidomethyl)benzoyl
BARF   tetrakis[3,5-bis(trifluoro-
         methyl)phenyl]borate
BCN   N-benzyloxycarbonyl-oxy-
         5-norbornene-2,3-
         dicarboximide
BDPP   (2*R*, 4*R*) or (2*S*, 4*S*) 2,4-
         bis(diphenylphosphino)-
         pentane
BER   borohydride exchange
         resin
BICP   (R,R) 2(R)-2'(R)-bis-
         (diphenylphosphino)-
         1(R),1'(R)-dicyclopentane
BIH   bis(*sym*-collidine)iodine(I)
         hexafluorophosphate
BINAL-H LiAlH4/ethanol/1,1'-
         bis-2-naphthol complex

BINAP = DINAP 2,2'-bis-
         (diphenylphosphino)-1,1'-
         binaphthyl
BINAS   6-fold sulfonated 2,2'-
         bis(diphenylphosphino-
         methyl)-1,1'-binaphthyl
Bip   biphenyl-4-sulphonyl
BLA   Bronsted acid assisted
         chiral Lewis Acid
bmin   1-butlyl-3-methyl-
         imidazolium cation
Bn   benzyl
BNAH   1-benzyl-1,4-dihydro-
         nicotinamide
BOB   4-benzyloxybutryl
Boc   *t*-butyloxycarbonyl
BOM   benzyloxymethyl
BPO   benzoyl peroxide
bpy   bipyridyl
BQ   benzoquinone
BSA   bovine serum albumin
BSA   *N,O*-bis-silylacetamide =
         O,N-bis-trimethylsilyl
         acetamide
Bt   1- or 2-benzotriazolyl
Bth   benzothiazole
BTI   PhI(O₂CCF₃)₂
BTI   [bis(trifluoroacetoxy)
         iodo]benzene
BTEAC   benzyl triethyl-
         ammonium chloride
BTFP   2-bromotrifluoroisoprene
BTMA   benzyltrimethyl
         ammonium
BTS   bis(trimethylsilyl)sulfate
BTSA   bis-trimethylsilyl chromate
BTSP   bis(trimethylsilyl) peroxide
Bu   butyl
Bus   *tert*-butylsulfonyl
Bz   benzoyl
CAN   ceric ammonium nitrate
cat.   catalyst

xv

Cbz   benzyloxycarbonyl
CCE   constant current
       electrolysis
CHD   cyclohexadiene
CHP   cumene hydroperoxide
Chx2BI dicyclohexyl iodoborane
CMHP   cumene hydroperoxide
CMMP   cyanomethylenetrimethyl
       phosphorane
cod   1,5-cyclooctadiene
cot   cyclooctatriene
Cp   cyclopentadienyl
CPTS   collidinium-p-toluene-
       sulfonate
Cr-PILC chromium-pillared clay
       catalyst
CRA   complex reducing agent
CSA   camphor sulfonic acid
CTAB   cetyl trimethyl-ammonium
       bromide
CTACl cetyl trimethylammonium
       chloride
CTMS = TMCS chlorotrimethyl-
       silyl
Cy   cyclohexyl
Δ   heat
D   day
DABCO 1,4-diazabicyclo[2.2.2]-
       octane
DAMFA (diethylaminoethylene)
       hexafluoroacetylacetone
DAST   diethylaminosulfur-
       trifluoride
DATMP diethylaluminum
       2,2,6,6-tetramethyl-
       piperidide
dba   dibenzylidene acetone
DBAD = DTBAD di-tert-
       butylazodi-carboxylate
DBH   di-tert-butyl hyponitrite
DBI   dibromoisocyanuric acid
DBn   p-dodecyloxybenzyl
dbpp   1,3-bis(dibenzo-
       phospholyl)propane

DBS   dibenzosuberyl
DBSA   dodecylbenzylsulfonic acid
DBU   1,5-diazabicyclo[5.4.0]-
       undec-5-ene
DCA   9,10-dicyanoanthracene
DCB   dichlorobenzene
DCC   dicyclohexylcarbodiimide
DCE   1,2-dichloroethane
DCN   1,4-dicyanonaphthalene
Dcpm   dicyclopropylmethyl
DDQ   2,3-dichloro-5,6-dicyano-
       benzoquinone
de = d.e. diastereomeric excess
DEAD   diethyl azodicarboxylate
DEPC   diethyl cyano-
       phosphoridate
DET   diethyl tartrate
DHAP   dihydroxyacetone
       phosphate
DHP   dihydropyran
DHQD   dihydroquinidine
DIAD   diisopropylazodi-
       carboxylate
DIB   (diacetoxyiodo)benzene
DIBAH = DIBAL diisobutyl-
       aluminum hydride
DIM   2-(hydroxyethyl)-1,3-
       dithiane
DIOP   2,3-O-isopropylidene-2,3-
       dihydroxy-1,4-bis-
       (diphenylphosphino)-
       butane
dippp 1,3-bis(diisopropyl-
       phosphino)propane
DIPT   diisopropyl tartrate
DLP   dilauroyl peroxide
DMA   N,N-dimethylacetamide
DMAD   dimethyl acetylene
       dicarboxylate
DMAP   4-(N,N-dimethyl)-
       aminopyridine
DMB   m-dimethoxybenzene
Dmb   2,4-dimethoxybenzyl
DMB   2,3-dimethylbuta-1,3-diene

DMD   dimethyl dioxirane
DME   dimethoxyethane
DMF   dimethylformamide
DMI   1,3-dimethylimidazolidin-
        2-one
DMM   dimethoxymethane
DMN   1,5-dimethoxynaphthalene
DMP   2,6-dimethylphenol
DMPS   dimethylphenylsilyl
DMPU N,N'-dimethylpropylene-
        urea
DMSO   dimethylsulfoxide
DMT   4,4'-dimethoxytrityl
DMTMM   4-(4,6-dimethoxy[1,3,5]
        triazin-2-yl)-4-methyl-
        morpholinium chloride
DMTr   dimethyltrityl
dpba   2-(diphenylphosphino)-
        benzoic acid
DPC   diphenylphosphoro
        chloridate
DPDC   diisopropyl peroxydi-
        carbonate
DPDM   diphenyl diazomethane
DPEDA   1,2-diphenylethane-1,2-
        diamine
DPPA   diphenylphosphorazidate
dppb   bis(1,4-diphenyl-
        phosphino)butane
dppe = DPPE bis(diphenyl-
        phosphino)ethane
dppf   dichloro[1,1'-bis-
        (diphenylphosphino-
        ferrocene)]
dppp   1,3-(diphenylphosphino)-
        propane
DPS t-butyldiphenylsilyl
DPTC   O,O'-di(2-pyridyl)
        thiocarbonate
dr   diastereomeric ratio
ds   diastereoselectivity
DTBAD = DBAD di-$tert$-
        butylazodi-carboxylate
DTBB   4,4'-di-$tert$-butylbiphenyl

DTBP   2,6-di-t-butylpyidine
d^{t}BPF 1,1'-bis(di-$tert$-
        butylphosphino)ferrocene
DTE   dithioerythritol
DTPM   [1,3-dimethyl-2,4,6-(1H,
        3H,5H)-trioxopyrimidin-5-
        ylidine]methyl
E   general electrophile
EDAC   ethyldimethylamino-
        propylcarbodiimide
EDC   1-ethyl-3-(3-dimethyl-
        aminopropyl)carbodiimide
EDCP   ethylene dicarboxylic
        diphosphonic acid
EDTA   ethylenediamine
        tetraacetic acid
ee = e.e. enantiomeric excess
en   ethylene diamine
EPHP   N-ethylpiperidine
        hydrophosphorous acid
Et   ethyl
EWG   electron withdrawing
        group
F_C ferrocenyl
FDP   fructose-1,6-diphosphate
Fe(dbm)$_3$ tris(dibenzoylmethido)
        iron(III)
FePHEN tris(1,10-phenan-
        throline)iron(III)hexa-
        fluorophosphate
fl   flavin
flosyl = Fs fluorosulfonate
Fmoc   9-fluorenylmethoxy-
        carbonyl
fod   6,6,7,7,8,8,8-heptafluoro-
        2,2-dimethyl-3,5-octane-
        dione
FSM   mesoporous silica
Fs = flosyl fluorosulfonate
FTT   1-fluoro-2,4,6-trimethyl-
        pyridinium triflate
FVP   flash vapor pyrolysis
GaSO   chiral Ga heterometallic
        complex

GEBC gel entrapped base catalyst
Gr   graphite
GSNO nitrosoglutathione
h   hours
Hap   hydroxyapatite
HDNIB [hydroxy(2,4-dinitro-
        benzenesulfonyloxy)iodo]b
        enzene
hfacac hexafluoroacetylacetone
HFIP   hexafluoro-2-propanol
HGK   4-hydroxy-2-ketoglutarate
HLE   horse liver esterase
Hmb   2-hydroxy-4-methoxy-
        benzyl
HMDS   1,1,1,3,3,3-
        hexamethyldisilazane
HMPA = HMPT hexamethyl-
        phosphoramide
HMS   hexagonal mesoporous
        silica
HMTAB hexamethylenetetramine
        bromide
HNIB   [hydroxy(p-nitrobenzene-
        sulfonyloxy)iodo]benzene
hv   irradiation with light
HTIB   [hydroxy(p-tolyl-
        sulfonyloxy)iodo]benzene
IBDA   iodobenzene diacetate
IBX   o-iodoxybenzoic acid
IDCP   iodonium dicollidine
        perchlorate
INOC   Intramolecular Nitrile
        Oxide Cycloaddition
Ipc2   diisopropylcamphyl
IPr   1,3-bis(2,6-diisopropyl-
        phenyl)imidazol-2-ylidene
K-10   a Montmorillonite clay
KMBA   potassium N-
        methylbutyramide
KSF   a Montmorillonite clay
LA   Lewis acid
L-selectride" lithium tri-
        ^{s}butylborohydride
L.R.   Lawesson's reagent

LAH   lithium aluminum hydride
LDA   lithium diisopropylamide
LDBB   lithium 4,4'-^{t}butylbi-
        phenylide
LDE   lithium diethylamide
LDPE   lithium perchlorate-diethyl
        ether
LICKOR BuLi/tBuOK
liq.   liquid
LR   Lawesson's reagent
LTMP   lithium 2,2,6,6-tetra-
        methylpiperidide
MABR   methylaluminum bis(4-
        bromo-2,6-di-^{t}butyl-
        phenoxide)
MAD   methylaluminum bis-(2,6-
        di-^{t}butyl-4-methyl-
        phenoxide)
MAPh   methylaluminumbis(2,6-
        diphenoxide)
MBDA   magnesium bis-
        (diisopropylamide)
MBT   2-mercaptobenzothiazole
MCPBA   m-chloroperbenzoic acid
Me   methyl
Mek   methyl ethyl ketone
MEM   β-methoxyethoxymethyl
MEPY   methyl 2-pyrrolidone-
        5(S)-carboxylate
Mes = mesityl 2,4,6-trimethyl-
        phenyl
MMPP   magnesium mono-
        peroxyphthalate
Mmt   4-methoxytrityl
MOM   methoxymethyl
MPD   1-methylpyrrolidone
MPM   methoxy(phenylthio)-
        methyl
Mpm = PMB p-methoxybenzyl
MPPC   N-methyl piperidine
        chlorochromate
MS   molecular sieves
Ms   methanesulfonyl
MSA   methanesulfonic acid

MSH   o-mesitylenesulfonyl
      hydroxylamine
Mspoc 2-methylsulfonyl-3-
      phenyl-1-prop-2-enyloxy-
      carbonyl
MTO   methyltrioxorhenium
      (MeReO$_3$)
MTPA methoxy-α-trifluoro-
      methylphenylacetyl
MV^{2+} methyl viologen
MVK   methyl vinyl ketone
mw = μW microwave
NaBMGS sodium butylmono-
      glycosulfate
NAP   naphthylmethyl
Naph = Np naphthyl
NBS   N-bromosuccinimide
NCS   N-chlorosuccinimide
Nf   nonaflate
N$_f$ nonafluorobutylsulfonyl
NFOBS N-fluoro-O-benzenedi-
      sulfonimide
NHPI N-hydroxyphthalimide
NIS   N-iodosuccinimide
NMM   N-methyl morpholine
NMO   N-methylmorpholine-N-
      oxide
NPM   N-phenylmaleimide
N-PSP N-phenylseleno-
      phthalimide
NR   no reaction
Nuc. general nucleophile
[O]   general oxidation
ONf   nonaflate= nonafluoro-1-
      butanesulfonyl
Oxone potassium peroxy-
      monosulfate
PBP   pyridinium bromide
      perbromide
PCC   pyridinium chloro-
      chromate
PDC   pyridinium dichromate
PEG   polyethylene glycol
Pf   9-phenylfluorenyl
pfb   perfluorobutyrate

PFC   pyridinium fluorochromate
Ph   phenyl
Ph-H benzene
Ph-Me toluene
PhTRAP 2,2'-bis[1-(diphenyl-
      phosphino)ethyl]-1,1'-
      biferrocene
pic   2-pyridinecarboxylate
PIDA phenyliodonium diacetate
PIFA phenyliodo bis-
      (trifluoroacetate)
PLAP porcine liver acetone
      powder
PLE PIG LIVER ESTERASE
PMHS polymethylhydrosiloxane
PMB = Mpm p-methoxybenzyl
PMP   1,2,2,6,6-pentamethyl-
      piperidine
PMP   p-methoxyphenyl
PNB   p-nitrobenzyl
PNP dimer cis, exo-2-phenyl-
      norbornylpalladium Cl$^-$
PNZ   p-nitrobenzyloxycarbonyl
PPA   polyphosphoric acid
PPHF pyridinium polyhydrogen
      fluoride
PPI   2-phenyl-2-(2-pyridyl)
      imidazole
PPL   pig pancreatic lipase
ppp   poly(p-phenylene)
PPSE polyphosphoric acid
      trimethylsilyl ester
PPTS pyridinium p-toluene-
      sulfonate
Pr   propyl
Proc propargyloxycarbonyl
PSCBH poly supported
      cyanoborohydride
PSE   phenylsulfonylethylidene
psi   pounds per square inch
PTAB phenyltrimethylammonium
      perbromide
PTC   phase transfer catalysis

PTMSE (2-phenyl-2-trimethyl-silyl)ethyl
PTS   *p*-tolylsulphonate
PTSA   *p*-toluenesulfonic acid
pyr   pyridine
GFC   quinolinium fluoro-chromate
rac   racemic
RaNi   Raney nickel
R*f*   perfluorinated alkyl
rt   room temperature
Salen *N*,*N*'-ethylenebis-(salicylideneiminato)
SAMP   (s)-1-amino-2-methoxymethyl-pyrrolidine
SBER   sulfurated borohydride exchange resin
SC CO2 super critical CO_2
SDS   sodium dodecylsulfate
SEM = TEOC β-trimethylsilyl-ethoxymethyl
SES   2-[(trimethylsilyl)ethyl]-sulfonyl
Sia   Siamyl
Si-BEZA O-(trisubstituted silyl)benzamide
SMEAH sodium bis(2-methoxy-ethoxy)aluminum hydride
SPB   sodium perborate
SPC   sodium percarbonate
T3p®   propane phosphonic acid anhydride
TADDOL 2,2,-dimethyl-α,α,α^1, α^1-tetraaryl-1,3-dioxolan-4,5-dimethanol
TASF   tris(dimethylamino)-sulfur(trimethylsilyl)-difluoride
TBAB   tetrabutylammonium bromide
TBAF   tetrabutylammonium fluoride

TBAHS tetra-*n*-butylammonium hydrogen sulfate
TBATB tetrabutlyammonium tribromide
TBATFA tetrabutylammonium trifluoroacetate
TBATFA tetrabutylammonium trifluoroacetate
TBCO   tetrabromocyclohexa-dienone
TBDMS = TBS *t*-butyldimethyl-silyl
TBDPS *t*butyldiphenylsilyl
Tbfmoc Tetrabenzo[a,c,g,i]fluorenyl-17-methyloxy-carbonyl
TBH   di-*tert*-butyl hypochlorite
TBHP   *t*butyl hydroperoxide
TBME   *t*butyl methyl ether
TBP   tributylphosphine
TBST   triphenyl silanethiol
Tbs   4-methoxy-3-*t*-butyl-benzenesulphonyl
TBSOP N-*t*butylcarbonyl-2-(*t*butyldimethylsiloxy)-pyrrole
TBTH   tributyltin hydride
TBTSP *t*-butyl trimethylsilyl peroxide
TCAA   trichloroacetyl anhydride
TCCA   trichloroisocyanuric acid
TCF   trichloromethyl chloro-formate
TCNE   tetracyanoethylene
TCNEO tetracyanoethylene oxide
TDTAP tetradecyl trimethyl-ammonium permanganate
TCPC^{TFE} (tetrakis(2,2,2-trifluoro-ethoxycarbonyl)palladium cyclopentadiene
TDS   dimethyl thexylsilyl
TEA   triethylamine
TEAA   tetraethylammonium acetate

I
CARBON-CARBON BOND FORMING REACTIONS

I.A. Carbon - Carbon Single Bonds

(see also: I.E., I.F., I.G., I.H.)

I.A.1. Alkylations of Aldehydes, Ketones and Their Derivatives

I.A.1-1 Normant, J.F. et al., *TL*, **42**, 1883.

49-90%
e.e. = 76-96%

(also trapping with E$^+$)

I.A.1-2 Palomo, C. et al., *OL*, **3**, 3249.

70-92%
d.r. = 49:1

I.A.1-3 Hou, X.-L., Dai, l.-X. et al., *OL*, **3**, 149.

93%
e.e. = 95%

1

I.A.1-4 Tamaru, Y. et al., *JACS*, **123**, 10401 and *TL*, **42**, 3113.

63-85%

I.A.1-5 Saicic, R.N. et al., *T*, **57**, 583.

(other examples given)   46%

I.A.1-6 Enders, D. et al., *S*, 1406.

1. ^{t}BuLi, THF, -78°C
2. R-X, -100°C→rt

76-99%
d.e. = 71-96%

I.A.2. Alkylations of Nitriles, Acids and Acid Derivatives

I.A.2-1 Davies, H.M.L., Ren., *JACS*, **123**, 2070.

Rh$_2$(S-DOSP)$_4$

65-66%
d.e. = 95%, e.e. = 71-84%

I.A.2-2 Tian, Q., Nayyar, N.K. et al., *TL*, **42**, 6807.

LHMDS, BrCH$_2$CN
-78C

90%

I.A.2-3 Ley, S.V., *AG(E)*, **40**, 2906; Tadano, K. et al., *SL*, 1772.

57-96%

I.A.2-4 Kazmaier, U., Zume, F.L., *EJOC*, 4067; Gong, L., Mi, A. et al., *TA*, **12**, 1567.

88%
d.s. = 96%

I.A.2-5 Trost, B.M., Lee, C., *JACS*, **123**, 12191; Konno, T. et al., *TA*, **12**, 2743.

70%
e.e. = 87%

I.A.2-6 Ma, D., Kozikowski, A.P. et al., *BMCL*, **11**, 99; Cahard, D. et al., *TA*, **12**, 983.

73%
e.e. = 75%

I.A.2-7 Uneyama, K. et al., *TA*, **12**, 1303.

81% d.e. = 64%

I.A.2-8 Oshima, K. et al., *TL*, **42**, 4535.

46-95%

I.A.2-9 Gibson, C.L. et al., *JCS(P1)*, 1538.

31-69%
d.r. = 10-65:1

I.A.2-10 Park, Y.S. et al., *SL*, 613.

65%
d.r. = 5.7:1

I.A.2-11 Badia, D. et al., *JOC*, **66**, 5801; Myers, A.G. et al., *JACS*, **123**, 7207.

85-93%
d.r. = 24:1

I.A.2-12 Boeckman, R.K., Jr., et al., *OL*, **3**, 3777; Young, D.W. et al., *JCS(P1)*, 2367.

50-92%
d.r. = 20-99:1

I.A.2-13 Papahatjis, D.P. et al., *CL*, 192.

58-94%

I.A.2-14 Lugtenburg, J. et al., *JOC*, 1269; Basavaiah, D. et al., *TL*, **42**, 85.

70-96%

I.A.3. Alkylations of β-Dicarbonyl, β-Cyanocarbonyl Systems and Other Active Methylene Compounds

I.A.3-1 Burgess, K. et al., *JOC*, **66**, 206; Santelli, M. et al., *JOC*, **66**, 1633; Iwao, M. et al., *TA*, **12**, 2793; Ding, K. et al., *TL*, **42**, 7659; Dieguez, M., Gomez, M. et al., *CC*, 1132; Pozzi, G., Sinou, D. et al., *CC*, 1220; Zhou, Q.-L. et al., *JOM*, **640**, 65; Buono, G. et al., *TA*, **12**, 1345; Anderson, J.C. et al., *TA*, **12**, 923; Nakano, H., Hongo, H. et al., *JOC*, **66**, 620; Mino, T., Yamashita, M. et al., *JOC*, **66**, 1795; Hou, X.-L. et al., *JACS*, **123**, 6508; Claver, C., van Leeuwen, P.W.N.M. et al., *JOC*, **66**, 8867; Hayashi, T. et al., *OM*, **20**, 3913; Gilbertson, S.R., Lan, P., *OL*, 2237; Gomez, M., Masdem-Bulto, A.M. et al., *TA*, **12**, 1469; Hallman, K., Moberg, C., *TA*, **12**, 1475; Mino, T. et al., *TA*, **12**, 1677.

≤96%
e.e. = ≤98%

I.A.3-2 Krafft, M.E. et al., *JACS*, **123**, 9174.

61-85% 19:1

I.A.3-3 Trost, B.M., Lee, C.B., *JACS*, **123**, 3671; Gilbertson, S.R. et al., *JOC*, **66**, 7240; Xiao, J. et al., *OM*, **20**, 138; Kocovsky, P. et al., *OM*, **20**, 673; Bolm, C. et al., *SL*, 1878.

58-94%

I.A.3-4 Evans, P.A., Kennedy, L.J., *JACS*, **123**, 1234; Hou, X.-L., Dai, L.-X. et al., *JACS*, **123**, 7471.

79-91% 47:1-1:99
d.s. = 3-39:1

I.A.3-5 Llera, J.M. et al., *TA*, **12**, 1089.

71-93%
d.e. = 96%

I.A.3-6 Yoshida, J. et al., *JACS*, **123**, 6957.

51-99% 94-100:1

I.A.3-7 Dehli, J.R., Gotor, V., *TA*, **12**, 1485.

44-69%
d.e. = 99:1 e.e. = ≤98%

I.A.3-8 Hayashi, T. et al., *JACS*, **123**, 2089; McLaughlin, M.L., Hammer, R.P. et al., *JOC*, **66**, 7118.

34-98%
e.e. = 52-89%

I.A.3-9 Snider, B.B., O'Hare, S.M., *SC*, **31**, 3753; Lavilla, R. et al., *JOC*, **66**, 1487.

51%

I.A.3-10 Henry, G.E., Jacobs, H., *T*, **57**, 5335; Argade, N.P. et al., *S*, 702; Dahan, A., Portnoy, M., *JOC*, **66**, 6480; Rodriquez, J. et al., *ECJ*, **7**, 1056.

75%

I.A.3-11 Pei, T., Widenhoefer, R.A., *JACS*, **123**, 11290.

32-89%

I.A.4. Alkylations of N-, P-, S-, Se and Similar Stabilized Carbanions

I.A.4-1 Ballini, R. et al., *T*, **57**, 4461.

80%

I.A.4-2 Coldham, I. et al., *OL*, **3**, 3799.

$$\xrightarrow[\text{(-)-sparteine, Et}_2\text{O}]{\text{R-X, }^s\text{BuLi}}$$

40-50%
e.r. = 1-15.7:1

I.A.4-3 Bertini Gross, K.M., Beak, P., *JACS*, **123**, 315; Quirion, J.-C. et al., *JOC*, **66**, 8744.

$$\xrightarrow{^s\text{BuLi, TMEDA, Me}_2\text{SO}_4}$$

68-77%

I.A.4-4 Dieter, R.K. et al., *JACS*, **123**, 5132.

$$\xrightarrow[\text{2. CuCN}\cdot\text{2LiCl, THF}]{\text{1. }^s\text{BuLi, (-)-sparteine, Et}_2\text{O}}$$

53-80%
e.r. = 4-19:1

I.A.4-5 Quirion, J.-C. et al., *JOC*, **66**, 5566.

+ R-X $\xrightarrow[\text{THF, -20°C}]{\text{BuLi}}$

60-85%
d.e. = 41-72%

I.A.4-6 Yoshida, J. et al., *TL*, **42**, 2173.

$$\xrightarrow[\text{2. R}^2\text{-MgBr, Et}_2\text{O, -20°C, 20min}]{\text{1. -2e, -H}^+\text{, -72°C}}$$

45-72%

I.A.5. Alkylations of Organometallic Reagents

(see also: I.B.3., I.B.4., I.F., I.G.)

I.A.5-1 Hoveda, A.H. et al., *AG(E)*, **40**, 1456.

I.A.5-2 Backvall, J.-E. et al., *SL*, 923; Alexakis, A. et al., *SL*, 927.

I.A.5-3 Kobayashi, Y. et al., *JOC*, **66**, 5881.

22-99% 13.3:1-1:24

I.A.5-4 Kulinkovich, O.G. et al., *SL*, 49.

I.A.5-5 Yamamoto, Y. et al., *JACS*, **123**, 6702.

79%

I.A.5-6 Yang, F., Zhao, G., Ding, Y., *TL*, **42**, 2839.

52-75%

I.A.5-7 Knochel, P. et al., *TL*, **42**, 6847.

$$\text{(structure)} \xrightarrow[\text{THF, -30°C}]{^i\text{PrMgCl, E}^+} \text{(structure)}$$

57-83%

I.A.5-8 Fu, G.C. et al., *JACS*, **123**, 10099.

$$\text{RCH}_2\text{-9-BBN} + \text{R}^1\text{—Br} \xrightarrow[\text{THF, rt, 16-24h}]{\text{Pd(OAc)}_2, \text{PCy}_3, \text{K}_2\text{PO}_4\cdot\text{H}_2\text{O}} \text{R}^1\text{∿∿R}$$

66-93%

I.A.5-9 Lapinsky, D.J., Bergmeier, S.C., *TL*, **42**, 8583.

$$\text{(structure)NTs} \xrightarrow[\substack{\text{aq K}_3\text{PO}_4, \text{DMF, rt}}]{\substack{\text{1. 9-BBN} \\ \text{2. R-X, PdCl}_2(\text{dppf})\cdot\text{CH}_2\text{Cl}_2}} \text{R(structure)NTs}$$

28-70%

I.A.5-10 Sarandeses, L.A. et al., *JACS*, **123**, 4155.

$$\text{R}_3\text{In} + \text{R}^1\text{-X} \xrightarrow[\text{THF}]{\text{PdCl}_2(\text{PPh}_3)_2} \text{R-R}^1$$

82-97%

I.A.5-11 Yadav, J.S. et al., *CL*, 18; Singh, G., Vankayalapati, H., *TA*, **12**, 1727.

$$\text{R}\overset{\text{OAc}}{\underset{\text{OAc}}{\diagdown}} + \text{TMS}\diagup\diagdown \xrightarrow[\text{CH}_2\text{Cl}_2, \text{rt}]{\text{InCl}_3} \text{R}\overset{\text{OAc}}{\diagup}\diagdown\diagup$$

70-92%

I.A.5-12 Uemura, S., Hidai, M. et al., *JACS*, **123**, 3393.

$$\text{R}\overset{\text{}}{\underset{\text{OH}}{\diagdown}} + \text{(acetone)} \xrightarrow[\text{NH}_4\text{BF}_4, \text{reflux}]{\text{Cp}^*\text{RuCl}(\mu_2\text{-SMe})_2\text{RuCp}^*\text{Cl}} \text{(product)}$$

55-82%

I.A.5-13 Yamamoto, Y. et al., *JACS*, **123**, 759.

$$\text{(structure)Cl} + \text{Bu}_3\text{Sn}\diagdown\diagup \xrightarrow[\text{Me}_2\text{CO, rt}]{\text{Pd}_2(\text{dba})_3\cdot\text{CHCl}_3, \text{PPh}_3} \text{(product)}$$

71-85%

I.A.5-14 Begtrup, M. et al., *JOC*, **66**, 8344.
**Synthesis of 4- and 5-Substituted 1-Hydroxyimidazoles
Through Directed Lithiation and Metal-Halogen Exchange.**

I.A.5-15 Fernandez-Mayoralas, A. et al., *JOC*, **66**, 1768.

1. MeLi·LiBr, Et$_2$O/THF
2. ^{t}BuLi, THF, -78°C
3. E$^+$

16-61%

I.A.5-16 Chong, J.M. et al., *JOC*, **66**, 8248.

1. BuLi, TMEDA, THF, 0°C
2. R-X, Et$_2$O, -78°C

49-94%

I.A.5-17 Natale, N.R. et al., **57**, 8039; Clayden, J. et al., *JCS(P1)*, 371.

1. LDA, THF, -78°C
2. E$^+$

10-95%

I.A.5-18 Dalpozzo, R. et al., *TL*, **42**, 8833.

1. LiH
2. R^3-Li, CeCl$_3$

14-94%

I.A.5-19 Hosom, A. et al., *OM*, **20**, 5014.

1. "Bu$_6$CrLi$_3$", THF
2. R^3-X, -78°C→rt

62-96%

I.A.5-20 Nenajdenko, V.G. et al., *TA*, **12**, 2517.

+ R^{1}MgBr

CuI
Et$_2$O, -78→0°C

53-89%

I.A.5-21 Sato, F. et al., *JACS*, **123**, 4857.

73%

I.A.5-22 Nakamura, E. et al., *OL*, **3**, 3137.

82-96%

I.A.6. Other Alkylation Procedures

I.A.6-1 Kim, J.N. et al., *TL*, **42**, 9023.

41-74%   14-32%

I.A.6-2 Katoh, T. et al., *OL*, **3**, 2701.

94%

I.A.6-3 Liu, H.-J. et al., *SL*, 214.

51-99%

I.A.6-4 Jiang, B. et al., *T*, **57**, 1581.

$$R\text{-}X \xrightarrow[K_3PO_4]{Pd(PPh_3)_4, PPh_3} R\text{-}CN \quad 10\text{-}89\%$$

I.A.6-5 Jain, R. et al., *BMCL*, **11**, 1133.

$$R^2\text{-}CO_2H + \quad \xrightarrow[aq\ H_2SO_4,\ 70°C,\ 15min]{AgNO_3,\ (NH_4)_2S_2O_8} \quad 12\text{-}45\%$$

I.A.7. Nucleophilic Addition to Electrophilic Carbon

I.A.7.a.1. Aldol-Type 1,2-Additions

I.A.7.a.1-1 Denmark, S.E., Ghosh, S.K., *AG(E)*, **40**, 4759.

1. R^1-CHO, cat., CHCl$_3$/CH$_2$Cl$_2$, -78°C
2. MeOH

cat. =

69-99%
anti:syn = 32-99:1
e.e. = 19-90%

I.A.7.a.1-2 Fernandez, R., Lassaletta, J.M. et al., *JOC*, **66**, 5201 and *SL*, 1158.

CH$_2$Cl$_2$, rt

38-82%
d.r. = ≤49:1

I.A.7.a.1-3 Li, G. et al., *OL*, **3**, 823.

cat., EtCN, -78°C

60-75%
E:Z = 1:20
e.e. = 61-98%

I.A.7.a.1-4 Sakthivel, K., Barbas, C.F., III, *JACS*, **123**, 5260; Barbas, C.F., III, et al., *TL*, **42**, 199; Trost, B.M. et al., *OL*, **3**, 2497; Kitahara, T. et al., *T*, **57**, 4107; Fenc, J.-C. et al., *JCR(S)*, 414; Bell, T.W. et al., *JOC*, **66**, 1525; Yamamoto, H. et al., *SL*, 69, 1245 and *BCJ*, **74**, 1477.

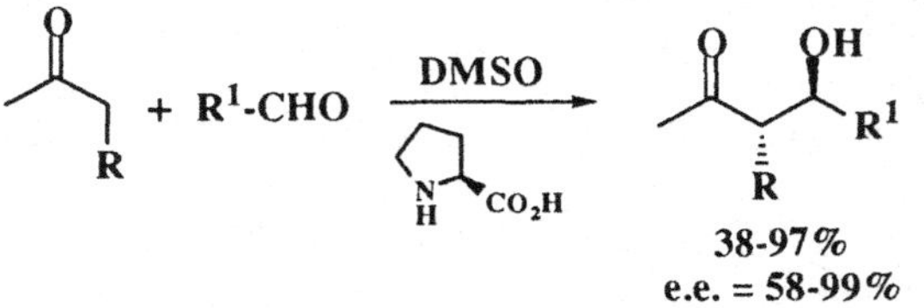

+ R1-CHO
DMSO
R
CO2H
38-97%
e.e. = 58-99%

I.A.7.a.1-5 Shibasaki, M. et al., *OL*, **3**, 1539 and *JACS*, **123**, 2466; Sheldon, R.A. et al., *JOC*, **66**, 4559; Marco, J.A. et al., *OL*, **3**, 901.

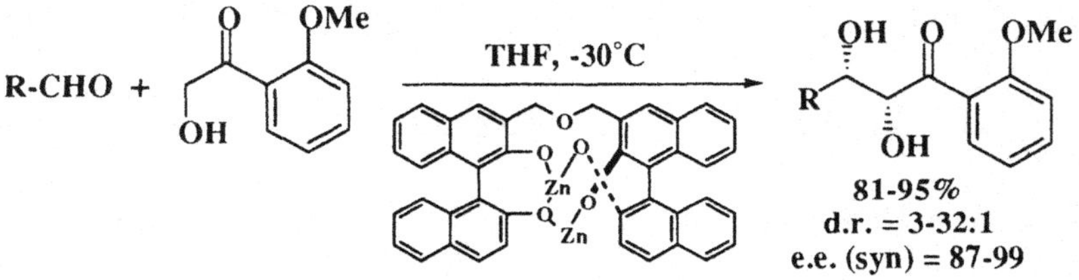

R-CHO +
OMe
OH
THF, -30°C
Zn
81-95%
d.r. = 3-32:1
e.e. (syn) = 87-99

I.A.7.a.1-6 Denmark, S.E., Pham, S.M., *OL*, **3**, 2201.

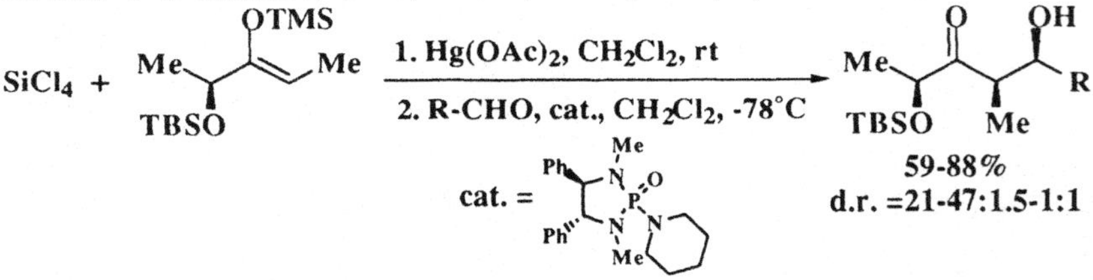

SiCl4 +
OTMS
Me
TBSO
1. Hg(OAc)2, CH2Cl2, rt
2. R-CHO, cat., CH2Cl2, -78°C
cat. =
59-88%
d.r. =21-47:1.5-1:1

I.A.7.a.1-7 Evans, D.A. et al., *OL*, **3**, 3133; Yamauchi, S. et al., *JCS(P1)*, 2158; Battaglia, A. et al., *TA*, **12**, 1015; Andrus, M.B. et al., TL, **42**, 7197.

+
CbzN
Me
LDA
HN
80%
d.r. = 15:1

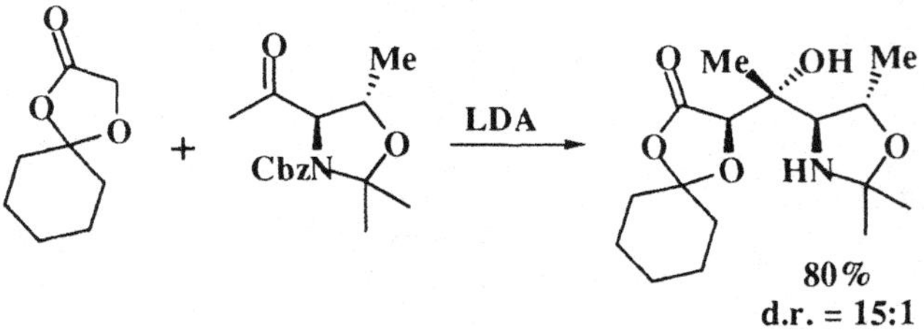

I.A.7.a.1-8 Chiu, P. et al., *TL*, **42**, 4091 and *OL*, **3**, 1901; Romo, D. et al., *JACS*, **123**, 7945.

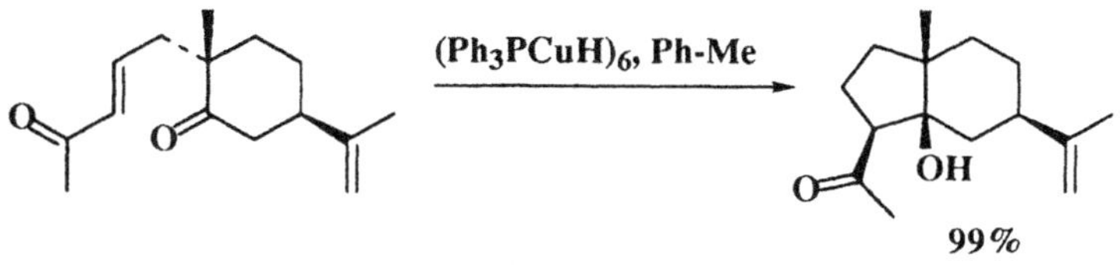

$$(Ph_3PCuH)_6, \ Ph\text{-}Me$$

99%

I.A.7.a.1-9 Veenstra, S.J., Kinderman, S.S., *SL*, 1109.

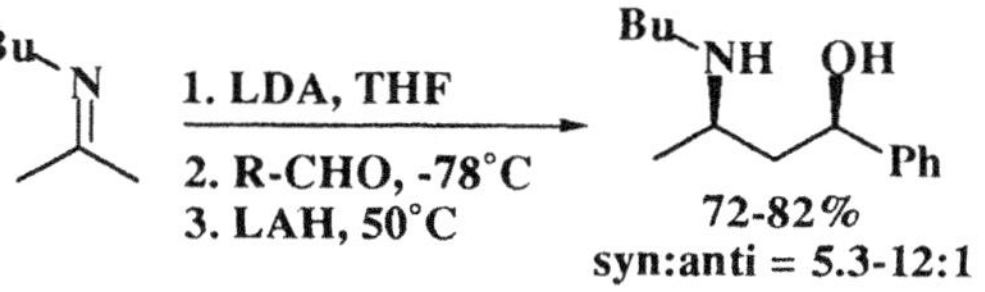

1. LDA, THF
2. R-CHO, -78°C
3. LAH, 50°C

72-82%
syn:anti = 5.3-12:1

I.A.7.a.1-10 Ghosh, A.K., Kim. J.H., *TL*, **42**, 1227; Kurosu, M., Lorca, M., *JOC*, **66**, 1205; Crimmins, M.T. et al., *OL*, **3**, 949.

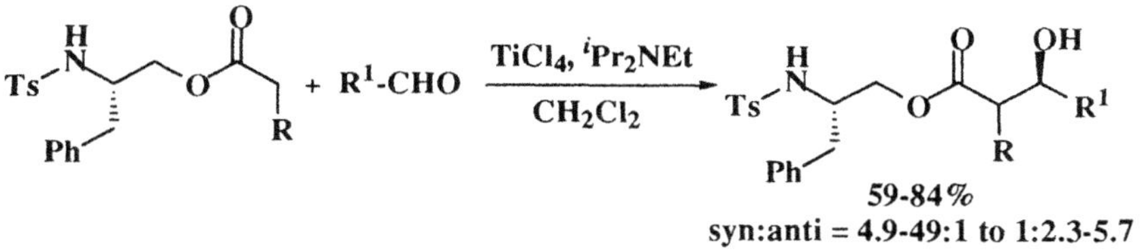

$$+ \ R^1\text{-}CHO \quad \xrightarrow{\ TiCl_4, \ ^iPr_2NEt \ }{CH_2Cl_2}$$

59-84%
syn:anti = 4.9-49:1 to 1:2.3-5.7

I.A.7.a.1-11 Romea, P., Urpi, F. et al., *TL*, **42**, 4629.

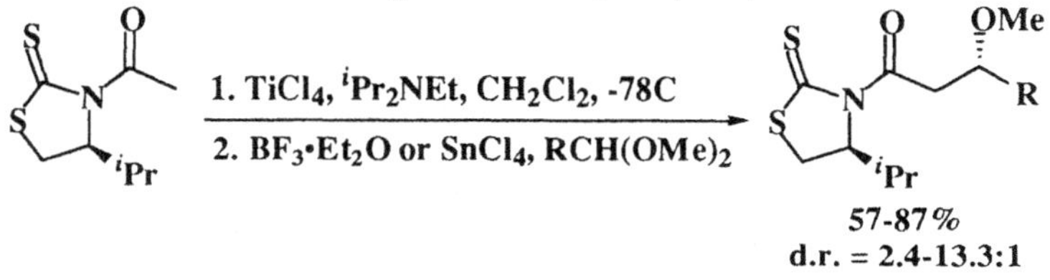

1. TiCl$_4$, ^{i}Pr$_2$NEt, CH$_2$Cl$_2$, -78C
2. BF$_3$•Et$_2$O or SnCl$_4$, RCH(OMe)$_2$

57-87%
d.r. = 2.4-13.3:1

I.A.7.a.1-12 Evans, D.A. et al., *OL*, **3**, 3133; Yamauchi, S. et al., *JCS(P1)*, 2158; Battaglia, A. et al., *TA*, **12**, 1015; Andrus, M.B. et al., TL, **42**, 7197.

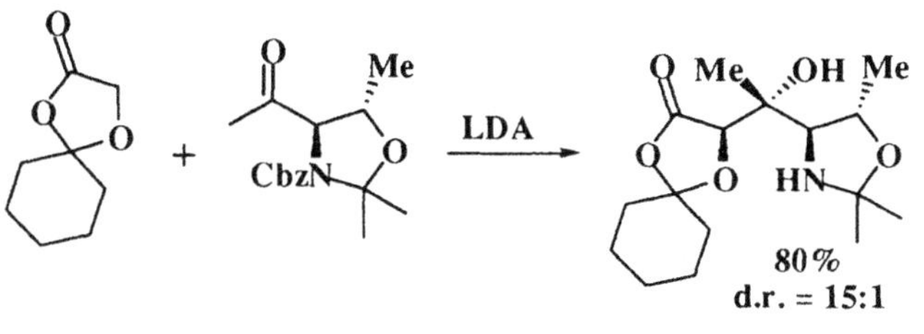

LDA

80%
d.r. = 15:1

I.A.7.a.1-13 DeMong, D.E., Williams, R.M., *TL*, **42**, 183, 3529; Katritzky, A.R. et al., *JOC*, **66**, 4041.

Ph, O O + O NHCbz → Bu$_2$BOTf, TEA / CH$_2$Cl$_2$ → Ph, O O ... OH

Ph, N Me

Ph, N Me NHCbz

69%

I.A.7.a.1-14 Imashiro, R., Kuroda, T., *TL*, **42**, 1313; Guidon, Y. et al., *JACS*, **123**, 8496; Bellasoued, M. et al., *JOC*, **66**, 5054 and *SC*, **31**, 1007; Rassu, G., Casiraghi, G. et al., *JOC*, **66**, 8070.

MeO CHO + Cl Cl OTMS OMe → CH$_2$Cl$_2$, -78°C, 7h → MeO Cl Cl CO$_2$Me OH

X S O$_2$ ^{i}Pr B H O

30-89%
e.e. =77-96%

I.A.7.a.1-15 Tanabe, Y. et al., *SL*, 1959.

R^1 O + R^2 O SAr → TiCl$_4$, Bu$_3$N / CH$_2$Cl$_2$, -78°C → R^1 OH O SAr

R   R^3 R R^2 R^3

76-99%
syn:anti = 2.3-15.7:1

I.A.7.a.1-16 Lefebvre, O., Brigaud, T., Portella, C., *JOC*, **66**, 1941; Portella, C. et al., *JOC*, **66**, 4543; Shen, Y., Zhang, Y., *OL*, **3**, 2805.

O TMS → 1. CF$_3$-TMS, Bu$_4$N$^+$ Ph$_3$SnF$_2^-$ / 2. R^1-CHO, BF$_3$·Et$_2$O → O OH R^1

R   R F F

43-78%

I.A.7.a.1-17 Wulff, W.D. et al., *AG(E)*, **40**, 2271; Akiyama, T. et al., *TL*, **42**, 4025; Badia, D. et al., *JOC*, **66**, 9030.

OH Ph N + Me Me OTMS OMe → NMI, Ph-Me, rt → OH Ph N H Me CO$_2$Me Me

Ph Ph OH HO

78-99%
e.e. = 47-99%

I.A.7.a.1-18 Campagne, J.-M., Moreau, X., *TL*, **42**, 4467; Kalesse, M. et al., *OL*, **3**, 3561; Christmann, M., Kalesse, M., *TL*, **42**, 1269; Mukaiyama, T. et al., *CL*, 190; Soriente, A. et al., *TA*, **12**, 959; Scettri, A. et al., *TA*, **12**, 1529; Bluet, G., Campagne, J.-M., *JOC*, **66**, 4293.

I.A.7.a.1-19 Kobayashi, S. et al., *JACS*, **123**, 12511 and *JOC*, **66**, 809; Blond, G., Billard, T., Langlois, B.R., *JOC*, **66**, 4826; Cooke, J.W.B., *JOC*, **66**, 334.

I.A.7.a.1-20 Evans, D.A. et al., *AG(E)*, **40**, 1884.

I.A.7.a.1-21 Hu, L. et al., *JOC*, **66**, 5413.

I.A.7.a.1-22 Shi, M., Feng, Y.-S., *JOC*, **66**, 406; Oshima, K. et al., *JOC*, **66**, 7854.

I.A.7.a.1-23 Morken, J.P. et al., *OL*, **3**, 1829, 2839; Baba, A. et al., *JOC*, **66**, 8690.

$$BnO\diagup CHO \; + \; \diagdown CO_2Me \xrightarrow[\text{indane-pybox, rt}]{[Ir(cod)Cl]_2, Et_2MeSiH} BnO\diagup\overset{OH}{\diagup}\underset{Me}{\diagup}CO_2Me$$

59%
syn:anti = 9.5:1
e.e. =96%

I.A.7.a.1-24 Balan, D., Adolfsson, H., *JOC*, **66**, 6498.

$$R\text{-}CHO + TsNH_2 + \diagup CO_2Me \xrightarrow[\text{IPA, base}]{La(OTf)_3, 4Å MS} \underset{NHTs}{Ph}\diagup CO_2Me$$

12-83%

I.A.7.a.1-25 Pan, Y. et al., *OPP*, **33**, 351; Tanabe, Y., *CC*, 1674.

$$Ar\text{-}CO_2Et + RCH_2CN \xrightarrow[\text{Ph-Me}]{K, Np, (((\bullet} \underset{O}{Ar}\overset{R}{\diagup}CN$$

59-87%

I.A.7.a.1-26 Nokami, J. et al., *JOC*, **66**, 1228; Siebenhaar, B. et al., *CJC*, **79**, 566.

$$R\diagup CHO + \underset{O}{Ar}\overset{}{\diagup}S\overset{O}{\diagup}R^1 \xrightarrow[\text{EtCN, rt}]{Et_2NH} R\overset{OH}{\diagup}\diagup\underset{O}{\diagup}R^1$$

18-84%

I.A.7.a.1-27 Gree, R. et al., *TL*, **42**, 3069.

$$R\text{-}CHO + \diagup\diagup\underset{OH}{\diagup}C_5H_{11} \xrightarrow[\text{THF, reflux}]{Ru(PPh_3)_3HCl} \underset{OH \; O}{Ph}\overset{Me}{\diagup}C_5H_{11}$$

26-72%

I.A.7.a.1-28 Trost, B.M., Oi, S., *JACS*, **123**, 1230.

$$R\diagup\diagup\underset{OH}{\overset{Ph}{\diagup}} + R^1\text{-}CHO \xrightarrow[\text{DCE, 180°C}]{O=Vn(OSiPh_3)_3} \underset{O \quad Ph}{R}\overset{R^1 \diagdown OH}{\diagup} + \underset{O \quad Ph}{R}\overset{R^1}{\diagup}\diagdown OH$$

42-95% 9-99:1

I.A.7.a.1-29 Jauch, J., *JOC*, **66**, 609; Bauer, T., Tarusiuk, J., *TA*, **12**, 1741; Basavaiah, D. et al., *SC*, **31**, 2987; Krishna, P.R. et al., *TA*, **12**, 829; Alcaide, B. et al., *JOC*, **66**, 1612; Hatakeyama, S. et al., *CC*, 2030; Bosanac, T., Wilcox, C.S., *CC*, 1618.

PhSeLi, R-CHO

THF, -60C

65-89%
d.e. = 99%

I.A.7.a.1-30 Schneider, C., Hansch, M., *CC*, 1218; Morken, J.P. et al., *AG(E)*, **40**, 601.

R-CHO +

Zr(O^{t}Bu)$_4$

THF, -20°C

70-89% 1-22:1

I.A.7.a.2 Addition of N-, P-, S-, Se and Similar Stabilized Carbanions

I.A.7.a.2-1 Shibasaki, M. et al., *SL*, 980; Qian, C., et al., *TL*, **42**, 4673.

+ RCH$_2$NO$_2$

Al-Li(R-binaphthoxide)$_2$

CH$_2$Cl$_2$, -40°C

68-98%
e.e. = 60-83%

I.A.7.a.2-2 Knudsen, K.R., Jorgensen, K.A., *JACS*, **123**, 5843; Jorgensen, K.A. et al., *AG(E)*, **40**, 2992, 2995.

+

cis-DiPh-BOX-Cu

THF, -100°C

67-94%
erythro:threo = 5-39:1
e.e. = 83-98%

I.A.7.a.2-3 Deshpande, V.H. et al., *SC*, **31**, 3623; Sartori, G. et al., *TL*, **42**, 2401; Yao, C.-F. et al., *JOC*, **66**, 1984.

$$R\text{-CHO} + R^1CH_2NO_2 \xrightarrow[\text{rt, 4-15min}]{\text{BnNMe}_3\text{OH}} \begin{array}{c} HO \quad R^1 \\ \diagdown \\ R \quad NO_2 \end{array}$$

78-98%

I.A.7.a.2-4 Seebach, D. et al., *JOC*, **66**, 3059; Caddick, S. et al., *T*, **57**, 6295.

25-85%
d.r. = 2.5-19:1

I.A.7.a.3. Addition of Organometallic and Related Species

I.A.7.a.3-1 Marshall, J.A. et al., *OL*, **3**, 3369; Marshall, J.A., Schaaf, G.M., *JOC*, **66**, 7825; Roush, W.R. et al., *OL*, **3**, 3057.

69-89%
anti:syn = 49:1
e.r. = 99:1

I.A.7.a.3-2 Carreira, E.M. et al., *OL*, **3**, 3017; Sasaki, Boyall, D., Carreira, E.M., *HCA*, **84**, 964; Cozzi, P.G., Umani-Ronchi, A. et al., *TA*, **12**, 1063.

54-95%
e.e. = 88-97%

I.A.7.a.3-3 Yang, F., Zhao, G., Ding, Y., *TL*, **42**, 2839.

52-76%

I.A.7.a.3-4 Tanaka, T. et al., *JOC*, **42**, 1867.

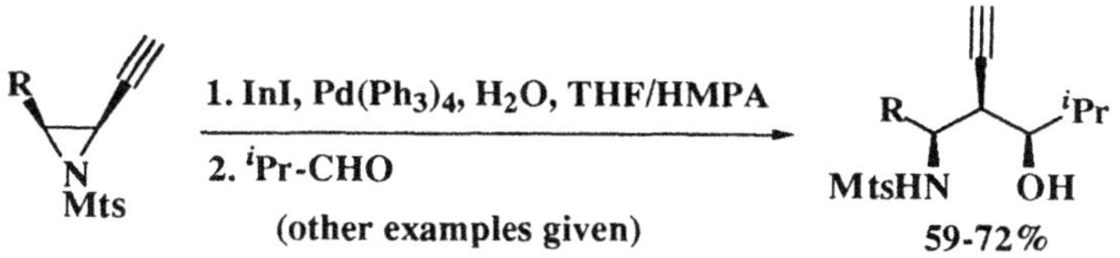

1. InI, Pd(Ph₃)₄, H₂O, THF/HMPA

2. *i*Pr-CHO

(other examples given)

59-72%

I.A.7.a.3-5 Bolm, C. et al., *AG(E)*, **40**, 1488; Hu, Q.-S. et al., *TL*, **42**, 7725; Long, J., Ding, K., *AG(E)*, **40**, 544; Nevalainen, M., Nevalainen, V., *TA*, **12**, 1771; Fan, Q.-H. et al., *TA*, **12**, 1559; Cobb, A.J.A., Marson, C.M., *TA*, **12**, 1547; Rosini, C. et al., *TA*, **12**, 1235; Dimitrov, V. et al., *TA*, **12**, 1313, 1323; Walsh, P.J. et al., *OL*, **3**, 2161; Wang, J.-X. et al., *CL*, 174; Gau, H.-M. et al., *CC*, 1546; Soai, K. et al., *AG(E)*, **40**, 1096.

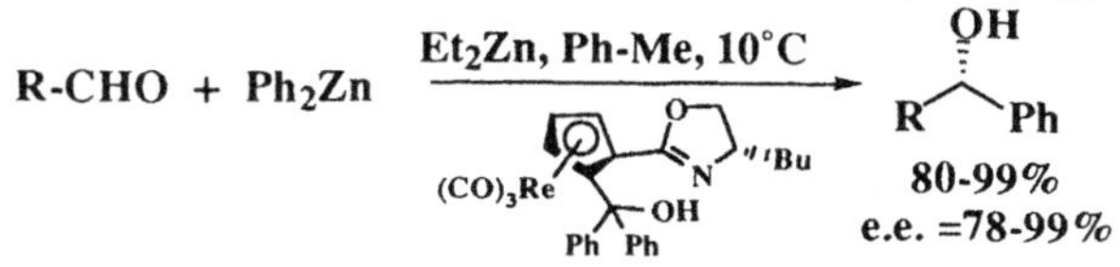

R-CHO + Ph₂Zn

Et₂Zn, Ph-Me, 10°C

80-99%

e.e. =78-99%

I.A.7.a.3-6 Zercher, C.K. et al., *OL*, **3**, 4169; Hilgenkamp, R., Zercher, C.K., *T*, **57**, 8793.

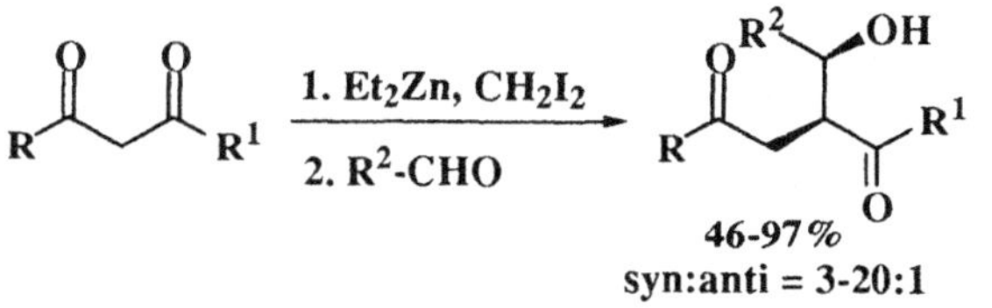

1. Et₂Zn, CH₂I₂

2. R²-CHO

46-97%

syn:anti = 3-20:1

I.A.7.a.3-7 Oshima, K. et al., *T*, **57**, 987; Thebtaranonth, Y. et al., *JOC*, **66**, 4692.

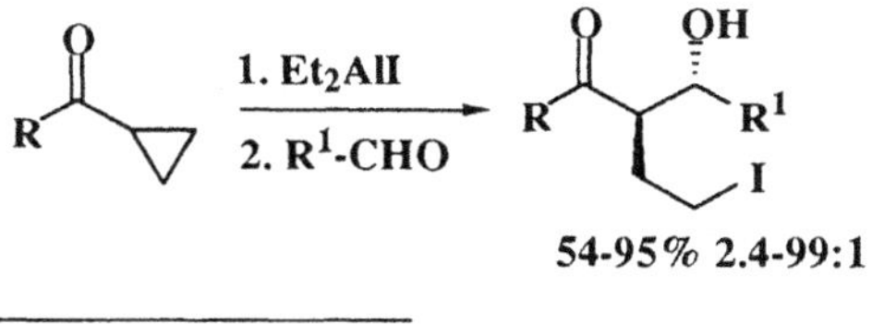

1. Et₂AlI

2. R¹-CHO

54-95% 2.4-99:1

I.A.7.a.3-8 Smyj, R.P., Chong, J.M., *OL*, **3**, 2903.

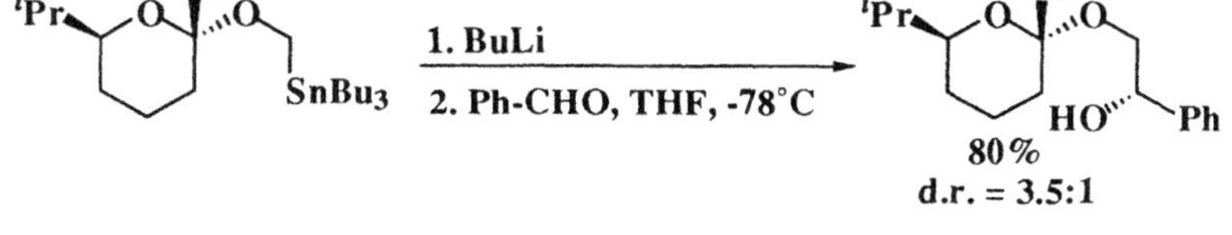

1. BuLi

2. Ph-CHO, THF, -78°C

80%

d.r. = 3.5:1

I.A.7.a.3-9 Righi, G. et al., *T*, **57**, 10039.

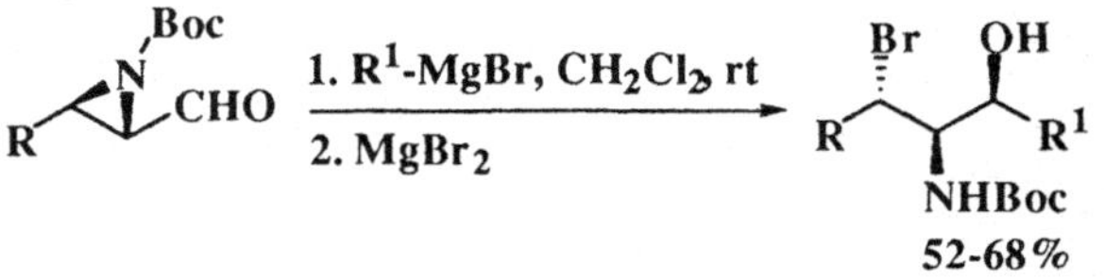

$$\underset{R}{\overset{\text{Boc}}{\bigtriangleup}}\text{-CHO} \xrightarrow[\text{2. MgBr}_2]{\text{1. R}^1\text{-MgBr, CH}_2\text{Cl}_2, \text{ rt}} \underset{\underset{\text{NHBoc}}{}}{R}\overset{\overset{\text{Br OH}}{}}{\diagdown}R^1$$

52-68%

I.A.7.a.3-10 Li, Z., Zhang, Y., *JCR(S)*, 522.

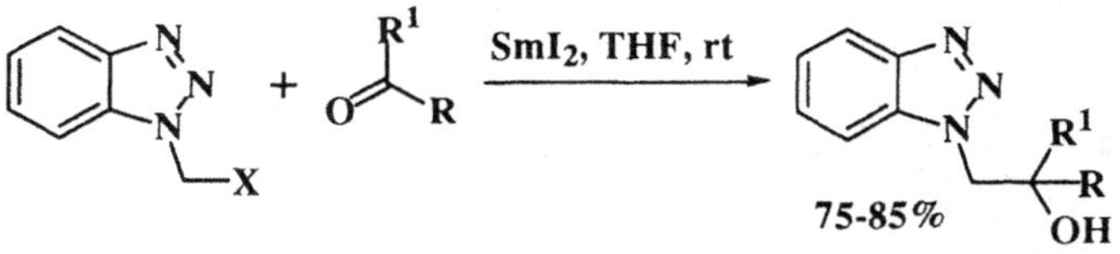

75-85%

I.A.7.a.3-11 Toru, T. et al., *T*, **57**, 8469.

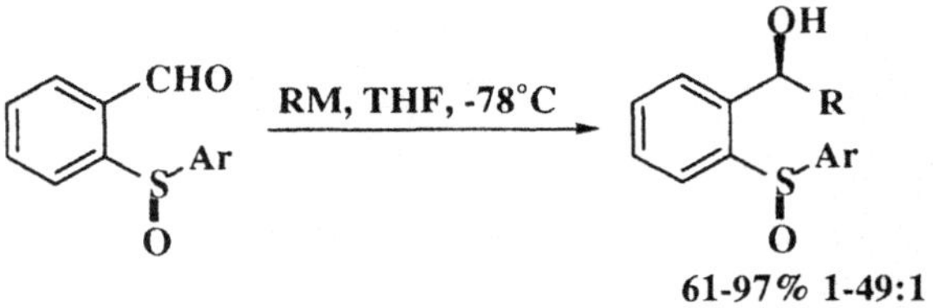

61-97% 1-49:1

I.A.7.a.3-12 Kabalka, G.W. et al., *T*, **57**, 1663.

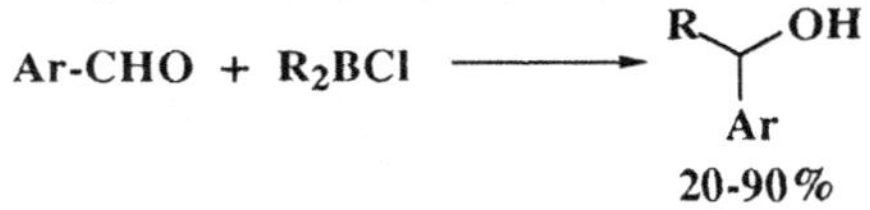

$$\text{Ar-CHO} + \text{R}_2\text{BCl} \longrightarrow \underset{\text{Ar}}{R}\diagdown\text{OH}$$

20-90%

I.A.7.a.3-13 Durandetti, M. et al., *OL*, **3**, 2073; Dahmen, S., Brase, S., *OL*, **3**, 4119; Oi, S. et al., *OM*, **20**, 1036; McCluskey, A., Young, D.J. et al., *JOC*, **66**, 7811.

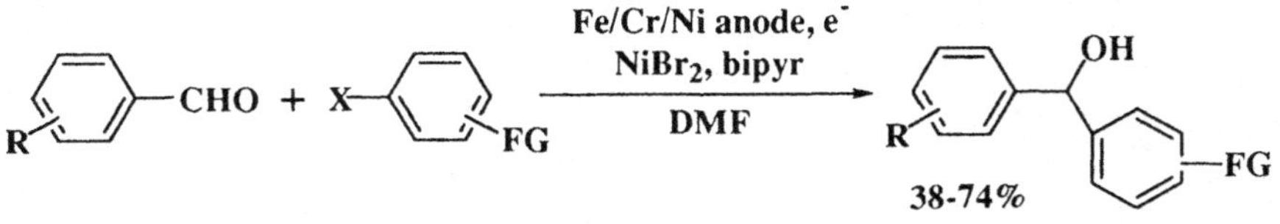

Fe/Cr/Ni anode, e⁻
NiBr₂, bipyr
DMF

38-74%

I.A.7.a.3-14 Basu, M.K., Banik, B.K., *TL*, **42**, 187; Cho, Y.S. et al., *TL*, **42**, 1957; Micskei, K. et al., *TL*, **42**, 7711.

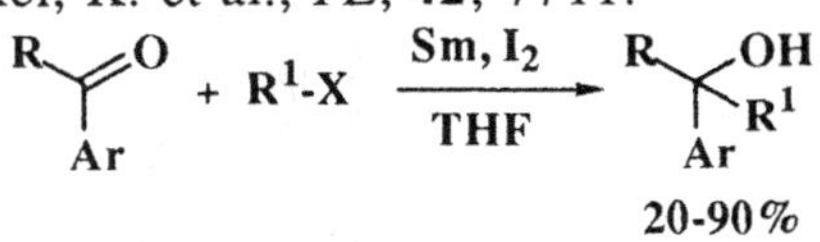

$$\underset{\text{Ar}}{\overset{\text{O}}{R\diagup}} + \text{R}^1\text{-X} \xrightarrow[\text{THF}]{\text{Sm, I}_2} \underset{\text{Ar}}{\overset{\text{OH}}{R\diagdown}}R^1$$

20-90%

I.A.7.a.3-15 Degl'Innocenti, A. et al., *TL*, **42**, 4557.

12-83%

I.A.7.a.3-16 Cushman, M. et al., *JOC*, **66**, 4405.

50%

I.A.7.a.3-17 Miokowski, C., Falck, J.R. et al., *OL*, **3**, 4237; Cozzi, P.G., Umani-Ronchi, A. et al., *OL*, **3**, 1153; Takai, K. et al., *CC*. 1128.

$$R\text{-}CH_2CCl_3 + R^1\text{-}CHO \xrightarrow{\text{CrCl}_2, \text{THF, rt}}$$

48-97%

I.A.7.a.3-18 Nishiyama, H. et al., *OM*, **20**, 1580; Kii, S., Maruoka, K., *TL*, **42**, 1935; Denmark, S.E., Wynn, T., *JACS*, **123**, 6199; Yamamoto, H. et al., *CPB*, **74**, 1129; Loh, T.-P. et al., *TL*, **42**, 8701, 8705; Kundig, E.P. et al., *JOC*, **66**, 1852; Renaud, P. et al., *S*, 1573.

45-99%
e.e. =≤80%

I.A.7.a.3-19 Li, Z., Zhang, Y., *TL*, **42**, 8507.

70-89%
syn:anti = 1.1-19:1

I.A.7.a.3-20 Dias, L.C., Ferreira, E., *TL*, **42**, 7159; Andrade, C.K.Z., Azevedo, N.R., *TL*, **42**, 6473.

70-89%
syn:anti = 1.1-19:1

I.A.7.a.3-21 Lombardo, M., Trombini, C. et al., *OL*, **3**, 2981; Canac, Y., Levoirier, E., Lubineau, A., *JOC*, **66**, 3206; Baba, A. et al., *SL*, 1659.

70-96%
syn:anti 1:9-9:1

I.A.7.a.3-22 Delgado, A. et al., *TA*, **12**, 1625.

0-72%
d.r. = 6-121:1

I.A.7.a.3-23 Cozzi, F. et al., *JOC*, **66**, 3160; Yamada, T. et al., *OL*, **3**, 1937.

96%
e.e. = 95%

I.A.7.a.3-24 Lin, G.Q. et al., *JOC*, **66**, 3953.

58-84% 1.6-3.3:1
e.e. (trans) = 93-99%   e.e. (cis) = 0-75%

I.A.7.a.3-25 El-Sayed, E. et al., *JOC*, **66**, 4766.

14-56%
e.e. = ≤91%

I.A.7.a.3-26 Tang, C.-C. et al., *TA*, **12**, 1579; Shibasaki, M. et al., *TL*, **42**, 691; Cozzi, P.G., Umani-Ronchi, A. et al., *TL*, **42**, 3041; North, M. et al., *T*, **57**, 771; Brussee, J., *TA*, **12**, 1109; Marcus, J. et al., *TA*, **12**, 971; Groger, H., Vorlop, K.-D. et al., *OL*, **3**, 1969; Lin, G. et al., *TA*, **12**, 843; Chen, P., Lin, G. et al., *TA*, **12**, 3273.

≤99%
e.e. = 73%

I.A.7.a.3-27 Petit, G.R., Grealish, M.P., *JOC*, **66**, 8640.

70%

I.A.7.a.3-28 Izawa, K. et al., *TL*, **42**, 6337, 5887; Richards, N.G.J. et al., *JOC*, **66**, 6381; Pedrosa, R. et al., *T*, **57**, 8521.

31-67%

I.A.7.a.3-29 Araki, S. et al., *JOC*, **66**, 7919; Liu, R.-S. et al., *JOC*, **66**, 1781.

(other examples given(

41-78%
E:Z = 1:1.7-3.7

I.A.7.a.3-30 Tamaru, Y. et al., *OL*, **3**, 2181.

I.A.7.a.3-31 Takai, K.et al., *AG(E)*, **40**, 1116; Kumar, S. et al., *SL*, 1431.

49-99%
anti:syn = 99:1-1:9

I.A.7.a.3-32 Vandewalle, M. et al., *EJOC*, 3779.

74%

I.A.7.a.3-33 Minami, T. et al., *JOC*, **66**, 3924.

38-88%

I.A.7.a.3-34 Bailey, W.F. et al., *OL*, **3**, 1865; Chung, K.-H. et al., *JOC*, **66**, 2484; Carda, M., Marco, J.A. et al., *TA*, **12**, 1417.

78-96%
d.r. = 1.8-13.3

I.A.7.a.3-35 Barbero, A., Pulido, F.J. et al., *AG(E)*, **40**, 2161.

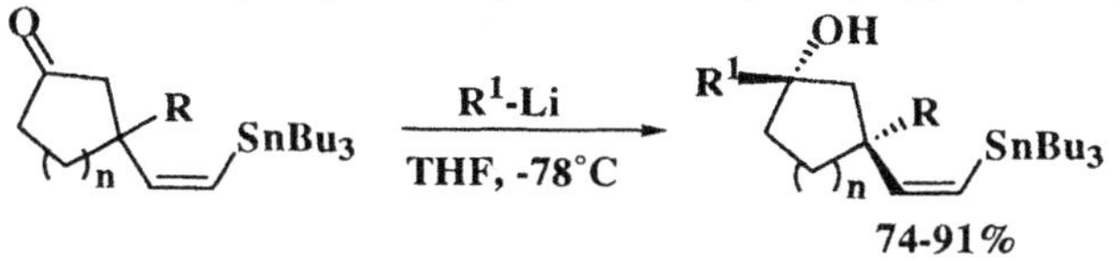

R^1-Li, THF, -78°C

74-91%

I.A.7.a.3-36 Procter, D.J. et al., *OL*, **3**, 2001.

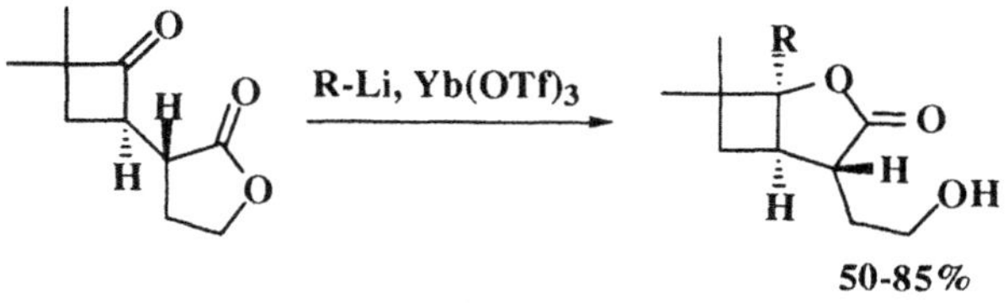

R-Li, Yb(OTf)$_3$

50-85%

I.A.7.a.3-37 Kang, H.-Y., Cho, Y.S. et al., *TL*, **42**, 5489.

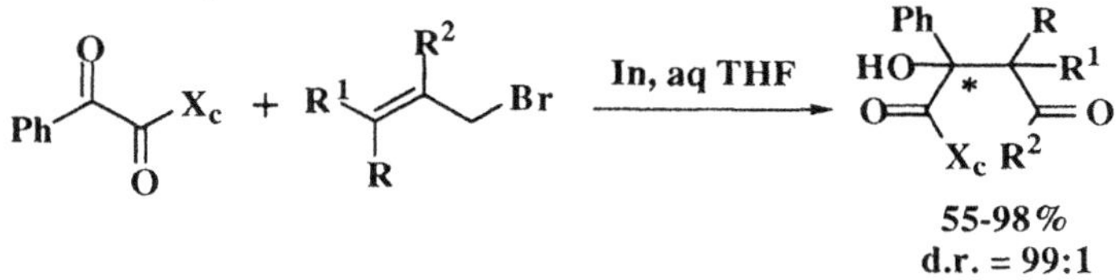

In, aq THF

55-98%
d.r. = 99:1

I.A.7.a.3-38 Alcaide, B. et al., *JOC*, **66**, 5208; Nair, V. et al., *T*, **57**, 9453 and *JCR(S)*, 551.

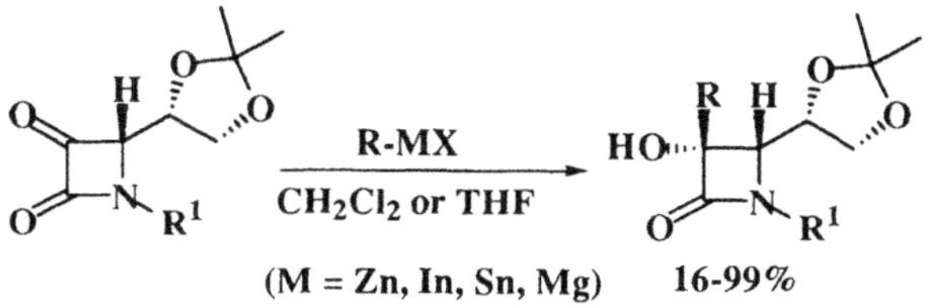

R-MX, CH$_2$Cl$_2$ or THF

(M = Zn, In, Sn, Mg)   16-99%

I.A.7.a.3-39 Pulido, F.J. et al., *JOC*, **66**, 7723.

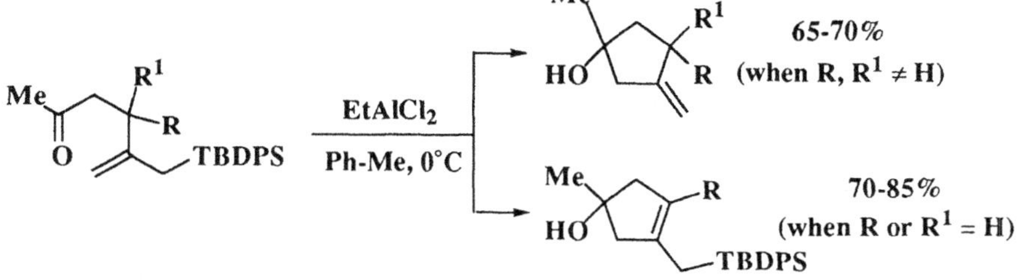

EtAlCl$_2$, Ph-Me, 0°C

65-70% (when R, R^1 ≠ H)

70-85% (when R or R^1 = H)

I.A.7.a.3-40 Wang, Y., Zhu, S., *TL*, **42**, 5441; Peppe, C. et al., *TL*, **42**, 4745.

$$R\text{-}C(=O)\text{-}R^1 + Cl\text{-}C(=O)\text{-}CF_2\text{-}CO_2Et \xrightarrow[\text{Et}_2\text{O, 0°C'rt}]{\text{Zn, CuCl}} R\text{-}C(OH)(R^1)\text{-}CF_2\text{-}C(=O)\text{-}CO_2Et$$

30-91%

I.A.7.a.3-41 Bartoli, G. et al., *TL*, **42**, 6093.

$$Me_2N\text{-}C(=O)\text{-}CH(Me)\text{-}C(=O)\text{-}R \xrightarrow[\text{2. } R^1\text{MgX, CeCl}_3]{\text{1. TiCl}_4} Me_2N\text{-}C(=O)\text{-}CH(Me)\text{-}C(OH)(R^1)(R)$$

55-99%
d.e. = 40-98%

I.A.7.a.3-42 Ranu, B.C. et al., *JOC*, **66**, 7519.

$$(EtO)_2P(=O)\text{-}CH_2\text{-}C(=O)\text{-}R + R^1\text{-}CH=CH\text{-}CH_2\text{-}Br \xrightarrow[\text{THF, rt}]{\text{In}} (EtO)_2P(=O)\text{-}CH(R)(C(OH)(R^1)\text{-}CH=CH_2)$$

65-95%

I.A.7.a.3-43 Vedejs, E., Zajac, M.A., *OL*, **3**, 2451.

LDA
THF, -23°C

57%

I.A.7.a.3-44 Pettus, T.R.R. et al., *JOC*, **66**, 3435.

1. R^1-M, Et_2O or THF
2. R^2-M, THF

50-97%

I.A.7.a.3-45 Burger, K. et al., *S*, 281.

$$F_3C\text{-}C(=N\text{-}R)\text{-}CF_3 + CH_2=CH\text{-}CO_2R^1 \xrightarrow[\text{THF}]{\text{DABCO, CaH}_2} F_3C\text{-}C(CF_3)(NH\text{-}R)\text{-}C(=CH_2)\text{-}CO_2R^1$$

27-65%

I.A.7.a.3-46 Rauter, A.P. et al., *TA*, **12**, 1131; Cho, Y.S., Lee, E. et al., *JCS(P1)*, 2079.

(many other examples)   65%

I.A.7.a.3-47 Molander, G.A. et al., *JOC*, **66**, 4511 and *OL*, **3**, 2257; Dudley, G.A., Danishefsky, S.J., *OL*, **3**, 2399.

21-94%

I.A.7.a.3-48 Goossen, L.J., Ghosh, K., *CC*, 2084 and *AG(E)*, **40**, 3458; Yamamoto, A. et al., *BCJ*, **74**, 371; Shimizu, T., Seki, M., *TL*, **42**, 429; **see also:** Frost, C.G., Wadsworth, K.J., *CC*, 2316; Giacomelli, G. et al., *OL*, **3**, 1519.

$$Ar\text{-}B(OH)_2 + R\text{-}CO_2H \xrightarrow[]{Pd(F_6\text{-}acac)_2, PCy_3}$$

37-95%

I.A.7.a.3-49 Fischer, C., Carreira, E.M., *OL*, **3**, 4319; Chang, S., Lee, M., Kim, S., *SL*, 1557.

54-85%

I.A.7.a.3-50 Vallee, Y. et al., *TA*, **12**, 1147; Shibasaki, M. et al., JACS, **123**, 6801.

20-95%
e.e. = 45-95%

I.A.7.a.3-51 Otaka, A. et al., *TL*, **42**, 5443.

1. $R_2Cu(CN)_2Li_2 \cdot LiX$
2. O_2

45-64%

I.A.7.a.3-52 Ukaji, Y., Inomata, K. et al., *CL*, 254.

1. Et_2Zn, cat.
2. $ICH_2CO_2{}^tBu$

26-80%
e.e. = 74-98%

I.A.7.a.3-53 Ellman, J.A. et al., *JOC*, **66**, 8772; Prakash, G.K.S., Mandal, M., Olah, G.S., *SL*, 77 and *OL*, **3**, 2847; Mabic, S., Cordi, A.A., *T*, **57**, 8861; Lu, W., Chan, T.H., *JOC*, **66**, 3467; Shaw, A.W., deSolms, S.J., *TL*, **42**, 7173; Pinho, P., Andersson, P.G., *T*, **57**, 1615.

R^2MgX or R^2Li, $AlMe_3$
Ph-Me, -78°C

20-93%
d.r. = 3.5-49:1

I.A.7.a.3-53 Kobayashi, S. et al., *AG(E)*, **40**, 1896; Yanada, R. et al., *JOC*, **66**, 1283, 7516; Kellogg, R.M., Broxterman, Q.B. et al., *OL*, **3**, 3943; Kobayashi, S. et al., *JACS*, **123**, 9493; Vilaivan, T. et al., *TL*, **42**, 9073; Sugimoto, Y. et al., *SL*, 1747; Yamagouchi, R. et al., *T*, **57**, 109; Kibayashi, C. et al., *TL*, **42**, 5029; Yamamoto, H. et al., *SL*, 1859; Friestad, G.K., Qin, J., *JACS*, **123**, 9922; Takeuchi, R. et al., *JACS*, **123**, 9525; Tokioka, K. et al., *JOC*, **66**, 7051; Chrzanowska, M., Sokolowska, J., *TA*, **12**, 1435; Soai, K. et al., *JCS(P1)*, 217; Hoveyda, A.H., Snapper, M.L. et al., *JACS*, **123**, 10409; Saidi, M.R. et al., *T*, **57**, 6829.

Ph-Me, 0°C

74-85%
syn:anti = 19:1
e.e. = 55-99%

I.A.7.a.3-55 Pilli, R.A. et al., *TL*, **42**, 5605; Royer, J. et al., *TA*, **12**, 1219; Liu, R.-S. et al., *JOC*, **66**, 6193; Kokayashi, S. et al., *SL*, 1225; Petrini, M. et al., *JOC*, **66**, 8264; Beifuss, U. et al., *T*, **57**, 1005; Kibayashi, C. et al., *TL*, **42**, 3013.

89-95%
cis:trans = 1-2.5:1

I.A.7.a.3-56 Lombardo, M., Fabbroni, S., Trombini, C., *JOC*, **66**, 1264; Prajapati, D. et al., *TL*, **42**, 7883; Shuto, S. et al., *JCS(P1)*, 599; Nelson, D.W. et al., *JOC*, **66**, 2572.

90% 13.3:1

I.A.7.a.3-57 Kakuuchi, A., Taguchi, T., Hanzawa, Y., *TL*, **42**, 1547.

37-64%

I.A.7.a.4. Addition of N-, P-, S-, Se and Similar Stabilized Carbanions

I.A.7.a.4-1 Riant, O. et al., *OL*, **3**, 3863; Mukaiyama, T. et al., *T*, **57**, 2499; Groth, U., Jeske, M., *SL*, 129; Yu, M., Zhang, Y., *OPP*, **33**, 187; Hekmatshoar, R. et al., *M*, **132**, 689; Uemura, S. et al., *BCJ*, **74**, 1497; Yoshimura, N., Mukaiyama, T., *CL*, 1334; Kise, N., Ueda, N., *TL*, **42**, 2365; Shimizu, M., Niwa, Y., *TL*, **42**, 2829.

73-95
dl:meso = 4-99:1
e.e. = 7-71%

I.A.7.a.4-2 Kakiuchi, K. et al., *TL*, **42**, 7595.

43-82%

I.A.7.a.4-3 Shimizu, M. et al., *CL*, 1196; Nicolaou, K.C. et al., *AG(E)*, **40**, 4705.

11-87%
anti:syn = 3.5:1-1:99

I.A.7.a.4-4 Quan, L.G., Cha, J.K., *TL*, **42**, 8567.

75%

I.A.7.a.4-5 Dolbier, W.R., Jr. et al., *OL*, **3**, 4271; Petrov, V.A., *TL*, **42**, 3267.

19-93%

I.A.7.a.4-6 Miles, W.H. et al., *T*, **57**, 9925.

90%
e.e. = 95%

I.A.7.a.4-7 Overman, L.E., Wolfe, J.P., *JOC*, **66**, 3167.

83% 5:1

I.A.7.a.4-8 Muller, M. et al., *JCS(P1)*, 633.
Enantioselective Synthesis of Hydroxy Ketones Through Cleavage and Formation of Acyloin Linkage. Enzymatic Kinetic Resolution *via* C-C Bond Cleavage.

I.A.7.a.4-9 Kim, S.S. et al., *TL*, **42**, 8315;

$$Ar\text{-}CHO + Me_2N\text{-}Ph \xrightarrow{h\nu}$$

66-79%

I.A.7.a.4-10 Tietze, L.F. et al., *ECJ*, **7**, 161; 1304.

≤51% 32.3:1

I.A.7.a.4-11 deArmas, P. Garcia-Tellado, F., Marrero-Tellado, J.J., *OL*, **3**, 1905.

0-85%

10-95%

I.A.7.a.4-12 Beifuss, W. et al., *AG(E)*, **40**, 568.

35-99%

I.A.7.b. Conjugate Additions

I.A.7.b.1. Enolate-Type Carbanions

I.A.7.b.1.-1 Enders, D. et al., *EJOC*, 4463; Palomo, C. et al., *TL*, **42**, 4829.

70-90%
d.e. = 85-96%

I.A.7.b.1.-2 Takasu, K., Ueno, M., Ihara, M., *JOC*, **66**, 4667; Shindo, M. et al., *OL*, **3**, 2029.

85%
d.e. = 99%

I.A.7.b.1.-3 Cai, C., Soloshonok, V.A., Hruby, V.J., *JOC*, **66**, 1339; O'Donnell. M.J. et al., *TA*, **12**, 821.

0-98%
d.r = 2.4-26:1

I.A.7.b.1.-4 Passarella, D. et al., *SL*, 132; Ishikawa, T., Saito, S. et al., *JOC*, **66**, 8000.

52%

I.A.7.b.1.-5 Evans, D.A. et al., *JACS*, **123**, 4480; **see also:** Harada, T. et al., *OL*, **3**, et al., 2101; Zhang, F.-Y., Corey, E.J., *OL*, **3**, 639; Kumaraswamy, G. et al., *TL*, **42**, 8515.

"Enantioselective and Diastereoselective Mukaiyama-Michael Reactions Catalyzed by Bis(oxazoline)Copper(II) Complexes."

I.A.7.b.1.-6 Ley, S.V. et al., *OL*, **3**, 3753 and *AG(E)*, **40**, 4763.

35-98%
d.r. = 1.4-65:1

I.A.7.b.1.-7 Suzuki, T., Torii, T., *TA*, **12**, 1077; Nakajima, M. et al., *CC*, 1596.

30-99%
e.e. = 10-75%

v

I.A.7.b.1.-8 Jahn, U., *CC*, 1600; **see also:** Collin, J. et al., *TL*, **42**, 9157.

53-86%

I.A.7.b.1.-9 Pfau, M. et al., *TA*, **12**, 1683; Jabin, I. et al., *JOC*, **66**, 256.

neat, 110°C, 7d

38%

I.A.7.b.1.-10 Hagiwara, H. et al., *JCS(P1)*, 316.

20-96%

I.A.7.b.2. Organometalic and Related Reagents

I.A.7.b.2-1 Smith, A.B., III. et al., *OL*, **3**, 3971.

1. CH$_2$=CHMgBr, THF
2. MeI, HMPA, -10°C→rt
3. aq citric acid

59%

I.A.7.b.2-2 Hoveyda, A.H. et al., *JACS*, **123**, 755; Alexakis, A. et al., *JACS*, **123**, 4358; Waldmann, H. et al., ECJ, **7**, 671; Kihanan, T.C., Tye, H., *TA*, **12**, 1255.

+ R$_2$Zn $\xrightarrow{\text{(CuOTf)}_2\text{·C}_6\text{H}_6\text{, cat.}}$

71-98%
e.e. = 72-98%

I.A.7.b.2-3 Piers, E. et al., *OL*, **3**, 3245.

'BuLi, TMS-Cl
HMPA, -78°C→rt

72-96%

I.A.7.b.2-4 Song, Z.J. et al., *OL*, **3**, 3357; Alexakis, A. et al., *SL*, 1375; Alexakis, A., Benhaim, C., *TA*, **12**, 1151.

Ar-Br, BuLi, -50°C

65-81%
e.e. = 57-92%

I.A.7.b.2-5 Kilburn, J.D. et al., *TL*, **42**, 347; Murakami, M., Miyamoto, Y., Ito, Y., *JACS*, **123**, 6441; Plumet, J. et al., *JOC*, **66**, 9026; Ichikawa, J. et al., *OL*, **3**, 2345; Dieter, R.K. et al., *JOC*, **66**, 2302.

59-95%

I.A.7.b.2-6 Kim, Y.H. et al., *SL*, 627; Merlic, C.A., Walsh, J.C., *JOC*, **66**, 2265; **see also:** Shibasaki, M. et al., *OL*, **3**, 4251.

62-74%

I.A.7.b.2-7 Sikorski, W.H., Reich, H.J., *JACS*, **123**, 6527.

"The Regioselectivity of Addition of Organolithium Reagents to Enones and Enals: The Role of HMPA."

I.A.7.b.2-8 Kabir, S.M.H., Rahman, M.T., *JOM*, **619**, 31.

"Preparations and Selectivities of Magnesium-Based Mixed Cuprates."

I.A.7.b.2-9 Sibi, M.P., Chen, J., *JACS*, **123**, 9472; Arai, Y. et al., *SL*, 529 and *CPB*, **49**, 1609; Dechoux, L. et al., *T*, **57**, 195.

79-93%

I.A.7.b.2-10 Han, G., Hruby, V.J., *TL*, **42**, 4281; Taber, D.F., *JOC*, **66**, 5911; William, A.D., Kobayashi, Y., *OL*, **3**, 2017; Tadano, K. et al., *JOC*, **66**, 5965.

80%

I.A.7.b.2-11 Lopes, C.C. et al., *S*, 845.

$$\text{Ar-Li} + \underset{\text{NO}_2}{\text{(cyclohexenyl)}} \xrightarrow{\text{THF, -78}\rightarrow\text{0°C}} \underset{\text{Ar} \quad \text{NO}_2}{\text{(cyclohexyl)}}$$

32-99%

I.A.7.b.2-12 Maddaford, S.P. et al., *OL*, **3**, 2571; Hayashi, T. et al., *JOC*, **66**, 6852; Reetz, M.T. et al., *OL*, **3**, 4083; Knochel, P. et al., *TL*, **42**, 8829; **see also:** Wang, D., Li, C.-J. et al., *SL*, 1470; Li, C.-J. et al., *TL*, **42**, 4459.

$$\text{Ar-B(OH)}_2 + \underset{\text{OAc, OAc}}{\text{(pyranone)}} \xrightarrow[\text{dioxane/H}_2\text{O, 100°C}]{\text{Rh(I)(cod)}_2\text{BF}_4} \underset{\text{Ar, OAc, OAc}}{\text{(pyranone)}}$$

50-81%

I.A.7.b.2-13 Hailes, H.C. et al., *TL*, **42**, 7325; Lee, P.H. et al., *OL*, **3**, 3205 and *JOC*, **66**, 8646; Yamamoto, Y. et al., *JACS*, **123**, 372.

$$\underset{\text{Me}}{\overset{\text{Ph}}{\text{(oxazaphospholidine-allyl)}}} + \underset{\text{R}}{\text{(cyclopentenone)}} \xrightarrow[\text{THF, -78°C}]{\text{BuLi}} \underset{\text{Me}}{\overset{\text{Ph}}{\text{(product)}}}$$

76-86%
d.r. = 11.5-19:1

I.A.7.b.2-14 Sato, F. et al., *OL*, **3**, 3543; Chang, S. et al., *OL*, **3**, 2089.

$$\underset{(\text{EtO})_2\text{P(O)}}{\overset{\text{TMS}}{\text{(alkyne)}}}\overset{R}{\underset{H}{}} + \underset{\text{CO}_2\text{Et}}{\overset{R^1}{\text{(alkene)}}}\text{CO}_2\text{Et} \xrightarrow{\text{Ti(O}^i\text{Pr)}_4 \ ^i\text{PrMgCl}} \underset{R \quad \text{CO}_2\text{Et}}{\overset{\text{TMS} \quad R^1}{\text{(product)}}}\text{CO}_2\text{Et}$$

82-97%
anti:syn = 32.3-100:1
e.r. = 92%

I.A.7.b.2-15 Hanzawa, Y., Taguchi, T. et al., *TL*, **42**, 1737; **see also:** Willis, M.C., Sapmaz, S., *CC*, 2558; Nishiguchi, I. et al., *OL*, **3**, 3439.

$$\underset{R}{\overset{O}{\text{ZrClCp}_2}} + \underset{O}{\overset{R^1}{\text{(alkyne)}}}R^2 \xrightarrow[\text{THF/Et}_2\text{O, rt}]{\overset{\text{PdCl}_2(\text{PPh}_3)_2 \text{ or}}{\text{Pd(PPh}_3)_4}} \underset{R^1 \quad O}{\overset{O \quad R^2}{\text{(product)}}}$$

40-95%

I.A.7.b.2-16 Arjona, O., Menchaca, R., Plumet, J., *OL*, **3**, 107.

I.A.7.b.3. Other Conjugate Additions

I.A.7.b.3-1 Sibi, M.P. et al., *AG(E)*, **40**, 1293 and *OL*, **3**, 3679; Jang, D.O. et al., *SL*, 1923.

54-76%
e.e. = 27-85%

I.A.7.b.3-2 Enholm, E.J. et al., *OL*, **3**, 145; Vidari, G. et al., *TA*, **12**, 1785.

85% 100:1

I.A.7.b.3-3 List, B. et al., *OL*, **3**, 2433; Betancort, J.M., Barbas, C.F., III, *OL*, **3**, 3737; List, B., Castello, C., *SL*, 1687; Christoffers, J., Mann, A., *ECJ*, **7**, 1014.

92-94%
d.r. = 20:1
e.e. = 10-23%

I.A.7.b.3-4 Kim, K., Jimenez, L.S., *TA*, **12**, 999.

1. NaH, THF
2. NaN$_3$, H$_2$O

35%
e.e. = 43%

I.A.7.b.3-5 Tius, M.A. et al., *TL*, **42**, 2419.

46-92%

I.A.8. Other Carbon-Carbon Single Bond Forming Reactions

I.A.8-1 Miyashita, M. et al., *JOC*, **66**, 5388; Mordini, A. et al., *JOC*, **66**, 3201.

R$_3$Al
CH$_2$Cl$_2$, -30°C

72-97%

I.A.8-2 Clive, D.L.J. Cheng, H., *CC*, 605; Clive, D.L.J. et al., *JOC*, **66**, 1966; Takasu, K., Ihara, M. et al., *TL*, **42**, 2157; Ikeda, M. et al., H, 54, 747; Gonzalez-Sierra, M. et al., *TL*, **42**, 1811; Bennasar, M.-L. et al., *JOC*, **66**, 7547; Marco-Conteues, J., Rodriquez-Fernandez, M., *JOC*, **66**, 3717; Yang, D. et al., **123**, 8612.

Bu$_3$SnH, AIBN
Ph-H, reflux

51-85%

I.A.8-3 Barrero, A.F. et al., *JOC*, **66**, 4074; Taylor, E.C., Liu, B., *JOC*, **66**, 3726.

TiCl$_2$Cp$_2$, Mn
THF, rt

44%

I.A.8-4 Yao, C.-F., *JOC*, **66**, 6021; Kim, S. et al., *AG(E)*, 2524.

57-99%

I.A.8-5 Correia, R., Deshong, P., *JOC*, **66**, 7159.

$$Ar\text{-}Si(OEt)_3 \xrightarrow[\text{2. R-OBz, 50-60°C}]{\text{1. TBAF, THF}} Ar\text{-}R$$

34-88%

I.A.8-6 Breit, B., Zahn, S.K., *AG(E)*, **40**, 1910.

41-71%
syn:anti = 24-99:1

I.A.8-7 Marcaccini, S. et al., *JCR(S)*, 465.

44-61%

I.A.8-8 Oshima, K. et al., *SL*, 293; Sha, C.-K. et al., *CC*, 39.

34-99%
E:Z = 1:1-99

I.A.8-9 Helquist, P. et al., *JOC*, **66**, 3449; Hashimoto, S. et al., *CC*, 1604.

31-96%

I.A.8-10 Ishii, Y. et al., *JOC*, **66**, 6425.

$$R\text{-}H + R^1\diagup\diagdown EWG \xrightarrow[\text{MeCN, 75°C}]{\text{NHPI, Co(acac)}_2} R\diagup\overset{OH}{\diagdown}EWG + R\diagup\overset{O}{\diagdown}EWG$$

42-98% 1-9:1

I.B. Carbon-Carbon Double Bonds

(see also: I.E.1)

I.B.1. Wittig-Type Olefination Reactions

I.B.1-1 Ando, K., *SL*, 1272.

$$R\text{-}CHO + \underset{O}{\overset{O}{\diagdown}}\overset{NHPh}{\underset{P(OPh)_2}{}} \xrightarrow[\text{THF, -78°C}]{{}^t\text{BuOK}} R\diagup\diagdown\overset{O}{\diagdown}NHPh$$

78-98%
Z:E = 5.7-24:1

I.B.1-2 Petroski, R.J., Weisleder, D., *SC*, **31**, 89; Verkade, J. et al., *T*, **57**, 8047.

$$R\text{-}CHO + (EtO)_2\overset{O}{\underset{R^1}{\overset{\|}{P}}}CO_2Et \xrightarrow[\text{hex, rt}]{{}^t\text{BuOLi}} R\diagup\overset{CO_2Et}{\underset{R^1}{\diagdown}}$$

86-98%
E:Z = 1.7-140:1

I.B.1-3 Qing, F.-L., Zhang, X., *TL*, **42**, 5929.

$$(CF_3CH_2O)_2\overset{O}{\overset{\|}{P}}CO_2Et \xrightarrow[\substack{\text{3. NaH, -78°C}\\\text{4. R-CHO, -78 to -5°C}}]{\substack{\text{1. NaH, THF, -30°C}\\\text{2. Br}_2\text{, -30 to -15C}}} R\diagup\overset{CO_2Et}{\underset{Br}{\diagdown}}$$

44-87%
E:Z = 2.4-100:1

I.B.1-4 Dai, W.-M., Lau, C.W., *TL*, **42**, 2541.

$$\xrightarrow[\text{THF/Me}_2\text{CO, -60°C}]{Ph_3As=CHCO_2R^*}$$

93%
d.r. = 7.3:1

I.B.1-5 Harcken, C., Martin, S.F., *OL*, **3**, 3591; Rein, T., Helquist, P., Norrby, P.-O. et al., *JACS*, **123**, 9738; Hillier, M.C., Meyers, A.I., *TL*, **42**, 5145; Comins, D.L., Ollinger, C.G., *TL*, **42**, 4115; Motoyoshiya, J. et al., *T*, **57**, 1715; Madsen, R. et al., *JOC*, **66**, 4625; Costantino, V. et al., *TL*, **42**, 8185; Markidis, T., Kokotos, G., *JOC*, **42**, 1919; Westman, J., *OL*, **3**, 3745.

64-95%
E:Z = 3-19:1

I.B.1-6 Chapleur, Y. et al., *TL*, **42**, 7265; Martin, V.S. et al., *JOC*, **42**, 7231.

40-87%

I.B.1-7 Iyengar, D.S. et al., *TL*, **42**, 531; Tsunoda, T. et al., *TL*, **41**, 235.

92-97%

I.B.1-8 Fukuyama, T. et al., *SL*, 1403; Kung, H.F. et al., *JMC*, **44**, 2270; Verkade, J.G. et al., *JOC*, **66**, 3521.

78-93%

I.B.1-9 Charette, A.B. et al., *TL*, **42**, 5149; Balas, L. et al., *TL*, **42**, 3709.

44-75%
E,Z:E,E = 7.3-23:1

I.B.1-9 Shen, Y. et al., S, 389 and *JCS(P1)*, 519.

I.B.1-10 Warren, S. et al., *JCS(P1)*, 118.

I.B.1-11 Doxsee, K.M. et al., *TL*, **42**, 1411; Lebel, H. et al., *AG(E)*, **40**, 2887; Brookhart, M. et al., *JACS*, **123**, 2442.

I.B.1-12 Langer, P., Kracke, B., *SL*, 1790.

I.B.2. Eliminations

I.B.2.a. Eliminations of Alcohols and Derivatives

I.B.2.a-1 Chung, K.-H. et al., *JOC*, **66**, 5937; Lees, W.J. et al., *JOC*, **66**, 1914.

I.B.2.a-2 Yamamoto, H. et al., *SL*, 1690; Concellon, J.M. et al., *ECJ*, 7, 3062.

42-99%

I.B.2.a-3 Moltrasio, G., *JCR(S)*, 508.

80-99%

I.B.2.b. Eliminations of Halides

I.B.2.b-1 Ando, R. et al., *JOC*, **66**, 3617.

70%

I.B.2.c. Other Eliminations

I.B.2.c-1 Farcas, S., Namy, J.-L., *TL*, **42**, 879.

55-95%

I.B.2.c-2 Tokuda, M. et al., *TL*, **42**, 3893.

73-99%
Z:E = 3-100:1

I.B.2.c-3 Furstner, A. et al., *OL*, **3**, 3955.

56-79%

I.B.2.c-4 Yadav, J.S. et al., *TL*, **42**, 6385 and *SL*, 1608; Kim, K. et al., *JOC*, **66**, 2149; Chao, B., Dittmer, D.C., *TL*, **42**, 5789.

$$R^1 \underset{R}{\overset{O}{\diagdown}}N \diagdown O \diagdown I \xrightarrow[\text{MeOH, reflux}]{\text{In}} R^1 \diagdown \underset{R}{\overset{NH}{\diagdown}} \diagup$$

81-96%

I.B.2.c-5 Carretero, J.C. et al., *OL*, **3**, 2957.

$$\xrightarrow[\text{2. Na/Hg, Na}_2\text{HPO}_4\text{, MeOH}]{\text{1. MeOTf, CH}_2\text{Cl}_2\text{, rt}}$$

64%

I.B.2.c-6 Fleming, F.F. et al., *JOC*, **66**, 2171.

$$R^2 \underset{R^1 \quad R}{\overset{O \quad R^3}{\diagup}} CN \xrightarrow[\text{THF, -78°C}]{\text{LDA}} R^2 \underset{R^1 \quad R}{\overset{HO \quad R^3}{\diagup}} CN$$

60-94%

I.B.2.c-7 Hodgson, D.M. et al., *OL*, **3**, 3401.

$$O \underset{n}{\diagup} O \xrightarrow[\text{THF, -78C}]{\text{R-Li}} HO \underset{n}{\diagup} \overset{R}{\diagup} OH$$

54-93%

I.B.3. Olefin Metathesis

I.B.3-1 Louie, J., Grubbs, R.H., *AG(E)*, **40**, 247; **see also:** de Clerc, B., Verpoort, F., *TL*, **42**, 8959; Hoveyda, A.H., Schrock, R.R. et al., *AG(E)*, **40**, 1452; Furstner, A., Leitner, W. et al., *JACS*, **123**, 9000.
   "Highly Active Metathesis Catalysts Generation in situ from Inexpensive and Air-Stable Precursors."

I.B.3-2 Grubbs, R.H. et al., *JACS*, **123**, 6543; **see also:** Hoveyda, A.H., Schrock, R.R., *ECJ*, **7**, 945; Grubs, R.H. et al., *OM*, **20**, 5314; Furstner, A. et al., *ECJ*, **7**, 4811.
   "Mechanism and Activity of Ruthenium Olefin Metathesis Catalysis."

I.B.3-3 Georg, G.I. et al., *OL*, **3**, 1411.
"A Convenient Method for the Efficient Removal of Ruthenium Byproducts Generated during Olefin Metathesis Reactions."

I.B.3-4 Madsen, R. et al., *JOC*, **66**, 4630; Hanna, I., *JOC*, **66**, 4094; Callum, C.S., Lowary, T.L., *JOC*, **66**, 8961; Wood, J.L. et al., *OL*, **3**, 1563; Meyers, A.I. et al., *JOC*, **66**, 5545; Jeong, L.S. et al., *JOC*, **66**, 6490; Nevalainen, M., Koskinen, A.M.P., *AG(E)*, **40**, 4060; Lin, C.-C. et al., *TL*, **42**, 4079; Jacobsen, K.A. et al., *OL*, **3**, 597; Paquette, L.A. et al., *JOC*, **66**, 6695; Negishi, E. et al., *EJOC*, 3039; Praly, J.-P. et al., *EJOC*, 2939; Mori, K. et al., *EJOC*, 4395; de Armas, P., Garcia-Tellado, F., Marrero-Tellado, J.J., *EJOC*, 4423; Burke, T.R., Jr. et al., *OL*, **3**, 1617; O'Leary, D.J., Grubbs, R.H. et al., *JOC*, **66**, 5291; Furstner, A. et al., *ECJ*, **7**, 3236; Mol, J.C. et al., *ECJ*, **7**, 2842; **see also:** Shibassaki, M. et al., *TL*, **42**, 8023; Agrofoglio, L.A. et al., *TL*, **42**, 8817; Young, D.G.J. et al., *TL*, **42**, 5363.

72-96%

I.B.3-5 Yao, Q., *OL*, **3**, 2069; Harrison, B.A., Verdine, G.L., *OL*, **3**, 2157; Eustache, J. et al., *TL*, **42**, 239; Denmark, S.E., Yang, S.-M., *OL*, **3**, 1749; **see also:** Hanson, P.R. et al., *OL*, **3**, 3939.

82%

I.B.3-6 Grubbs, R.H. et al., *AG(E)*, **40**, 1277 and *SL*, 1034 and *JACS*, **123**, 10417; Taylor, R.E. et al., *OL*, **3**, 2209; Blechert, S. et al., CC, 1692, 1796; Bouz Bouz, S., Cossy, J., *OL*, **3**, 1451; Smith, A.B., III, et al., **123**, 990; Sakamoto, Y., Nakata, T. et al., *TL*, **42**, 7633.

77-99%

I.B.3-7 Nicolaou, K.C. et al., *AG(E)*, **40**, 4441; Liu, L., Postema, M.H.D., *JACS*, **123**, 8602.

80%

I.B.3-8 Hanna, I. et al., *OL*, **3**, 3095; Granja, J.R. et al., *OL*, **3**, 1483; Hoveyda, A.H. et al., *JACS*, **123**, 7767; Undheim, K. et al., *JCS(P1)*, 2697; Blechert, S. et al., *TL*, **42**, 5245.

79-96%

I.B.3-9 Kozmin, S.A. et al., *AG(E)*, **40**, 4274; Clark, S.J. et al., *TL*, **42**, 3235; Dixneuf, P.H. et al., *SL*, 397; Mori, M. et al., *S*, 654; Smulik, J.A., Diver, S.T., *TL*, **42**, 171.

50-95%

I.B.4. Other Carbon-Carbon Double Bond Forming Reactions

I.B.4-1 Kozikowski, A.P. et al., *TL*, **42**, 5359; Hopf, H., Kruger, A., *ECJ*, **7**, 4378; **see also:** Chattopadhyay, S. et al., *JOC*, **66**, 2990.

62%

I.B.4-2 Shindo, M. et al., *TL*, **42**, 8357; Jong, S.-J., Fang, J.-M., *JOC*, **66**, 3533; **see also:** Tang, Y. et al., *CC*, 1384.

$$\underset{Ph}{\overset{Ph}{>}}\!\!=\!\!O \;+\; \underset{Br\ \ Br}{\overset{R\quad CO_2Et}{\diagdown\diagup}} \xrightarrow[-78°C,\ 3h]{Li,\ C_{10}H_8} \underset{Ph}{\overset{Ph\quad R}{\diagdown\diagup}}\!\!=\!\!\underset{CO_2H}{}$$

72-99%

I.B.4-3 Kabalka, G.W. et al., *TL*, **42**, 4759; Wicha, J. et al., *JOC*, **66**, 6994; Nenajdenko, V.G. et al., *S*, 2081.

$$X\text{—}\!\!\diagup\!\!\diagdown\text{—}CH\!\!=\!\!NNHTs \xrightarrow[\text{THF, reflux}]{^tBuOK,\ B(OMe)_3} X\text{—}\!\!\diagup\!\!\diagdown\text{—}CH\!\!=\!\!CH\text{—}\!\!\diagup\!\!\diagdown\text{—}X$$

79-88%

I.B.4-4 Liu, J.-T., Yao, C.-F., *TL*, **42**, 6147.

$$Ar\text{-}CHO + MeNO_2 \xrightarrow[\text{2. Et}_3B,\ Et_2O/H_2O]{\text{1. AcOH, NH}_4OAc,\ 70\text{-}100°C} \underset{Et}{\overset{Ar}{\diagdown\diagup}}$$

51-85%

I.B.4-5 Dujardin, G. et al., *SL*, 147.

$$R^1\text{-}CHO + \underset{O}{\overset{Me}{RO_2C\!\!-\!\!C}} \xrightarrow[\text{CH}_2Cl_2,\ \text{reflux}]{Cu(OTf)_3,\ HC(OMe)_3} \underset{RO_2C\quad O}{\overset{R^1}{\diagdown\diagup}}$$

26-91%

I.B.4-6 Alcaraz, L. et al., *OL*, **3**, 4051.

$$\underset{R}{\overset{R^1}{>}}\!\!\diamondsuit\!\!O + Me_2S^+\!\!=\!\!CH_2 \xrightarrow[\text{THF, -10°C→rt}]{} \underset{R\quad OH}{\overset{R^1}{\diagdown\diagup}}$$

68-93%

I.B.4-7 Reissig, H.-U. et al., *SL*, 1293; **see also:** Henderson, K.W., Kerr, W.J., Moir, J.H., *SL*, 1253; Frauenrath, H. et al., *AG(E)*, **40**, 177.

$$^tBu\text{—}\!\!\bigcirc\!\!=\!\!O \xrightarrow[\substack{\text{2. Nf-F, Bu}_4NF,\ KF,\ THF \\ \text{3. CH}_2=CHCO_2Me,\ Pd(OAc)_2,\ LiCl}]{\substack{\text{Li·LiCl} \\ Me_{\diagdown}N_{\diagdown}Me \\ \text{1.}\quad Ph\ \ Ph\quad,\ THF,\ -105°C}} {}^tBu\text{—}\!\!\bigcirc\!\!=\!\!CH\text{—}CO_2Me$$

82%
e.e. = 89%

I.B.4-8 Matsubara, S. et al., *SL*, 513.

Me—C(OBn)=... $CH_2(ZnI)_2$ / THF, rt → product, 73%

I.B.4-9 Nair, V. et al., *OL*, **3**, 3495.

R—C$_6$H$_4$—CHO + alkyne(CO_2Me)(CO_2Me) → pyr / DME, -10°C→rt → product, 43-85%

I.B.4-10 Franck, R.W. et al., *OL*, **3**, 197.

$CBrF_2CBrF_2$ / KOH/Al_2O_3, ^tBuOH → 70%

I.B.4-11 Horvath, A., Backvall, J.-E., *JOC*, **66**, 8120.

R—C(AcO)—R^1 $Pd(OAc)_2$, LiBr / AcOH, 40°C → product, 77-94%

I.B.4-12 Trost, B.M. et al., *JACS*, **123**, 12466, 12504, 2897.

R—allene + methyl vinyl ketone $CeCl_3\cdot7H_2O$, DMF, 60°C → product, 55-74%

Ru catalyst, Et—$\equiv$—OH

I.B.4-13 Tanaka, K., Fu, G.C., *JACS*, **123**, 11492.

R^2—alkyne...CHO $Rh(dppe)_2(BF_4)_2$ / $MeCN/Me_2CO$, 100°C → product, 67-88%

I.B.4-14 Ren, H.-J., Wang, Y.-G., *SC*, **31**, 1204.

I.B.4-15 Oh, C.H. et al., *TL*, **42**, 8669; Widenhofer, R.A. et al., *OL*, 3, 385; Kang, S.-K. et al., *CC*, 1306.

I.B.4-16 Zhang, R. et al., *JCS(P1)*, 2958.

I.B.4-17 Hayashi, T. et al., *JACS*, **123**, 9918.

I.B.4-18 Koketsu, M., Ishihara, H. et al., *JOC*, **66**, 4099.

I.B.5. Vinylations

I.B.5-1 Mootoo, D.R. et al., *OL*, 3, 1323.

I.B.5-2 Yao, C.F. et al., *TL*, **42**, 3613.

$$Ar\diagdown\diagup NO_2 + R\text{-}I \xrightarrow[\text{Et}_2\text{O, rt}]{\text{AlEt}_3, (\text{PhCO}_2)_2} Ar\diagdown\diagup R$$

60-99%

I.B.5-3 Lee, K., Cha, J.K., *JACS*, **123**, 5590.

$$\xrightarrow{\text{Pd(OAc)}_2, \text{HCO}_2\text{K, Bu}_4\text{NBr}}$$

79%

I.B.5-4 Feuerstein, M., Doucet, H., Santelli, M., *SL*, 1980; Velasco, D. et al., *OL*, **3**, 541; **see also:** Beletskaya, I.P. et al., *JOM*, **622**, 89; Iyer, S., Jayanthi, A., *TL*, **42**, 7877; Livingston, A.G. et al., *TL*, **42**, 8219; Gron, L.U. et al., *TL*, **42**, 8555.

$$\xrightarrow[\text{DMF, 130°C}]{[\text{PdCl}(C_3H_5)]_2, K_2CO_3}$$

7-99%

I.B.5-5 Gilbertson, S.R., Fu, Z., *OL*, **3**, 161.

$$\xrightarrow{\text{Pd}_2(\text{dba})_3, \,^{i}\text{Pr}_2\text{NEt, 70°C}}$$

78-100% conversion
e.e. = 40-94%

I.B.5-6 Percy, J.M. et al., *OL*, **3**, 2859; Buynak, J.D. et al., *OL*, **3**, 2953.

$$\xrightarrow[\substack{\text{2. PdCl}_2(\text{PPh}_3)_2 \\ \text{TEA, 90°C} \\ \equiv\!-\!C_8H_{17}}]{\substack{\text{1. Pd(OAc)}_2, \text{PPh}_3 \\ \text{CuI, 50°C}}}$$

39%

I.B.5-7 Danion, D. et al., *SC*, **31**, 249; Jiang, B. et al., *TL*, **42**, 4083; see also: Falck, J.R. et al., *TL*, **42**, 7213.

$$(HO)_2B \diagup \diagdown \diagup \overset{CO_2Me}{\underset{NHBoc}{|}} + R\text{-}Br \xrightarrow[\text{DME, 80°C}]{Pd(PPh_3)_4, CsF} R \diagup \diagdown \diagup \overset{CO_2Me}{\underset{NHBoc}{|}}$$

37-64%

I.B.5-8 Ide, M., Nakata, M., *SL*, 1511; Cahiez, G., Knochel, P. et al., *SL*, 1901.

$$R^2\text{-MgX} + \underset{R^1 \quad R}{\overset{TfO \quad CO_2Me}{\diagup\diagdown}} \xrightarrow[\text{DMPU/THF, -30°C→rt}]{ZnCl_2, CuCl_2, LiCl} \underset{R^1 \quad R}{\overset{R^2 \quad CO_2Me}{\diagup\diagdown}}$$

83-98%

I.B.5-9 Hanamoto, T. et al., *SL*, 281; Mori, A., Suguro, M., *SL*, 845.

$$R \text{—} \bigcirc \text{—} I + \overset{SiPh_2Me}{\underset{F}{=}} \xrightarrow[\text{DMI, rt}]{PdPPh_3)_4, CuI, CsF} R \text{—} \bigcirc \text{—} \overset{}{\underset{F}{=}}$$

49-91%

I.B.5-10 Panek, J.S. et al., *OL*, **3**, 3281; Miller, M.W. et al., *SL*, 254; Uenishi, J., Matsui, K., *TL*, **42**, 4353.

53-54%

I.B.5-11 Marshall, J.A. et al., *OL*, **3**, 4107.

73-97%

I.B.5-123 Piers, E., Coish, P.D.G., *S*, 251.

57%

I.B.5-13 Mori, A., Kosugi, M. et al., *OL*, **3**, 3313; **see also:** Lautens, M. et al., *JACS*, **123**, 5358.

$$Ph\text{-}B(OH)_2 + \overset{}{\diagup}\!\!\!\!\diagdown R \xrightarrow[\text{DMF, 100°C}]{\text{Pd(OAc)}_2,\ \text{Cu(OAc)}_2,\ \text{LiOAc}} Ph\diagdown\!\!\!\!\diagup\!\!\!\!\diagdown R$$

58-84%

I.B.6. Allene Forming Reactions

I.B.6-1 Shibuya, S. et al., *SL*, 287.

$$XO\overset{R^1}{\underset{R}{\diagdown}}\!\!\!\equiv \xrightarrow{\text{BrZnCF}_2\text{PO}_3\text{Et, CuBr}} \overset{R^1}{\underset{R}{\diagdown}}\!\!\!=\!\!\!=\!\!\!\overset{}{CF_2PO_3Et}$$

72-98%

I.B.6-2 Kamigata, N. et al., *JOC*, **66**, 1787; **see also:** Reissig, H.-U. et al., *S*, 1377.

$$\overset{Ph}{\diagdown}\!\!\!\equiv\!\!\!-Ph \xrightarrow[\text{2. R-SeX, TMEDA, Ph-H}]{\text{1. BuLi, hex, rt}} \overset{Ph}{\underset{RSe}{\diagdown}}\!\!\!=\!\!\!=\!\!\!\overset{Ph}{\underset{SeR}{\diagup}}$$

64-70%

I.B.6-3 Tius, M.A., Pal, S.K., *TL*, **42**, 2605.

$$\overset{R}{\underset{HO}{\diagup}}\!\!\!\overset{SiEt_3}{\diagdown}\!\!\!{}_{R^1} \xrightarrow[\text{CCl}_4/\text{pyr, 0°C}]{\text{SOCl}_2} \overset{R}{\diagdown}\!\!\!=\!\!\!=\!\!\!\overset{}{R^1}$$

55-90%

I.B.6-4 Sato, F. et al., *JACS*, **123**, 7937.

$$Bu\!-\!\!\!\equiv\!\!\!-Bu \xrightarrow[\substack{\text{2. }\equiv\!-\text{CH}_2\text{OCO}_2\text{Et} \\ \text{3. H}^+ \text{ or D}^+\text{, -50°C}}]{\text{1. (}\Pi\text{-C}_3\text{H}_5\text{)Ti(O}^i\text{Pr)}_2}$$

(other examples given)

Bu, Bu, H (D)

65%

I.B.6-5 Sato, F. et al., *TL*, **42**, 4147.

$$\xrightarrow[\text{Et}_2\text{O, -50}\rightarrow\text{-20°C}]{{}^i\text{Pr-MgCl, Ti(O}^i\text{Pr)}_4}$$

25-45%

I.B.6-63 Langer, P. et al., *ECJ*, **7**, 573.

I.B.6-7 Esteruelas, M.A. et al., *OM*, **20**, 3202.

I.B.6-8 Prabharasuth, R., Van Vranken, D.L., *JOC*, **66**, 5256.

I.C. Carbon-Carbon Triple Bonds

I.C-1 Zeni, G. et al., *TL*, **42**, 8927, 7921; Braga, A.L., Zeni, G. et al., *TL*, **42**, 8563.

I.C-2 Anastasia, L., Negishi, E., *OL*, **3**, 3111; Mori, A. et al., *SL*, 649; **see also:**, Dai, W.-M. et al., *TL*, **42**, 81, 5275; Jones, G.B. et al., *JOC*, **42**, 3688; Tobe, Y. et al., *OL*, **3**, 2419.

I.C-3 Savarin, C., Srogi, J., Liebeskind, L.S., *OL*, **3**, 91.

I.C-4 Kang, S.-K. et al., *OL*, **3**, 2697; Radhakrishnan, U., Stang, P.J., *OL*, **3**, 859.

I.C-5 Anand, N.K., Carreira, E.M., *JACS*, **123**, 9687.

R——— + R¹-CHO $\xrightarrow[\substack{Ph\quad Me \\ HO\quad NMe_2}]{Zn(OTf)_2, TEA, Ph-Me, 60°C}$ R———$\overset{R^1}{\underset{OH}{<}}$

55-94%
e.e. = 86-99%

I.C-6 Pale, P. et al., *TL*, **42**, 8641 and *OL*, **3**, 1661.

R^2R^1 C=C R^3 OTf + R_3Si———R^4 $\xrightarrow[H_2O, DMF, rt]{Pd(PPh_3)_4, AgI, Bu_4NF}$ R^2 R^3 C=C with R¹ and ———R⁴

66-99%

I.C-7 Buck, M., Chong, J.M., *TL*, **42**, 5825.

R——— + R¹-X $\xrightarrow[THF, reflux]{BuLi}$ R———R¹

52-99%

I.C-8 Kabalka, G.W. et al., *SL*, 676, 108.

R——— + $(CH_2O)_n$ + R^1R^2NH $\xrightarrow[\substack{CuI, Al_2O_3}]{\mu W}$ R———$CH_2NR^1R^2$

40-90%

I.C-9 Gimeno, J. et al., *OM*, **20**, 3175.
"Efficient Routes to Terminal γ-Keto-Alkynes and Unsaturated Cyclic Carbene Complexes Based on Regio- and Diastereo-Selective Nucleophilic Addition of Enolates on Ru(II) Indenyl Allenylindenes."

I.C-10 Furstner, A., Mathes, C., *OL*, **3**, 221.

EtO_2C-C₆H₄-C≡C-TMS + Bu———Bu $\xrightarrow[\substack{Me \\ (\quad N-)_3Mo \\ ^tBu \\ Me}]{CH_2Cl_2/Ph-Me, 80°C}$ EtO_2C-C₆H₄-C≡C-Bu

65%

I.C-11 Tykwinski, R.R. et al., *TL*, **42**, 8575.

$\substack{Br \\ Br}$ C=C $\substack{Ar}$ ———TMS $\xrightarrow[hex, -78\rightarrow -40°C]{BuLi}$ Ar———===—TMS

28-93%

I.C-12 Krafft, M.E. et al., *TL*, **42**, 7733.

$$R\!\!-\!\!\equiv \quad \xrightarrow[\text{Co}_2(\text{CO})_8\text{phenanthroline}]{\text{CO, MeCN}} \quad R\!\!-\!\!\equiv\!\!-\!\!\equiv\!\!-\!\!R$$

53-91%

I.C-13 Bonin, M., Micouin, L. et al., *TL*, **42**, 3171.

66-92%
d.e. = 66-97%

I.C-10 Ishihara, T. et al., *JOC*, **42**, 3442.

$$CF_3CF_2CH_2NR_2 \quad \xrightarrow[\text{DMPU/THF, 0°C}]{\text{LDA}} \quad F_3C\!\!-\!\!\equiv\!\!-\!\!NR_2$$

0-99%

I.D. Cyclopropanations

I.D.1. Carbene or Carbenoid Additions to a Multiple Bond

I.D.1-1 Katsuki, T. et al., *TL*, **42**, 2521 and *SL*, 114; Lahuerta, P. et al., *OL*, **3**, 3317.

10-93%
e.e. = 38-98%

I.D.1-2 Yamada, T. et al., *BCJ*, **74**, 2139; Shi, M. et al., *JCR(S)*, 375; Scott, P. et al., CC, 1638; Mezzeti, A. et al., *OM*, **20**, 2102; Luis, S.V., Mayoral, J.A. et al., *JOC*, **66**, 8893; Scott, P. et al., *TA*, **12**, 1055; Chelucci, G. et al., *T*, **57**, 1099; **see also:** Davies, H.M.L., Townsend, R.I., *JOC*, **66**, 6595; **see also:** Barrett, A.G.M. et al., *JOC*, **66**, 8260; Zheng, Z. et al., *TA*, **12**, 197.

$$Ph\diagdown\!\!\diagup \quad + \quad N_2CHCO_2{}^tBu \quad \xrightarrow{\text{Co-Salen}} \quad Ph\diagdown\!\!\triangle\!\!CO_2{}^tBu$$

99%
trans:cis = 10:1
e.e. (trans) = 96%

I.D.1-3 Livant, P. et al., *JOC*, **66**, 4945; Maquire, A.R. et al., *JOC*, **66**, 7166.

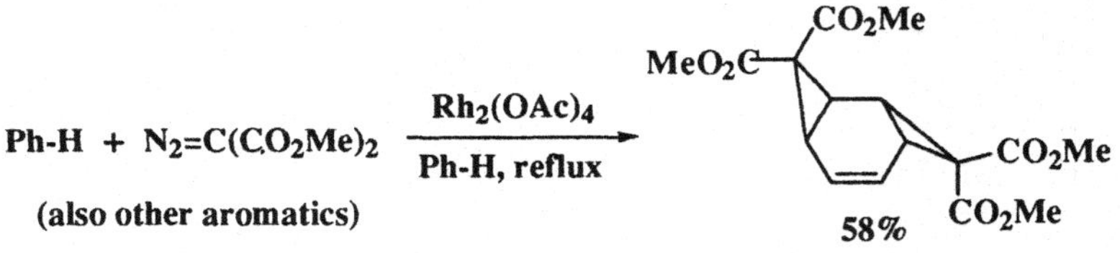

Ph-H + N2=C(CO2Me)2
(also other aromatics)
Rh2(OAc)4
Ph-H, reflux
CO2Me
MeO2C
CO2Me
CO2Me
58%

I.D.1-4 Aggararwal, V.K. et al., *OL*, **3**, 2785.

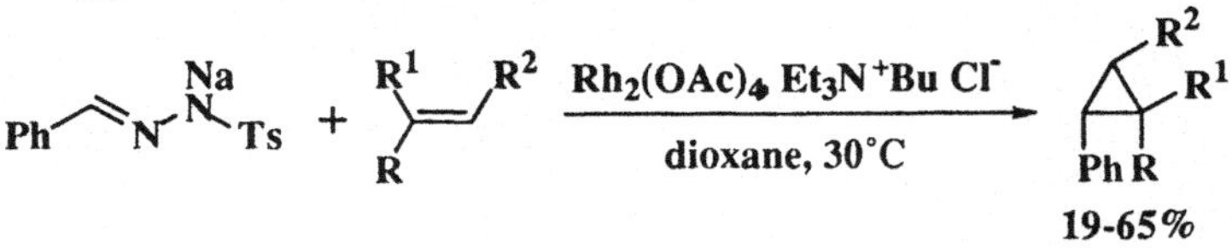

Na
Ph N N Ts
+
R1 R2
R
Rh2(OAc)4, Et3N+Bu Cl-
dioxane, 30°C
R2
R1
Ph R
19-65%

I.D.1-5 Mohapatra, D.K., *JCS(P1)*, 1851; Evans, D.A., Burch, J.D., *OL*, **3**, 503; **see also:** Charette, A.B. et al., *OL*, **3**, 3293 and *JACS*, **123**, 8139..

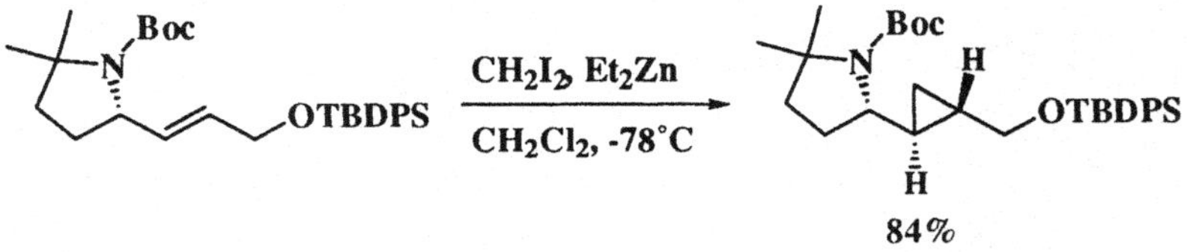

Boc
N
OTBDPS
CH2I2, Et2Zn
CH2Cl2, -78°C
Boc
N
H
OTBDPS
H
84%

I.D.1-6 Barluenga, J. et al., *JACS*, **123**, 10494 and *ECJ*, **7**, 4723; Mikolajczyk, M., Bertrand, C. et al., *JOC*, **66**, 8240.

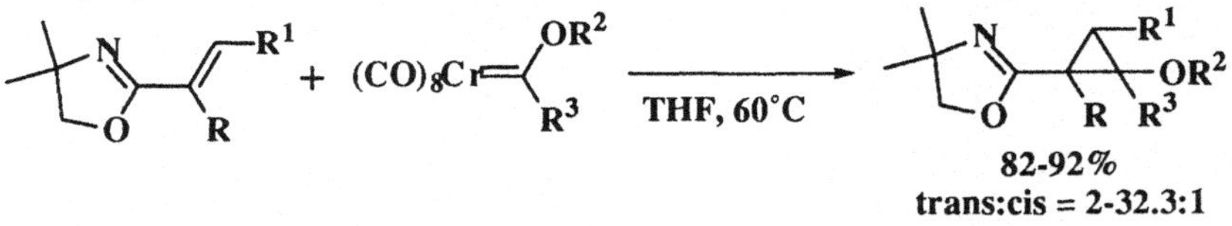

N
R1
O
R
+ (CO)5Cr
OR2
R3
THF, 60°C
N
R1
OR2
O
R R3
82-92%
trans:cis = 2-32.3:1

I.D.1-7 Cohen, T. et al., *OL*, **3**, 2121.

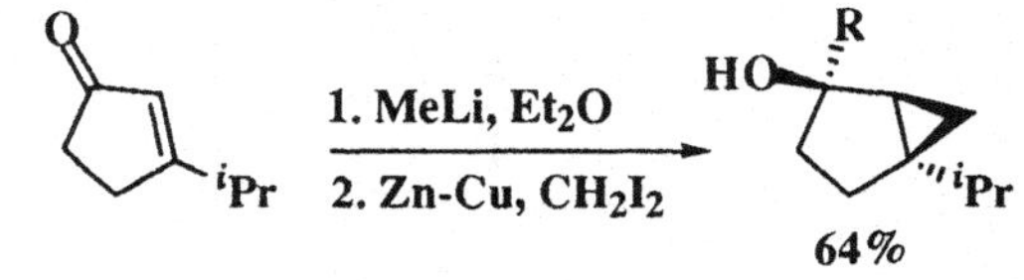

O
iPr
1. MeLi, Et2O
2. Zn-Cu, CH2I2
R
HO
iPr
64%

I.D.1-8 Dzhemilev, U.M. et al., *JOM*, **636**, 91.

$$R \underequal R^1 \xrightarrow[\text{hex, 20-25°C}]{\text{AlEt}_3,\ \text{CH}_2\text{I}_2} \begin{array}{c} R \quad R^1 \\ \triangle \\ Et \quad Et \end{array}$$

43-58%

I.D.2. Other Cyclopropanations

I.D.2-1 Tang, Y. et al., *JOC*, **66**, 5717.

$$\begin{array}{c} ^iBu_2Te \\ TMS \end{array} + \begin{array}{c} R \\ CO_2R^* \end{array} \xrightarrow[\text{THF, -78°C}]{\text{LiTMP}} \begin{array}{c} CO_2R^* \quad CO_2R^* \\ TMS \quad R \quad R \quad TMS \end{array}$$

72-99% ≤11.5:1

I.D.2-2 Hell, Z. et al., *TA*, **12**, 1287.

$$\begin{array}{c} R \quad O \quad O \\ Me \overset{\cdots}{} N \quad O \\ H \\ R^2 \end{array} \xrightarrow[\text{Ph-Me, reflux}]{\text{I}_2,\ \text{K}_2\text{CO}_3,\ \text{TCMC}} \begin{array}{c} O \quad Me \\ R^2 \quad O \quad N \quad R \\ H \\ O \end{array}$$

32-67%
cis:trans = 1:1

I.D.2-3 Taguchi, T. et al., *T*, **57**, 7487.

$$\begin{array}{c} X \overset{I}{\underset{n}{\bigcirc}} CO_2Et \\ F \end{array} \xrightarrow{\text{LHMDS}} \begin{array}{c} F \\ X \overset{}{\underset{n}{\bigcirc}} CO_2Et \end{array}$$

58-79%

I.D.2-4 Wipf, P. et al., *JACS*, **123**, 5122.

$$R \equiv \xrightarrow[\substack{\text{2. ArCH=NP(O)Ph}_2 \\ \text{CH}_2\text{I}_2,\ \text{CH}_2\text{Cl}_2,\ \text{reflux}}]{\text{1. Cp}_2\text{ZrHCl},\ \text{Me}_2\text{Zn}} \begin{array}{c} R \quad \triangle \quad Ar \\ NHP(O)Ph_2 \end{array}$$

45-84%

I.D.2-5 Taylor, R.E. et al., *JACS*, **123**, 2964.

$$\begin{array}{c} Ph \quad OH \\ R \end{array} \xrightarrow[\substack{\text{1. CH}_2=\text{CHCH}_2\text{TMS, CH}_2\text{Cl}_2,\ \text{reflux} \\ \text{Schrock's cat.} \\ \text{2. SOCl}_2,\ \text{pyr-2,6-Me}_2,\ \text{rt}}]{} \begin{array}{c} Ph \\ \triangle \\ R \end{array}$$

47-77%

I.D.2-6 Langer, P., Freifeld, I., *OL*, **3**, 3903; **see also:** Zhu, X.-Q., Liu, Y.C. et al., *JOC*, **66**, 344.

34-81%
cis:trans = 5-8:1

I.D.2-7 Taddei, M. et al., *OL*, **3**, 3273; Salaun, J. et al., *SL*, **42**, 4991; Cha, J.K. et al., *TL*, **42**, 2059; **see also:** Bertus, P., Szymoniak, J., *CC*, 1792.

64%

I.D.2-8 Chen, B.-C., Ahmad, S. et al., *TL*, **42**, 6227.

72-88%

I.D.2-9 Takeda, T. et al., *TL*, **42**, 3369.

37-73%

I.D.2-10 Wessig, P., Muhling, O., *AG(E)*, **40**, 1064.

61-86%

I.D.2-11 Ramnauth, J., Lee-Ruff, E., *CJC*, **79**, 114.

72%

I.D.2-12 Dechoux, L., Doris, E. et al., *EJOC*, 4107.

70-88%

I.E. Thermal and Photochemical Reactions

I.E.1. Cycloadditions

I.E.1-1 Chiba, K. et al., *JACS*, **123**, 11314.

73%

I.E.1-2 Taguchi, T. et al., *TL*, **42**, 2165; see also: Sugita, Y. et al., *CPB*, **49**, 657.

65-88%

I.E.1-3 Bungard, C.J., Morris, J.C., *S*, 741.

55-62%

I.E.1-4 Jordan, R.W., Tarn, W., *OL*, **3**, 2367; Cheng, C.-H. et al., *JOC*, **66**, 8804.

80-95%   68-97%

I.E.1-5 Ishikawa, T., Saito, S. et al., *JOC*, **66**, 3834.
    "Simultaneous Discrimination of Diastereotopic Groups and Faces: The First Example in Intramolecular [3+2] and [2+2+1] Cycloaddition Reactions."

I.E.1-6 Ito, Y. et al., *OL*, **3**, 2411.
"A Novel [2+2] Photodimerization of N-[(E)-3,4-Methylene-dioxycinnamoyl]dopamine in the Solid State."

I.E.1-7 Davies, H.M.L. et al., *JACS*, **123**, 7461; see also: Hodgson, D.M. et al., *ECJ*, **7**, 4465.

49-86%
e.e. = 70-99%

I.E.1-8 Krische, M.J. et al., *JACS*, **123**, 6716.

48-72% 7:1-1:100

I.E.1-9 Gevorgyan, V., Tsuboya, N., Yamamoto, Y., *JOC*, **66**, 2743, 2835.

20-85%

I.E.1-10 Oshima, K. et al., *JACS*, **123**, 4629.

66-98%

I.E.1-11 Ellman, J.A. et al., *JACS*, **123**, 1539; Fukuzawa, S. et al., *SL*, 709.

50-96%
d.r. = 19-99:1
e.e. = 32-98%

I.E.1-12 Tadano, K. et al., *OL*, **3**, 3029; Yakelis, N.A., Roush, W.R., *OL*, 3, 957; Harmata, M. et al., *TL*, **42**, 149; Fallis, A.G. et al., *OL*, 1021; Deslongchamps, P. et al., *T*, **57**, 4167; Padden-Row, M.N., Sherburn, M.S. et al., *JOC*, **66**, 3963.

I.E.1-13 Danishefsky, S.J. et al., *OL*, **3**, 2949; Wolter, M. et al., *EJOC*, 4051; Ishikawa, T. et al., T. **57**, 2717; Scheeren, H.W. et al., *EJOC*, 2869; Tome, F. et al., *TL*, **42**, 7233; Inokuchi, T. et al., *JOC*, **66**, 8059; Fujita, R. et al., *CPB*, **49**, 893, 900; Fujita, R., Hongo, H. et al., *CPB*, **49**, 407, 497; Burnell, D.J. et al., *JCS(P1)*, 2644, 2657; Singh, S.B. et al., *BMCL*, **11**, 3143; Sestelo, J.P., del Mar Real, M., Sarandeses, L.A., *JOC*, **66**, 1395.

I.E.1-14 Calverley, M.J., *ST*, **66**, 249; Dilley, A.S., Romo, D., *OL*, **3**, 1535; Davies, D.L. et al., *JCS(P1)*, 1500; Gelmi, M.L. et al., *JOC*, **66**, 4941, 6299; Jones, S., Atherton, J.C.C., *TA*, **12**, 1117 and *TL*, **42**, 8239; Breunung, M., Corey, E.J., *OL*, **3**, 1559; Fujita, R. et al., *CPB*, **49**, 601.

I.E.1-15 Ishikawa, T., Saito, S. et al., *JACS*, **123**, 4607; Sparks, S.M., Shea, K.J., *OL*, **3**, 2265; Tantillo, D.J., Houk, K.N., Jung, M.E., *JOC*, **66**, 1938; Chackalamannil, S., Davies, R., *OL*, **3**, 1427.

I.E.1-16 Vollhardt, K.P.C. et al., *JACS*, **123**, 9324; Malacria, M. et al., *EJOC*, 3491.

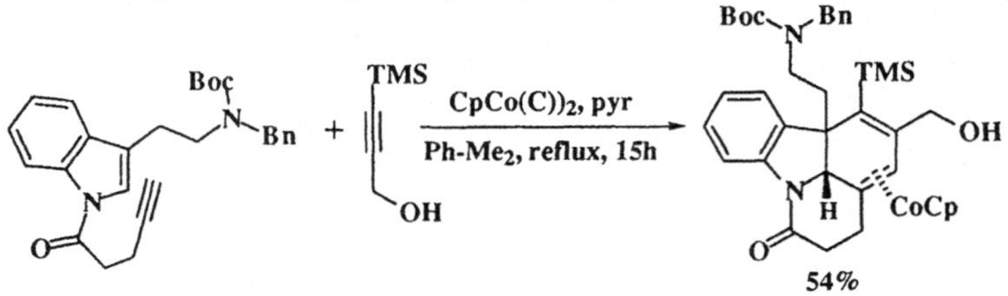

54%

I.E.1-17 Norret, M., Sherburn, M.S., *AG(E)*, **40**, 4198.

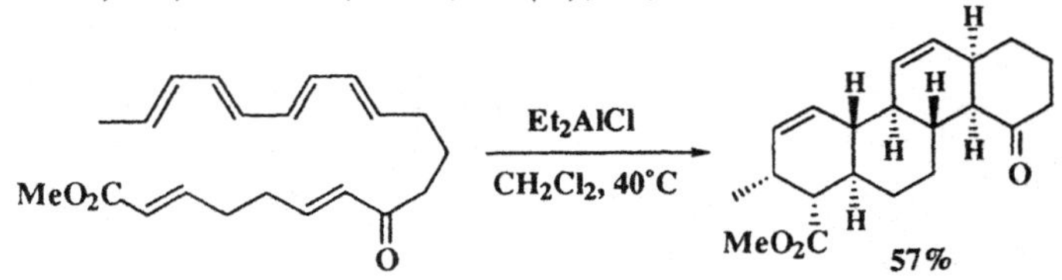

Et₂AlCl

CH₂Cl₂, 40°C

57%

I.E.1-18 Coelho, P.J., Blanco, L., *SL*, 1455; Cho, C.-G. et al., *TL*, **42**, 8193.

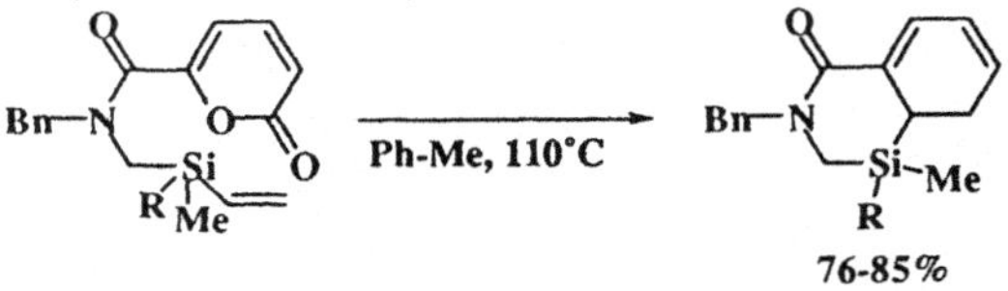

Ph-Me, 110°C

76-85%

I.E.1-19 Kita, Y. et al., *OL*, **3**, 4015.

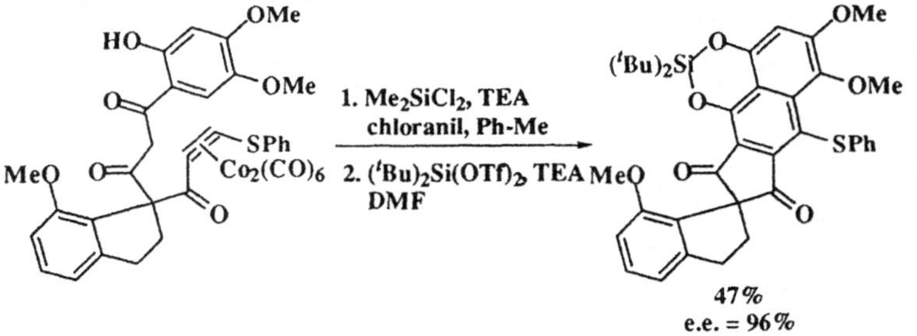

1. Me₂SiCl₂, TEA
chloranil, Ph-Me

2. (ᵗBu)₂Si(OTf)₂, TEA
DMF

47%
e.e. = 96%

I.E.1-20 Hoornaert, G.J. et al., *TL*, **42**, 3759; Kurth, M.J. et al., *JOC*, **66**, 5528.

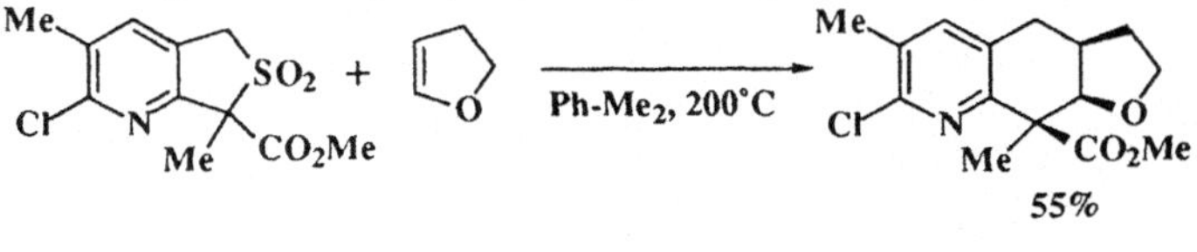

Ph-Me₂, 200°C

55%

I.E.1-21 Gauvry, N., Huet, F., *JOC*, **66**, 583.

I.E.1-22 Santelli, M. et al., *TL*, **42**, 843, 847; Pellissier, H., Santelli, M. et al., *JOC*, **66**, 115.

I.E.1-23 Jarosz, S., Szewczyk, K., *TL*, **42**, 3021.

I.E.1-24 Johnson, T.W., Corey, E.J., *JACS*, **123**, 4475.

I.E.1-25 Nicolaou, K.C. et al., *AG(E)*, **40**, 2145.

I.E.1-26 Faller, J.W., Parr, J., *OM*, **20**, 697.
  "Utility of Osmium (II) in the Catalysis of Asymmetric Diels-Alder Reactions."

I.E.1-27 Caster, K.C. et al., *JOC*, **66**, 2932.

86-97%

I.E.1-28 Harmata, M. et al., *JOC*, **66**, 5232.

69-90%
3-99:1

I.E.1-29 Kotha, S. et al., *T*, **57**, 6261; Beller, M. et al., *JACS*, **123**, 8398; Harmon, W.D. et al., *JACS*, **123**, 10756; **see also:** Passarella, D. et al., *JCS(P1)*, 127.

74-99%

I.E.1-30 Kishbaugh, T.L.S., Gribble, G.W., *TL*, **42**, 4783; Mancini, P.M.E., *JOC*, **66**, 3906.

35-85%

I.E.1-31 Grigg, R. et al., *T*, **57**, 10347.

$$Ar\text{-}I + CH_2{=}C{=}CMe_2 +$$

76-90%

I.E.1-32 Winkler, J.D., Axelsen, P.H. et al., *AG(E)*, **40**, 743; **see also:** Carrano, M.C. et al., *JACS*, **123**, 7929.
  "**Design and Synthesis of Foldamers Based on an Anthracene Diels-Alder Adduct.** "

I.E.1-33 Kochi, J.K. et al., *JACS*, **123**, 87.
"Diels-Alder Topochemistry *via* Charge-Transfer Crystals: Novel (Thermal) Single-Crystal-to-Single-Crystal Transformations."

I.E.1-34 Najera, C. et al., *TA*, **12**, 1811; Sibi, M.P. et al., *JACS*, **123**, 8444; **see also:** Wong, Y.-S. et al., *ECJ*, **7**, 2349; **see also:** Node, M. et al., *TL*, **42**, 9237.

51%
d.e. = 99%

I.E.1-35 A. Boeckmann, R.K., Jr., et al., *TA*, **12**, 205; **B.** Nenajdenko, V.G. et al., *T*, **57**, 201; **C.** Kobayashi, C., Kiyotsuka, Y., *TL*, **42**, 9229; **D.** Uneyama, K. et al., *OL*, **3**, 3103; **E.** Dibble, P.W. et al., *TL*, **42**, 789; **F.** Hiroi, K. et al., *TL*, **42**, 7617.

A   B   C

D.A. Dienophile

D   E

D.A. Diene

F

D.A. Chiral Ligand

I.E.1-36 Trost, B.M., Shen, H.C., *AG(E)*, **40**, 2313; Wender, P.A. et al., *OL*, **3**, 2105 and *JACS*, **123**, 179.

85-93%

I.E.1-37 Cha, J.K. et al., *OL*, **3**, 2891; Hsung, R.P. et al., *JACS*, **123**, 7174; Montana, A.M., Grima, P.M., *TL*, **42**, 7809.

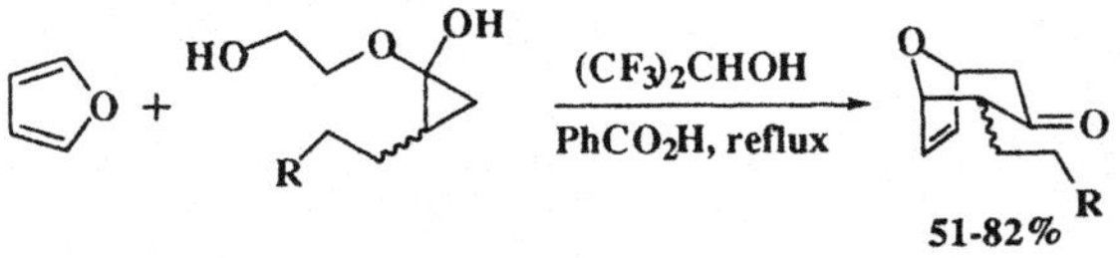

HO
OH
O
+
R
$(CF_3)_2CHOH$
$PhCO_2H$, reflux
O
R
51-82%

I.E.1-38 Magnus, P. et al., *TL*, **42**, 4947; Mascarenas, J.L. et al., *OL*, **3**, 623; Ohkata, K. et al., *CL*, 906.

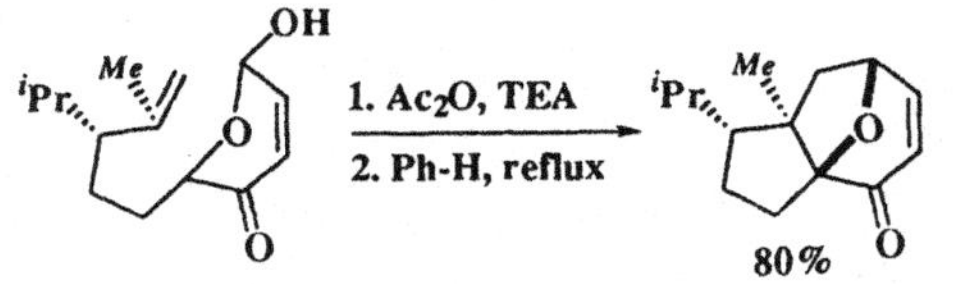

OH
^{i}Pr
Me
O
O
1. Ac_2O, TEA
2. Ph-H, reflux
80%

I.E.1-39 Wender, P.A. et al., *AG(E)*, **40**, 3895.

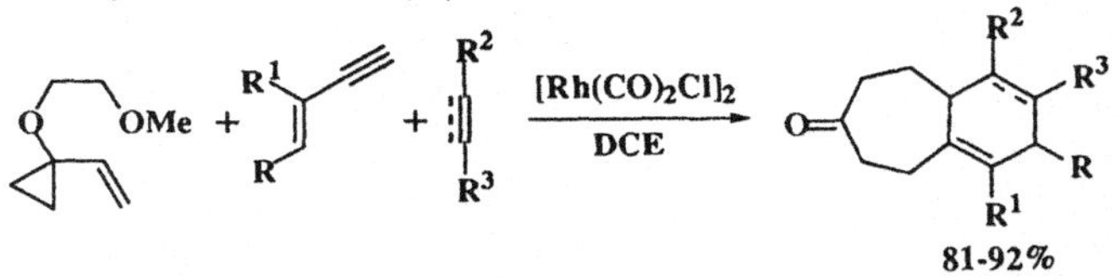

O
OMe
+
R^1
R
R^2
R^3
$[Rh(CO)_2Cl]_2$
DCE
81-92%

I.E.1-40 Gleason, J.L. et al., *OL*, **3**, 4189.

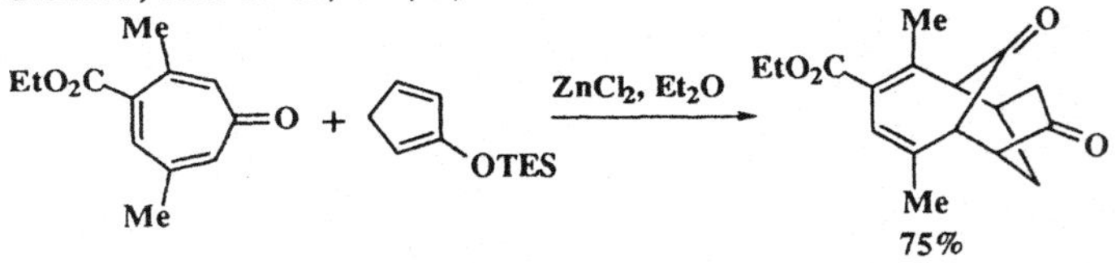

Me
EtO_2C
O
+
OTES
$ZnCl_2$, Et_2O
Me
75%

I.E.1-41 Hong, B.-C. et al., *BMCL*, **11**, 1981.

Me_2N
R
+
O
R^1
R^2
N
H
µW, DMF
30-65%
(without µW, 12-26%)

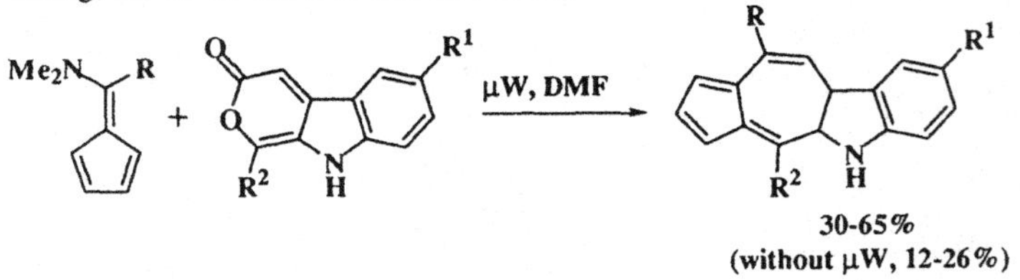

I.E.1-42 Snyder, J.K. et al., *JOC*, **66**, 6932; Chen, Y., Snyder, J.K., *JOC*, **66**, 6943.

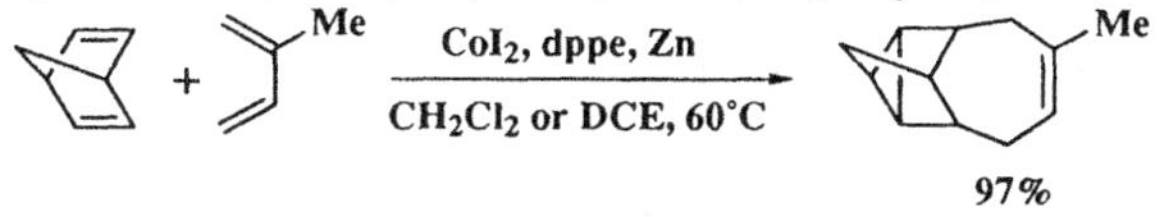

CoI$_2$, dppe, Zn

CH$_2$Cl$_2$ or DCE, 60°C

97%

I.E.2. Other Thermal Reactions

I.E.2-1 Snapper, M.L. et al., *JACS*, **123**, 5152.

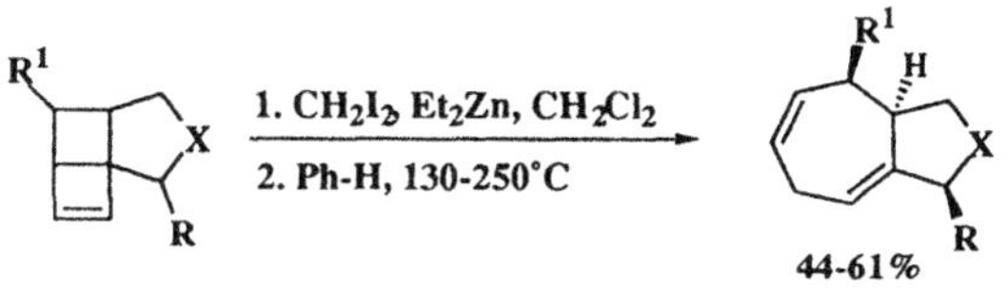

1. CH$_2$I$_2$ Et$_2$Zn, CH$_2$Cl$_2$

2. Ph-H, 130-250°C

44-61%

I.E.2-2 Mehta, G., Umarye, J.D., *TL*, **42**, 1991.

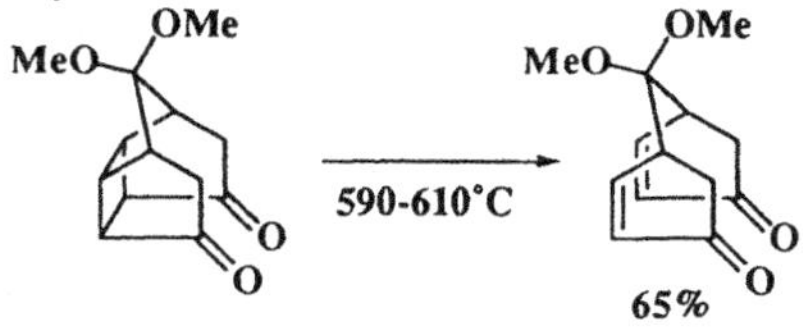

590-610°C

65%

I.E.2-3 Dixneuf, P.H. et al., *AG(E)*, **40**, 2912.

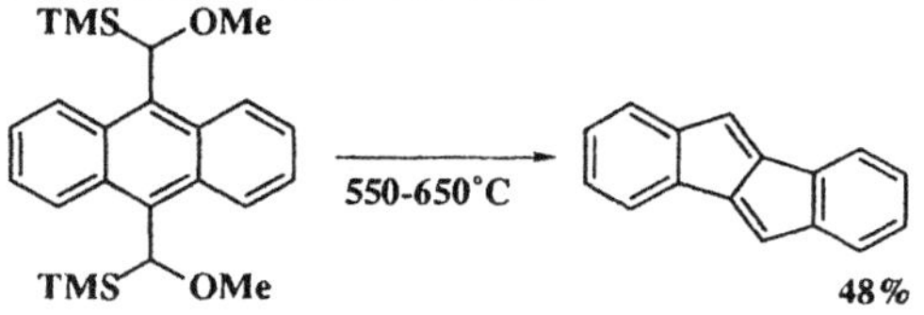

550-650°C

48%

I.E.2-4 Kendall, J.K., Shechter, H., *JOC*, **66**, 6643.

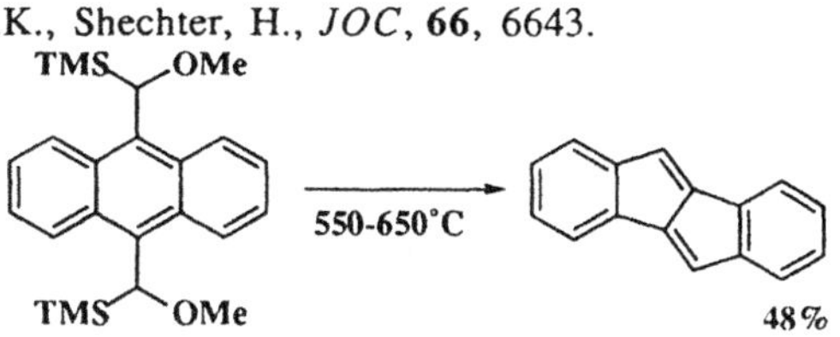

550-650°C

48%

I.E.3-5 Torrente, S., Alonso, R., *OL*, **3**, 1985.

(other alcohols and ethers used)

38-74%

I.E.3-6 Mariano, P.S. et al., *JACS*, **123**, 6433; Faure, S., Piva, O., *TL*, **42**, 255; **see also:** Ghosh, S. et al., *JCS(P1)*, 3013.

31-82%

I.E.3-7 Kwak, Y.-S., Winkler, J.D., *JACS*, **123**, 7429.

86%

I.E.3-87 Podlech, J., Linder, M.R., *OL*, **3**, 1849.

40-80%

I.E.3-9 Hakimelahi, G.H. et al., *JOC*, **66**, 7067.

70-76%

I.E.3-10 Ho, T.-I. et al., *TL*, **42**, 715.

63-92%

I.E.3-11 Carreira, E.M. et al., *AG(E)*, **40**, 2694.

I.E.3-12 Comins, D.L., Williams, A.L., *OL*, **3**, 3217.

I.E.3-13 Girard, J.-P. et al., *OL*, **3**, 3067.

I.E.3-14 Chang, D.J., Park, B.S., *TL*, **42**, 711.

I.E.3-15 Mizuno, K. et al., *OL*, **3**, 581 and *TL*, **42**, 3363.

I.E.3-16 Guo, Z., Schultz, A.G., *OL*, **3**, 1177.

I.E.3-17 Mizuno, K. et al., *TL*, **42**, 2361.

64-88% 2.3-32.3:1

I.E.3-18 Booker-Milburn, K.I. et al., *OL*, **3**, 3005.

61-73%

I.E.3-19 Mukiyaima, T. et al., *CL*, 633.

9-80% conversion

I.E.3-20 Tojo, G. et al., *OL*, **3**, 1343.

91%

I.E.3-21 Campos, P.J. et al., *TL*, **42**, 3575.

38-63%

I.E.3-22 Mattay, J. et al., *S*, 1275.

90%

I.E.3-23 Dopp, D. et al., *S*, 1223.

22-91%

I.F. Aromatic SubstitutionsForming a New Carbon-Carbon Bond

I.F.1. Friedel-Crafts Type Aromatic Substitution Reactions

I.F.1-1 Mi, A. et al., *S*, 1007; Davioud-Charvet, E. et al., *JOC*, **66**, 5616.

82%

I.F.1-2 Fukuyama, T. et al., *SL*, 1179.

83-98%
R:S = 4-100:1

I.F.1-3 Frost, C.G. et al., *TL*, **42**, 773; Furstner, A. et al.,*OL*, **3**, 417; Ottoni, O. et al., *OL*, **3**, 1005; Singh, V.K. et al., *T*, **57**, 241; Colquhoun, H.M. et al., *OL*, **3**, 2337; **see also:** Bagno, A., Kantlehner, W. et al., *EJOC*, 2947; Kantlehner, W. et al., *ZN*, **56B**, 105.

82-99%

I.F.1-4 Ogasawara, K. et al., *OL*, **3**, 1737.

99%

I.F.1-5 Brandange, S. et al., *JCS(P1)*, 2051; Basavaiah, D. et al., *S*, 919; **see also:** Yadav, J.-S. et al., *TL*, **42**, 8067

42-69%

I.F.1-6 Gray, A.D., Smyth, T.P., *JOC*, **66**, 7113; Malladi, R.R., Kabalka, G.W., *OPP*, **33**, 198.

37-76%

I.F.1-7 Avendano, C. et al., *JOC*, **66**, 5731; Deninno, M.P. et al., *JOC*, **66**, 6988.

65-80%

I.F.1-8 Bozell, J.J. et al., *JOC*, **66**, 3085.

R = Ac 0-60%
R = H 0-21%

0-50%

I.F.1-9 Gong, Y. et al., *BCJ*, **74**, 377.

83%

92%

I.F.2. Coupling Reactions to Form an Aromatic Carbon-Aromatic Carbon Bond

I.F.2-1 Feuerstein, M., Doucet, H., Santelli, M., *TL* , **42**, 5659, 6667; Doucet, H., Santelli, M. et al., *CC*, 325; Savignac, M., Genet, J.-P. et al., *TL*, **42**, 6523; Akiyama, R., Kobayashi, S., *AG(E)*, **40**, 3469; Mazet, C., Gade, L.H., *OM*, **20**, 4144; Xiao, J. et al., *TL*, **42**, 9085; Morris, G.A., Nguyen, S.T., *TL*, **42**, 2093; Coehlo, A., Ravina, E. et al., *S*, 871; Bedford, R.B., Cazin, C.S.J., *CC*, 1540; Havelkova, M., Hocek, M., Dvorak, D., *S*, 1704; Rival, Y. et al., *TL*, **42**, 2115; Cooke, G. et al., *T*, **57**, 2787; Schomaker, J.M., Delia, T.J., *JOC*, **66**, 7125; Aldous, D.J. et al., *SL*, 150; Monteiro, A.L. et al., *OL*, **3**, 3049; Zhuravel, M.A., Nguyen, S.T., *TL*, **42**, 7925; **see also:** Nolan, S.P. et al., *OL*, **3**, 1077; Kotha, S., Lahiri, K., *BMCL*, **11**, 2887; Uemura, M. et al., *OL*, **3**, 2033; **see also:** Sun, Y., Sowa, J.R., Jr. et al., *OL*, **3**, 1555; Enguehard, C., Gueffier, A. et al., *S*, 595; Ma, D., Wu, Q., *TL*, **42**, 5279.

12-98%

I.F.2-2 Harrowven, D.C. et al., *TL*, **42**, 2907; **see also:** Zhang, N. et al., *JOC*, **66**, 1500; **see also:** Gosmini, C. et al., *TL*, **42**, 267.

67%

I.F.2-3 Kauffmann, J.M., *S*, 197.

63%

I.F.2-4 Miller, J.A., *TL*, **42**, 6991.

$$\text{Ar-CN} + \text{Ar}^1\text{-MgX·LiZR} \xrightarrow[\text{THF, 60°C}]{\text{NiCl}_2(\text{PMe}_3)_2} \text{Ar-Ar}^1$$

69-97%

I.F.2-5 Murata, M., Masuda, Y. et al., *S*, 2231.

$$\text{Ar-Si(OEt)}_3 + \text{Ar}^1\text{-Br} \xrightarrow[\text{dioxane, 80°C}]{\text{PdCl}_2(\text{MeCN})_2, \text{ aq NaOH}} \text{Ar-Ar}^1$$

81-94%

I.F.2-6 Dai, C., Fu, G.C., *JACS*, **123**, 2719; Netherton, M.R., Fu, G.C., *OL*, **3**, 4295; Amat, M. et al., *S*, 267; **see also:** Bach, T., Bartels, M., *SL*, 1284; Gallagher, T. et al., *OL*, **3**, 835; **see also:**, Gosmini, C. et al., *T*, **57**, 1923.

$$\text{Ar-Cl + R-ZnCl} \xrightarrow[\text{THF/NMP, 100°C}]{\text{Pd(P'Bu}_3)_2} \text{Ar-R}$$
50-96%

I.F.2-7 Wong, M.S., Zhang, X.L., *TL*, **42**, 4087.

66-85%

I.F.2-8 Mc Clure, M.S. et al., *OL*, **3**, 1677; Mizufune, H. et al., *TL*, **42**, 437.

60-88%

I.F.2-9 Fort, Y. et al., *T*, **57**, 531; **see also:** Lin, G. et al., *SL*, 1527.

$$\text{Ar-X} \xrightarrow[\text{THF, 65°C}]{\text{Ni-Al, NaH, bipyr}} \text{Ar-Ar}$$
34-99%

I.F.2-10 Clive, D.L.J., Kang, S., *JOC*, **66**, 6083.

17-73%

I.F.2-11 Demir, A.S. et al., *JCS(P1)*, 3042; **see also:** Batey, R.A., Quach, T.D., *TL*, **42**, 9099.

(other oxidants used)

68-83%

I.F.2-12 Uang, B.-J. et al., *CC*, 980.

82%
e.e. = 51%

I.F.2-13 Ward, R.S., Hughes, D.D., *T*, **57**, 5633, 2057, 4015.

I.F.2-14 Albrecht, B.K., Williams, R.M., *TL*, **42**, 2755; Fort, Y. et al., *TL*, **42**, 1879; Clapham, B., Sutherland, A.J., *JOC*, **66**, 9033; see also: Kang, S.-K. et al., *SC*, **41**, 1021, 1027; Jung, K.W. et al., *TL*, **42**, 7729.

I.F.2-15 Bedford, R.B. et al., *JOM*, **633**, 173;
   "Silica-Supported Imine Palladacycles—Recyclable Catalysts for the Suzuki Reaction."

I.F.2-16 Doucet, H., Santelli, M. et al., *S*, 2320.
   "Palladium-Tetraphosphine Complex: An Efficient Catalyst for Allylic Substitution and Suzuki Cross-Coupling."

I.F.2-17 Ogoshi, S., Kurosawa, H. et al., *JCAS*, **123**, 8626; see also: Brinker, U.H. et al., *JOC*, **66**, 2874.

I.F.3. Other Aromatic Substitutions and Preparations

I.F.3-1 Knochel, P., et al., *OL*, **3**, 2871 and *SL*, 477; Oshima, K. et al., *JOC*, **66**, 4333; Iida, T., Mase, T. et al., *TL*, **42**, 4841; Hirao, T. et al., *JOC*, **66**, 300.

$$FG\text{-}\langle\text{aryl}\rangle\text{-I} + I\text{-}(CH_2)_n\text{-FG} \xrightarrow[\text{THF, -20°C}]{^iPrMgBr, CuCN\cdot 2LiCl} FG\text{-}\langle\text{aryl}\rangle\text{-}(CH_2)_n\text{-}FG^1$$

55-89%

I.F.3-2 Geneste, H., Schafer, B., *S*, 2259; Snieckus, V. et al., *JOC*, **66**, 3662; Weissensteiner, W. et al., *JOC*, **66**, 8912; Kowalik, J., Talbert, L.M., *JOC*, **66**, 3229; Townsend, L.G. et al., *JOC*, **66**, 4783.

$$CF_3\text{-}\langle\text{aryl}\rangle\text{-OTHP} \xrightarrow[\text{THF}]{E^+, BuLi, TMEDA} CF_3\text{-}\langle\text{aryl}\rangle\text{-OTHP (with E)}$$

53-90%

I.F.3-3 Jackson, R.F.W. et al., *JCS(P1)*, 1876; Yus, M., Gomis, J., *TL*, **42**, 5721; Rosowsky, A., Chen, H., *JOC*, **66**, 7522; Oshima, K. et al., *OL*, **3**, 1997; Lee, P.H. et al., *OL*, **3**, 3201.

$$IZn\text{-}\underset{CO_2{}^tBoc}{\overset{NHFmoc}{\diagup}} \xrightarrow[\text{DMF, 50°C}]{Ar\text{-}I, Pd_2(dba)_3, P(o\text{-tol})_3} Ar\text{-}\underset{CO_2{}^tBoc}{\overset{NHFmoc}{\diagup}}$$

21-59%

I.F.3-4 Andrus, M.B., Song, C., *OL*, **3**, 3761.

$$Ar\text{-}N_2{}^+ BF_4{}^- + (HO)_2B\text{-}R \xrightarrow[]{Pd(OAc)_2, THF} Ar\text{-}R$$

53-95%

I.F.3-5 Yang, C., Nolan, S.P., *SL*, 1539; Nolan, S.P., *OL*, **3**, 1511; Feuerstein, M., Doucet, H., Santelli, M., *JOC*, **66**, 5923; Muzart, J. et al., *JOM*, **634**, 153; Lautens, M. et al., *JOC*, **66**, 8127; **see also:** Alonso, I., Carretero, J.C., *JOC*, **66**, 4453.

$$R\text{-}\langle\text{aryl}\rangle\text{-Br} + \diagup\text{-}CO_2{}^tBu \xrightarrow[]{Pd(OAc)_2, Cs_2CO_3, DMA} R\text{-}\langle\text{aryl}\rangle\text{-CH=CH-}CO_2{}^tBu$$

16-99%

I.F.3-6 Furstner, A., Leitner, A., *SL*, 290; **see also:** Molander, G.A., Ito, T., *OL*, **3**, 393.

Pd(OAc)$_2$, THF, reflux

61-98%

I.F.3-7 Smith, A.B., III et al., et al., *OL*, **3**, 3967; Hirama, M. et al., *TL*, **42**, 5783; Parsons, P.J. et al., *TL*, **42**, 2209; **see also:** Ishibashi, H. et al., *OL*, **3**, 2427.

1. TMS-SnBu$_3$, Pd(PPh$_3$)$_4$
   Bu$_4$NBr, Li$_2$CO$_3$
   Ph-Me$_2$, reflux
2. N$_2$H$_4$·(H$_2$O)$_n$, EtOH

48%

I.F.3-8 Moradi, W.A., Buchwald, S.L., *JACS*, **123**, 7996; Hartwig, J.F. et al., *JACS*, **123**, 8410; Lee, S., Hartwig, J.F., *JOC*, **66**, 3402; **see also:** Goossen, L.J., *CC*, 669; **see also:** Konopelski, J.P., Deng, H., *OL*, **3**, 3001; Pigge, F.G., Fang, S., *TL*, **42**, 17.

Pd(OAc)$_2$, LiHMDS, Ph-Me

54-92%

I.F.3-9 Coudert, G. et al., *T*, **57**, 6969; Occhiato, E.G. et al., *JOC*, **66**, 2459; Wu, J. et al., *TL*, **42**, 159; Hallberg, A. et al., *JOC*, **66**, 544; **see also:** Hallberg, A. et al., *JACS*, **123**, 8217 and *JOC*, **66**, 4340.

ArB(OH)$_2$, Pd(PPh$_3$)$_4$

Na$_2$CO$_3$, DME

77-95%

I.F.3-10 Harrowven, D.C. et al., *T*, **57**, 4447.

NaCO^I salophen

THF, rt

61%

I.F.3-11 Sano, T. et al., *CPB*, **49**, 979.

I.F.3-12 Palmieri, G. et al., *JOC*, **66**, 4759; Barluenga, J. et al., *ECJ*, **7**, 5318.

I.F.3-13 Kim, J.N. et al., *TL*, **42**, 4195.

I.F.3-14 West, F.G. et al., *OL*, **3**, 3033.

I.F.3-15 Sato, F. et al., *JACS*, **123**, 7925; **see also:** Takeuchi, R. et al., *TL*, **42**, 2991.

I.F.3-16 Herndon, J.W. et al., *JOM*, **634**, 1.

86-92%

I.F.3-17 Indolese, A.F. et al., *JOC*, **66**, 4311.

34-89%

I.F.3-18 Serra, S. et al., *JOC*, **66**, 7883; Ruzziconi, R. et al., *JOC*, **66**, 617.

60-95%

(also using ClCO₂Et, TEA/aq NaOH)

I.F.3-19 Iyoda, M. et al., *JCS(P1)*, 159; Luh, T.-Y. et al., *JOC*, **66**, 7922.

46-70%

I.F.3-20 Kiselyov, A.S. et al., *TL*, **42**, 3053.

47-68%

I.G. Synthesis via Organometallics

I.G.1. Synthesis via Organoboranes

I.G.1-1 Zou, G., Falck, J.R., *TL*, **42**, 5817; **see also:**, Vice, S., McCombie, S. et al., *JOC*, **66**, 2487; **see also:** Pergament, I., Srebnik, M., *TL*, **42**, 8059.

$$RCH_2\text{-}\left(\begin{array}{c}O\\O\end{array}\right) + E\text{-}X \xrightarrow[\text{2. Pd}_2\text{(dba)}_3, \text{P(}^t\text{Bu)}_3, \text{NaOAc, 70-80°C}]{\text{1. }^s\text{BuLi, THF, -78°C}\rightarrow\text{rt}} \begin{array}{c}RCH_2\text{-E}\\58\text{-}90\%\end{array}$$

I.G.1-2 Sakuma, S., Miyaura, N., *JOC*, **66**, 8944.

$$R\diagdown\diagup\stackrel{NHR^1}{\underset{O}{\bigfrown}} + Ar\text{-B(OH)}_2 \xrightarrow[\text{dioxane/H}_2\text{O, 100°C}]{\text{RhX, (S)-BINAP, K}_2\text{CO}_3} \begin{array}{c}R,,\diagup\diagdown\stackrel{NHR^1}{O}\\H\text{ }Ar\end{array}$$

19-89%
e.e. = 77-93%

I.G.1-3 Maddaford, S.P. et al., *OL*, **3**, 2013.

$$\begin{array}{c}AcO\\AcO\\AcO\end{array} + Ar\text{-B(OH)}_2 \xrightarrow[\text{MeCN, rt}]{\text{Pd(OAc)}_2} \begin{array}{c}AcO\\AcO\end{array}\text{Ar}$$

55-83%

I.G.1-4 Collman, J.P. et al., *JOC*, **66**, 7892.

$$\begin{array}{c}Me\\B(OH)_2\end{array} + HN\diagdown N\text{-R} \xrightarrow[\text{CH}_2\text{Cl}_2, \text{rt}]{[\text{Cu(OH)L]}_2\text{Cl}_2, \text{O}_2} \begin{array}{c}Me\\N\diagdown N\end{array} + \begin{array}{c}Me\\N\diagdown N\end{array}$$

19-77% 1.9-8.7:1

I.G.1-5 Harrity, J.P.A. et al., *JOC*, **66**, 3525.

$$\begin{array}{c}R^2\\R^1\end{array}\stackrel{OMe}{=}\text{Cr(CO)}_5 + R\text{---}B\begin{array}{c}O\\O\end{array} \xrightarrow[\text{2. Ar-Br, PdCl}_2\text{(dppf)}]{\text{1. CAN, THF, 45°C}} \begin{array}{c}R^2\\R^1\end{array}\begin{array}{c}OMe\\Ar\\R\\OH\end{array}$$

K$_3$PO$_4$, dioxane, 85°C

0-83%

I.G.1-6 Savarin, C., Liebeskind, L.S., *OL*, **3**, 2149; Kang, S.-K. et al., *JCR(S)*, 283.

$$R^1\text{-I} + R\text{-B(OH)}_2 \xrightarrow[\text{thiophene-}CO_2Cu]{Pd(PPH_3)_4, THF, rt} R^1\text{-R}$$

72-96%

I.G.2. Carbonylations Reactions

I.G.2-1 Shibata, T. et al., *OL*, **3**, 1217; Chung, Y.K. et al., *OL*, **3**, 1065; Rausch, B.J., Gleiter, R., *TL*, **42**, 1651.

65-99%

I.G.2-2 Sugihara, T. et al., *ECJ*, **7**, 1589; **see also:** Gibson, S.E. et al., *TL*, **42**, 1183; Hyeon, T., Chung, Y.K. et al., *CC*, 2212; Cazes, B. et al., *TL*, **42**, 8153, 8157.

   "Advances in the Pauson-Khand Reaction: Development of Reactive Cobalt Complexes."

I.G.2-3 Takahashi, T. et al., *JOM*, **633**, 18.

50-89%

I.G.2-4 Rawal, V.H. et al., *OL*, **3**, 3615.

36%

I.G.2-5 Jiang, H. et al., *SC*, **31**, 199; **see also:** Gabriele, B. et al., *JOM*, **622**, 84.

$$R\text{≡} + CO + R^1\text{-OH} \xrightarrow[\text{BuOH, rt}]{PdBr_2, CuBr_2, \text{base}} R\text{≡}\text{-}CO_2R^1$$

40-83%

I.G.2-6 Stryker, J.M. et al., *JACS*, **123**, 8872; Sakakura, T. et al., *JOC*, **66**, 5262; **see also:** Alper, H. et al., *JOC*, **66**, 5424.

1. [(Me$_2$N-indenyl)$_2$TiCl]$_2$
   THF, -35°C→rt
2. R^1-I, SmI$_2$, -35°C→rt
3. CO, rt

45-99%

I.G.2-7 Carretero, J.C., Adrio, J., *S*, 1888 and *ECJ*, **7**, 2435; Ishizaki, M., Hoshino, O. et al., *T*, **57**, 2729; Tanimori, S. et al., *TL*, **42**, 4013; Hiroi, K., Watanabe, T., *H*, **54**, 73; Lovely, C.J. et al., *OL*, **3**, 2607; Pal, A., Bhattacharjya, A., *JOC*, **66**, 9071; Wilson, M.S.., Dake, G.R., *OL*, **3**, 2041; Perez-Castells, J. et al., *TL*, **42**, 3315; Krafft, M.E. et al., *TL*, **42**, 1427 and *JOC*, **66**, 3004; **see also:** Evans, P.A., Robinson, J.E., *JACS*, **123**, 4609; Chung, Y.K. et al., *CC*, 2440; Arjona, O., Plumet, J. et al., *TL*, **42**, 3085; Ishizaki, M., Hoshino, O. et al., *H*, **55**, 1439; Pericas, M.A., Riera, A. et al., *JOC*, **66**, 6400; Kerr, W.J. et al., *OL*, **3**, 2945.

1. Co$_2$(CO)$_8$, CH$_2$Cl$_2$, rt
2. MeCN, reflux

30-65%
d.r. = 49:1

I.G.2-8 Breit, B., Zahn, S.K., *JOC*, **66**, 4870; Leighton, J.L. et al., *JACS*, **123**, 11514; Beller, M. et al., *AG(E)*, **40**, 3408; Pereira, M.M., Bayon, J.C. et al., *TA*, **12**, 1083; Claver, C. et al., *CJC*, **79**, 560; **see also:** des Abbayes, H. et al., *TL*, **42**, 643; Bakos, J. et al., *CJC*, **79**, 725; Dahlenburg, L., Mertel, S., *JOM*, **630**, 221; Agboussou-Niedercorn, F. et al., *JOM*, **628**, 114; Diequez, M., Claver, C. et al., *ECJ*, **7**, 3086.

CO/H$_2$, Rh(CO)$_2$(acac), P(OPh)$_3$
Ph-Me, 70°C

79-80%
d.r. = 99:1

I.G.2-9 Kato, K., Akita, H. et al., *TL*, **42**, 4203.

CO, PdCl$_2$(MeCN)$_2$, BQ
MeOH, 0°C

69-91% CO$_2$Me

I.G.2-10 Beller, M. et al., *S*, 1098 and *AG(E)*, **40**, 2856; **see also:** Uozumi, Y. et al., *JOC*, **66**, 5272; **see also:** Lin, Y.-S., Alper, H., *AG(E)*, **40**, 779.

"**Efficient Palladium-Catalyzed Alkoxycarbonylation of N-Heteroaryl Chlorides—A Practical Synthesis of Building Blocks for Pharmaceuticals and Herbicides.**"

I.G.2-11 Grigg, R. et al., *T*, **57**, 7737; **see also:** Nieman, J.A., Ennis, M.D., *JOC*, **66**, 2175.

51-86%

I.G.2-12 Xiao, W.-J., Alper, H., *JOC*, **66**, 6229; **see also:** Takeuchi, S., Ukaji, Y., Inomata, K., *BCJ*, **74**, 955.

24-72%
e.e. = 0-89%

I.G.2-13 Mann, A. et al., *TL*, **42**, 265; Carpentier, J.-F., Castanet, Y. et al., *TL*, **42**, 3689; **see also:** Chatani, N., Murai, S. et al., *JACS*, **123**, 12686.

70-94%

I.G.2-14 Crowe, W.E. et al., *JACS*, **123**, 6457.

73-92%
e.e. = 0-90%

I.G.2-15 Huang, X. et al., *SC*, **31**, 311.

$$R^1 \longequal SnR_3 \quad \xrightarrow[\text{2. Br}_2, \text{R}^2\text{-OH, rt}]{\text{1. CO, Cp}_2\text{Zr(H)Cl, CH}_2\text{Cl}_2, \text{rt}} \quad \begin{array}{c} R^1 \quad SnR_3 \\ \diagdown \diagup \\ CO_2R^2 \end{array}$$

49-85%

I.G.3. Other Synthesis via Organometallics

I.G.3-1 Takahashi, T. et al., *OM*, **20**, 595; Trost, B.M., Pinkerton, A.B., *JOC*, **66**, 7714; **see also:**, Tsai, Y.-M. et al., *JOC*, **66**, 8983.

$$R \longequal R^1 + R^3\text{-N=C=O} + \begin{array}{c} CN \\ \diagup \\ R^3 \quad CN \end{array} \xrightarrow[\text{2. CuCl}]{\text{1. Cp}_2\text{ZrEt}_2} \begin{array}{c} R \quad R^1 \\ R^3 \diagdown \diagup O \\ NC \quad CN \end{array}$$

42-75%

I.G.3-2 Itoh, K. et al., *JACS*, **123**, 6372; Genet, J.-P. et al., *OL*, **3**, 2065; Widenhoefer, R.A. et al., *JOC*, **66**, 635, 1755; Pei, T., Widenhoefer, R.A., *JOC*, **66**, 7639; Studer, A. et al., *OL*, **3**, 2357; **see also:** Oh, C.H. et al., *T*, **57**, 1723; Oi, S. et al., *OM*, **20**, 3704; Mori, M. et al., *OM*, **20**, 1907; Montgomery, J. et al., *OM*, **20**, 370; Mikami, K. et al., *AG(E)*, **40**, 249; Lu, X. et al., *JOC*, **66**, 7676; **see also:** Shin, S., RajanBabu, T.V., *JACS*, **123**, 8416.

$$\begin{array}{c} R^1 \\ R \end{array} \xrightarrow[\text{}^i\text{PrOH, 90°C}]{\text{[Ru(cod)Cl}_2\text{]}_n} \begin{array}{c} R^1 \\ R \end{array}$$

62-98%

I.G.3-3 Tietze, L.F., Nordmann, G., *EJOC*, 3247.

$$\xrightarrow[\text{MeCN/H}_2\text{O, 80°C}]{\text{Pd(OAc)}_2, \text{PPh}_3, \text{TEA}}$$

23-89%

I.G.3-4 Trost, B.M., Surivet, J.-P., *AG(E)*, **40**, 1468; Ishii, T. et al., *AG(E)*, **40**, 2534; **see also:** Trost, B.M., Rudd, M.T., *JACS*, **123**, 8862; Hilt, G. et al., *AG(E)*, **40**, 387.

$$R \longequal TMS + \begin{array}{c} \| \\ \diagup \\ NHBoc \end{array} \xrightarrow[\text{Me}_2\text{CO/THF, rt}]{[\text{Ru}(\eta^5\text{-C}_5\text{H}_5)(\text{MeCN})_3]\text{PF}_6} \begin{array}{c} R \\ \diagdown \diagup \diagdown \diagup NHBoc \\ TMS \end{array}$$

81-97%

I.G.3-5 Shvo, Y., Arisha, A.H.I., *JOC*, **66**, 4921.

I.G.3-6 Saito, S., Yamamoto, T. et al., *JOC*, **66**, 796; Fu, Y.-S., Yu, S.J., *AG(E)*, **40**, 437; Jiang, H. et al., *JOC*, **66**, 3627; **see also:** Sen, A. et al., *JACS*, **123**, 7423.

(other examples given)

I.G.3-7 de Meijere, A. et al., *ECJ*, **7**, 4035; Miniere, S., Cintrat, J.-C., *JOC*, **57**, 7385; de Lera, A.R. et al., *JOC*, **66**, 8483; Venturello, P. et al., *JCS(P1)*, 437; Maycock, C.D., *CC*, 1662; Martinez, A.G., Barcina, J.O. et al., *OM*, **20**, 1020.

I.G.3-8 Srinivasan, K.V. et al., *CC*, 1544; Li, G.Y. et al., *JOC*, **66**, 8677; Littke, A.F., Fu, G.C., *JACS*, **123**, 6989; Faller, J.W., Crabtree, R.H. et al., *OM*, **20**, 5485; Carretero, J.C. et al., *ECJ*, **7**, 3890; Hoffmann, H.M.R. et al., *JCS(P1)*, 47; Zapf, A., Beller, M., *ECJ*, 2908; Moreno-Manas, M., Pleixats, R., Villaroya, S., *OM*, **20**, 4524; Shen, Q., Hammond, G.B., *OL*, **3**, 2213; Tanaka, H., *JOC*, **66**, 570; Tietze, L.F. et al., *ECJ*, **7**, 368; **see also:** Larock, R.C., Tian, Q., *JOC*, **66**, 7372; Yi, C.S. et al., *OM*, **20**, 802.

I.G.3-9 Huang, T.-S., Li, C.-J., *OL*, **3**, 2037.

I.G.3-10 Backvall, J.-E. et al., *JOC*, **66**, 8015.

I.G.3-11 Buchwald, S.L. et al., *OL*, **3**, 1897.

I.G.3-12 Denmark, S.E., Sweis, R.F., *JACS*, **123**, 6439; Naso, F. et al., *JOC*, **66**, 3878; Kang, S.-K. et al., *JCS(P1)*, 736.

I.G.3-13 Gottker-Schnetmann, I., Aumann, R., *OM*, **20**, 346; **see also:** Wulff, W.D. et al., *S*, 200.

I.G.3-14 Oshima, K. et al., *OL*, **3**, 1853; Clive, D.L.J. et al., *JOC*, **66**, 1233; **see also:** Nagashima, H. et al., *JOC*, **66**, 315; Hosomi, A. et al., *JOC*, **66**, 3348; Bailey, W.F., Longstaff, S.C., *OL*, **3**, 2217.

I.G.3-15 Araki, S. et al., *ECJ*, **7**, 2784.

≤77%
cis:trans = 8.1:1-13.3:1

I.G.3-16 de Armas, J., Hoveyda, A.H., *OL*, **3**, 2097; **see also:** Backvall, J.-E., Liepins, V., *OL*, **3**, 1861.

1. c-$C_6H_9CH_2CH_2MgBr$, $ZrCl_2Cp_2$
^{i}PrOTs, THF, 55°C

2. O_2, 0°C

73%

I.G.3-17 Lautens, M., Fagnou, K., *JACS*, **123**, 7170; **see also;** Nguyen, S.T. et al., *OL*, **3**, 2229.

"Effects of Halide Ligands and Protic Additives on Enantioselectivity and Reactivity in Rhodium-Catalyzed Asymmetric Ring-Opening Reaactions."

I.G.3-18 Doyle, M.P. et al., *JOC*, **66**, 8112; Lahuerta, P., Perez-Prieto, J. et al., *OM*, **20**, 950; Wardrop, D.J. et al., *OL*, **3**, 2261.

Rh_2L_4
CH_2Cl_2

58-90% 15.7:1-1:9

I.G.3-19 Matijoska, A. et al., *JCR(S)*, 324.

$(CH_2O)_n$, Cu_2Cl_2
Ph-H, 50°C, 10h

41-62%

I.G.3-20 Shilai, M., Kondo, Y., Sakamoto, T., *JCS(P1)*, 442.

Ph-CHO, ^{i}PrMgCl
THF, rt

60%

I.H. Rearrangements

I.H.1. Claisen, Cope and Similar Processes

I.H.1-1 Taguchi, T. et al., *TL*, **42**, 4865; **see also:** Uneyama, K. et al., *JOC*, **66**, 1026.

55-77%
e.e. = 79-95%

I.H.1-2 Moniz, G.A., Wood, J.L., *JACS*, **123**, 5095.

61%
e.e. = 90%

I.H.1-3 Itoh, T., Kudo, K., *TL*, **42**, 1317; **see also:** Etzkorn, F.A. et al., *OL*, **3**, 1789.

58%
e.e. = 82%

I.H.1-4 Yoon, T.P., MacMillan, D.W.G., *JACS*, **123**, 2911; Dong, V.M., MacMillan, D.W.C., *JACS*, **123**, 2448; **see also:** Kazmaier, U., Mues, H., *S*, 487; Natchus, M.G. et al., *BCML*, **11**, 627; **see also:** Trost, B.M., Lee, C.B., *JACS*, **123**, 3671; Loh, T.-P., Hu, Q.-Y., *OL*, **3**, 279.

74-95%
syn:anti = 11.5-99:1
e.e. = 86-97%

I.H.1-5 Garayt, M.R., Percy, J.M., *TL*, **42**, 6377; Sandford, G. et al., *JOC*, **66**, 4887; Tellier, F. et al., *TL*, **42**, 2665.

88-99%

I.H.1-6 Liao, C.-C. et al., *OL*, **3**, 263; Singh, V., Iyer, S., *CC*, 2578; Kuroda, C. et al., *TL*, **42**, 1915; Cook, J.M. et al., *OL*, **3**, 345.

70%

I.H.1-7 Kawasaki, T. et al., *JOC*, **66**, 1200.

44-87%

I.H.1-8 Wipf, P., Ribe, S., *OL*, **3**, 1503; **see also:** Ward, A.D. et al., *S*, 621.

39-78%
e.e. = 73-80%

I.H.1-9 Lindstrom, U.M., Somfai, P., *ECJ*, **7**, 94; Torwe, D. et al., *JOC*, **66**, 2884.

60-85%

I.H.1-10 Alayrac, C., Metzner, P. et al., *JOC*, **66**, 7841.

30-64%
d.r. = 2.33-100:1

I.H.1-11 Hiersemann, M. et al., *AG(E)*, **40**, 4700; Hiersemann, M., Abraham, L., *OL*, **3**, 49; **see also:** Majumdar, K.C., Roy, B., *JCR(S)*, 538.

97-99%
syn:anti = 15.7-99:1
e.e. = 82-84%

I.H.1-12 Ovaska, T.V. et al., *OL*, **3**, 115.

76%

I.H.1-13 Hinman, M.M., Heathcock, C.H., *JOC*, **66**, 7751; **see also:** Aumann, R. et al., *OM*, **20**, 2183.

70%

I.H.1-14 Grogan, G., Turner, N., Flitsch, S.L. et al., *AG(E)*, **40**, 1111. **"An Asymmetric Enzyme-Catalyzed Retro-Claisen Reaction for the Desymmetrization of Cyclic β-Diketones."**

I.H.1-15 Isobe, M. et al., *T*, **57**, 3875.

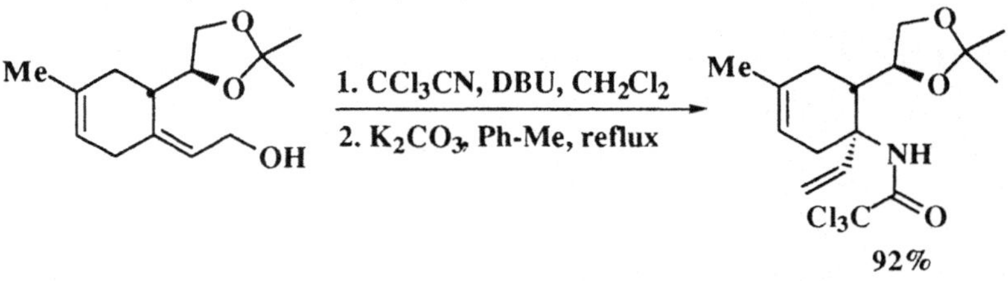

1. CCl$_3$CN, DBU, CH$_2$Cl$_2$
2. K$_2$CO$_3$, Ph-Me, reflux

92%

I.H.2. Other Rearrangements

I.H.2-1 Dake, G.R. et al., *OL*, **3**, 2109; Pearson, A.J., Dorange, I.B., *JOC*, **66**, 3140; **see also:** Paquette, L.A. et al., *JOC*, **66** 2828; **see also:** Doris, E. et al., *JOC*, **66**, 4450.

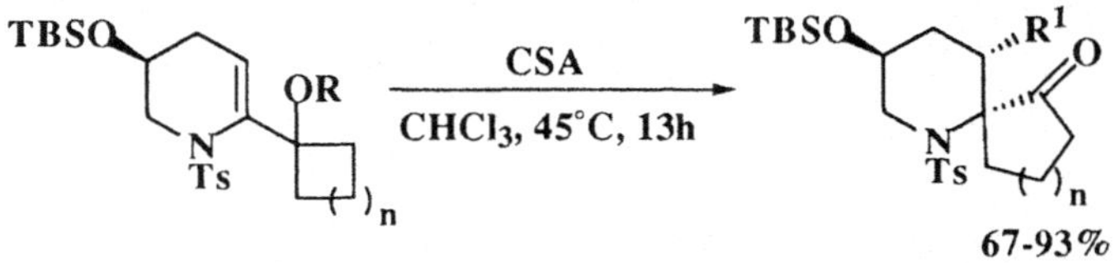

CSA
CHCl$_3$, 45°C, 13h

67-93%

I.H.2-2 Tu, Y.Q. et al., *TA*, **12**, 1459 and *SC*, **31**, 1613.

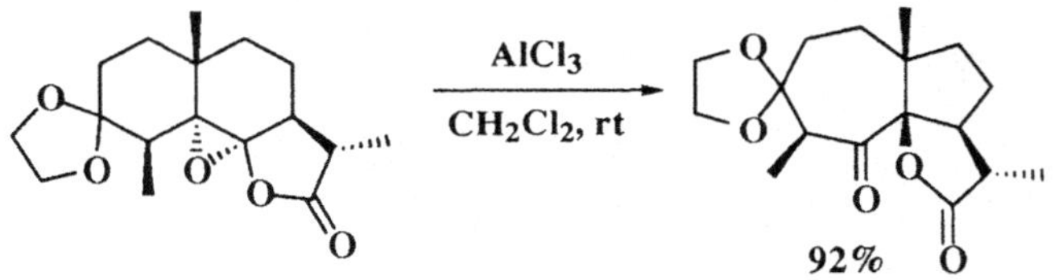

AlCl$_3$
CH$_2$Cl$_2$, rt

92%

I.H.2-3 Mohan, R.S. et al., *TL*, **42**, 8129; **see also:** Kita, Y. et al., *JOC*, **66**, 8779; Yamano, Y., Ito, M., *CPB*, **49**, 1662.

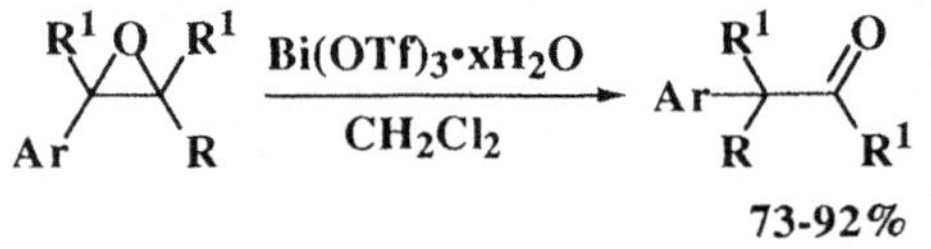

Bi(OTf)$_3$·xH$_2$O
CH$_2$Cl$_2$

73-92%

I.H.2-4 Wijnberg, J.B.P.A., de Groot, A. et al., *JOC*, **66**, 2350.

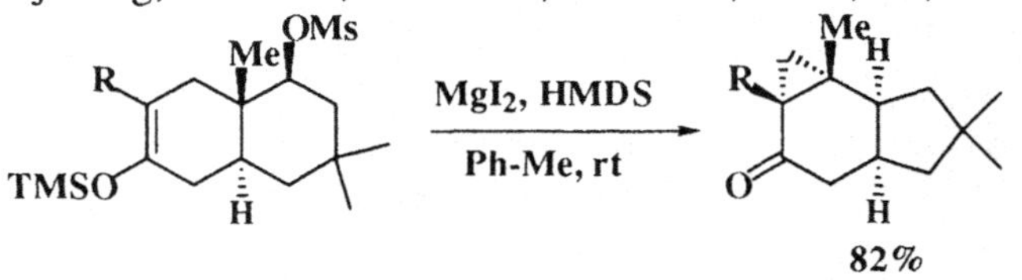

MgI$_2$, HMDS
Ph-Me, rt

82%

I.H.2-5 Rychnovsky, S.D. et al., *JOC*, **66**, 4679.

84%

I.H.2-6 Ila, H., Junjappa, H. et al., *JOC*, **66**, 1503.

61-82%

I.H.2-7 Bogdanowicz-Szwed, K. et al., *JOC*, **66**, 7205.

55-89%

I.H.2-8 Garner, C.M. et al., *TL*, **42**, 2261.

91%

I.H.2-9 Hodgson, D.M., Petroliagi, M., *TA*, **12**, 877.

81-93%

33-79%
e.e. = 12-62%

I.H.2-10 Trost, B.M., Yasukata, T., *JACS*, **123**, 7162.

58-99%
e.e. = 69-93%

I.H.2-11 Uemura, S. et al., *JOC*, **66**, 1455.

81-93%

I.H.2-12 Lees, W.J. et al., *JOC*, **66**, 4739.

75%

I.H.2-13 Ogasawa, R.A. et al., *TL*, **42**, 7587.

98%

I.H.2-14 Jung, M.E., Davidov, P., *OL*, **3**, 3025.

84%

I.H.2-15 Cha, J.K. et al., *TL*, **42**, 8769.
On the Stereochemistry of Anion-Accelerated [1,3]-Sigmatropic Rearrangement of Vinylcyclobutanols.

I.H.2-16 Cha, J.K. et al., *JACS*, **123**, 11322; **see also:** Mascarenas, J.L. et al., *OL*, **3**, 1181; Kim, Y.H. et al., *SL*, 1266.

I.H.2-17 Savoia, D. et al., *EJOC*, 2917.

I.H.2-18 Mc Allister, G.D., Taylor, R.J.K., *TL*, **42**, 1197.

II
OXIDATIONS

II.A. C-O Oxidations

II.A.1. Alcohols→Ketones, Aldehydes

II.A.1-1 Mukaiyama, T. et al., *CL*, 846; Itoh, A. et al., *OL*, **3**, 2653.

$$R\text{-}CH_2OH \xrightarrow[\text{CH}_2\text{Cl}_2,\ 0°C,\ 1h]{\text{NCS, Ph-SNH}^t\text{Bu. K}_2\text{CO}_3,\ 4\text{Å MS}} R\text{-}CHO$$
93-99%

[similar oxidation using FSM-16 mesoporous SiO$_2$, CsI, hv]

II.A.1-2 Oshima, K. et al., *SL*, 1421.

$$R\text{—}\!\!\equiv\!\!\text{—}CH_2OH \xrightarrow[\text{CH}_2\text{Cl}_2,\ 0°C]{\text{TiCl}_4,\ \text{TEA}} R\text{—}\!\!\equiv\!\!\text{—}CHO$$
80-98%

II.A.1-3 Giacomelli, G. et al., *JOC*, **66**, 7907 and *OL*, **3**, 3041.

$$\underset{R^1}{\overset{R}{\diagdown}}\!\!CH\text{-}OH + \text{(triazine)} \xrightarrow[\text{2. TEA, -30°C}]{\text{1. DMSO, THF, -30°C, 3min}} \underset{R^1}{\overset{R}{\diagdown}}C{=}O$$

90-94% (ketones)
20-93% (aldehydes)

II.A.1-4 Bolm, C. et al., *JOC*, **66**, 8154.

$$\underset{R^1}{\overset{R}{\diagdown}}\!\!CH\text{-}OH + \text{(TEMPO resin)} \xrightarrow[\text{CH}_2\text{Cl}_2/\text{H}_2\text{O},\ 0°C,\ 1h]{\text{NaOCl, KBr, pH 9.1}} \underset{R^1}{\overset{R}{\diagdown}}C{=}O$$

44-99%

II.A.1-5 Shaabani, A., Karimi, A.-R., *SC*, **31**, 759.

$$\underset{Ar}{\overset{R}{\diagdown}}\!\!CH\text{-}OTMS \xrightarrow[\text{MeCN/H}_2\text{O}]{\text{NaBrO}_3,\ \text{NH}_4\text{Cl}} \underset{Ar}{\overset{R}{\diagdown}}C{=}O$$

45-90%

II.A.1-6 Neumann, R. et al., *JOC*, **66**, 8650; Sheldon, R.A. et al., *JACS*, **123**, 6826; Galli, C. et al., *TL*, **42**, 7551.

$$\underset{R^1}{\overset{R}{\diagdown}}\text{OH} \xrightarrow[\text{Me}_2\text{CO}]{\text{H}_5(\text{PMo}_{10}\text{V}_2\text{O}_{40}),\ \text{TEMPO}} \underset{R^1}{\overset{R}{\diagdown}}\text{O}$$

94-99% conversion

[similar TEMPO oxidation with RuCl$_2$(PPh$_3$)$_3$ or Laccase in the presence of O$_2$]

II.A.1-7 Uemura, S. et al., *JOC*, **66**, 6620; Choudary, B.M. et al., *AG(E)*, **40**, 763.

$$\underset{R^1}{\overset{R}{\diagdown}}\text{OH} \xrightarrow[\text{pyr, Ph-Me, 65°C}]{\text{air, Pd(OAc)}_2,\ \text{hydrotalcite}} \underset{R^1}{\overset{R}{\diagdown}}\text{O}$$

74-99%

[smilarly with Ni/Al hydrotalcite/O$_2$]

II.A.1-8 Banik, B.K. et al., *SC*, **31**, 2691; Hirano, M. et al., *JCR(S)*, 274.

$$\underset{R^1}{\overset{R}{\diagdown}}\text{OH} \xrightarrow{\text{Montmorillonite, Bi(NO}_3)_3} \underset{R^1}{\overset{R}{\diagdown}}\text{O}$$

69-99%

[similar oxidation with moist Montmorillonite, (NH$_4$)$_2$S$_2$O$_8$/AgNO$_3$]

II.A.1-9 Hajipour, A.R., Mallakpour, S.E., Samimi, H.A., *SL*, 1735; Saidi, M.R. et al., *M*, **132**, 655.

$$\underset{R^1}{\overset{R}{\diagdown}}\text{OH} \xrightarrow[\text{MeCN, reflux}]{\text{BnP}^+\text{Ph}_3\ \text{IO}_4^-,\ \text{AlCl}_3} \underset{R^1}{\overset{R}{\diagdown}}\text{O}$$

78-99%

[benzylic oxidation also reported with CaOCl, moist Al$_2$O$_3$, μW]

II.A.1-10 Uemura, S. et AL., *TL*, **42**, 8877.

$$\underset{R}{\overset{\text{HO}}{\diagdown}}{=}\!\!=\!\!=\!\!R^1 \xrightarrow[\text{MeCN, 80°C, 3h}]{\text{O}_2,\ \text{VO(acac)}_2,\ 3\text{Å MS}} \underset{R}{\overset{\text{O}}{\diagdown}}{=}\!\!=\!\!=\!\!R^1$$

8-99%

II.A.1-11 Bruckner, C. et al., *TL*, **42**, 8793.

80-97%

II.A.1-12 Shirini, F. et al., *JCS(R)*, 476; Beller, M. et al., *TL*, **42**, 8447; Hashimoto, K. et al., *SL*, 922.

$$R \diagdown \diagup OH \xrightarrow[\text{wet SiO}_2, \text{hex or neat}]{(NH_4)_2Cr_2O_4, ZrCl_4} R \diagdown \diagup O$$

$$R^1 \qquad\qquad\qquad R^1$$

85-95%

[similar oxidations with K$_2$[OsO$_2$(OH)$_4$], O$_2$/DABCO
or with Ru(tmp)(O)$_2$, N$_2$O]

II.A.1-13 Crich, D., Neelamkavil, S., *JACS*, **123**, 7449.

$$R \diagdown \diagup OH + C_4F_9 \diagdown \diagup \overset{O}{\underset{\|}{S}} Me \xrightarrow[\text{CH}_2\text{Cl}_2, -30°C]{(COCl)_2} R \diagdown \diagup O$$

$$R^1 \qquad\qquad\qquad\qquad\qquad R^1$$

77-92%

II.A.2. Alcohols, Aldehydes→Acids, Esters

No Examples this year

II.B. C-H Oxidations

II.B.1 C-H→C-O

II.B.1-1 Li, Z. et al., *JOC*, **66**, 8424.

$$R\text{-}N\bigcirc \xrightarrow[\text{H}_2\text{O}]{O_2, H^+, \textit{sphingomonas sp.}} R\text{-}N\underset{*}{\bigcirc}OH$$

36-80% conversion
e.e. = 9-76%

II.B.1-2 Chaudhuri, M.K. et al., *T*, **57**, 2445; Li, P. et al., *JOC*, **66**, 4087.

$$\xrightarrow[\text{MeCN, reflux}]{\text{DmpzHFC}}$$

70-75%

(also examples of alcohol oxidations)
[similar ketone formation reported from alkynes using
CuCl$_2$·F$_2$H$_2$O/^{t}BuOOH]

II.B.1-3 Makosza, M. et al., *JOC*, **66**, 5022; Laali, K.K. et al., *JCS(P1)*, 578; Nicolaou, K.C. et al., *JACS*, **123**, 3183.

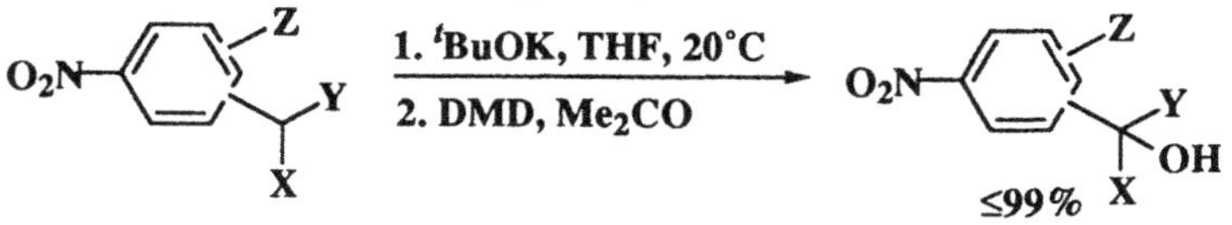

O_2N — (ring, Z, Y, X)

1. ^tBuOK, THF, 20°C
2. DMD, Me_2CO

→ O_2N — (ring, Z, Y, OH, X)

≤99%

[benzylic oxidation of aromatics also reported with cerium(II) triflate]

II.B.1-4 Ishii, Y. et al., *JOC*, **66**, 7889.

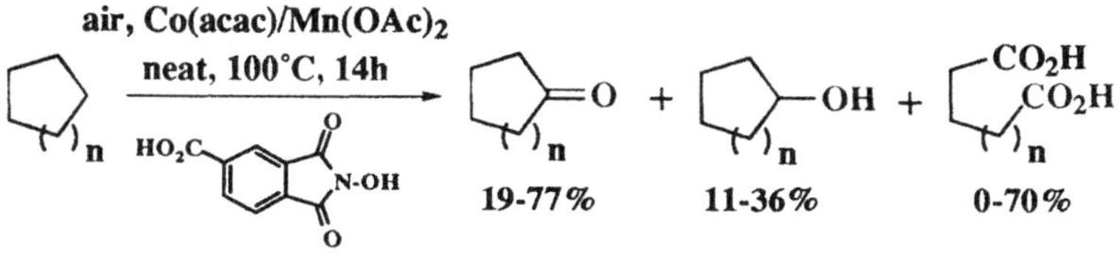

air, Co(acac)/Mn(OAc)$_2$

neat, 100°C, 14h

HO_2C — (N-OH imide)

→ (=O) 19-77% + (−OH) 11-36% + (CO_2H, CO_2H) 0-70%

II.B.1-5 Zhang, X. et al., *TL*, **42**, 5335

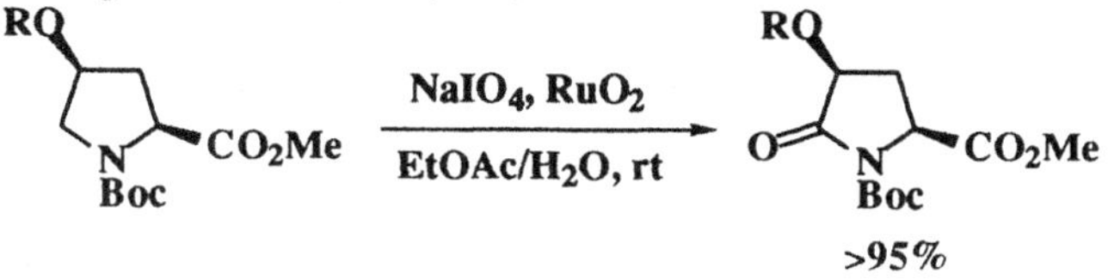

RO (pyrrolidine, N-Boc, CO_2Me)

$NaIO_4$, RuO_2

EtOAc/H_2O, rt

→ RO (O=, N-Boc, CO_2Me)

>95%

II.B.1-6 Koser, G.F. et al., *TL*, **42**, 5597; Togo, H. et al., *JOC*, **66**, 6174; Lee, J.C., Choi, J.-H., *SL*, 234.

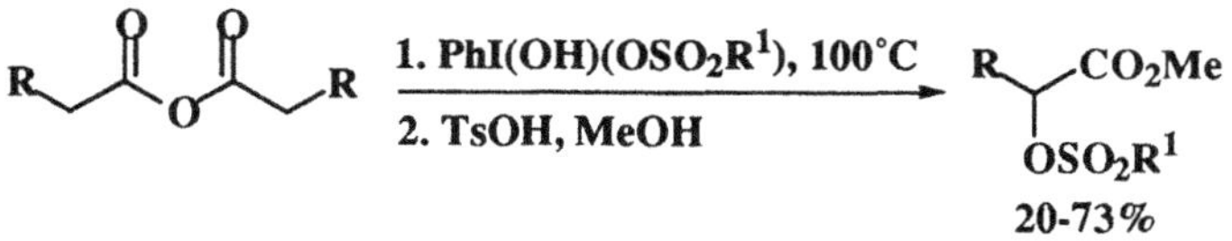

R (anhydride) R

1. $PhI(OH)(OSO_2R^1)$, 100°C
2. TsOH, MeOH

→ R CO_2Me, OSO_2R^1

20-73%

[similar tosylate formation reported using a polymer supported hypervalent iodine species or $PhI(OAc)_2$, RSO_3H]

II.B.1-7 Tanyeli, C. et al., *TL*, **42**, 6397; Li, Y. et al., *TL*, **42**, 3101.

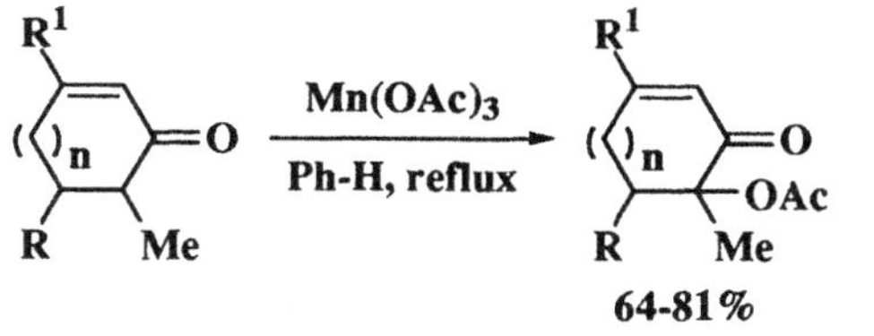

R^1 (cyclohexenone, =O, R, Me)

$Mn(OAc)_3$

Ph-H, reflux

→ R^1 (=O, OAc, R, Me)

64-81%

II.B.1-8 Kwong, H.-L. et al., *TA.*, **12**, 1007.

38-80%
e.e. = 0-52%

II.B.1-9 Ochiai, M. et al., *OL*, **3**, 2387.

54-84%

II.B.2 C-H→C-Hal

II.B.2-1 Schreimer, P.R., Fokin, A.A. et al., *ECJ*, **7**, 4996.
**Selective Radical Reactions in Multiphase Systems: Phase-
Transfer Halogenations of Alkanes.**

II.B.2-2 Braibante, M.E.F. et al., *S*, 1935.

NBS, Montmorillonite K-10

MeOH, rt

38-74%

II.C C-N Oxidations

II.C-1 Cicchi, S. et al., *TL*, **42**, 6503.

MnO$_2$

CH$_2$Cl$_2$, rt

85-96%

II.C-2 Reddy, P.S.N. et al., *SC*, **31**, 3447.

μW, Al$_2$O$_3$

48-67%

II.C-3 Mukaiyama, T. et al., *CL*, 712.

$$R\text{-}CH(NHMs)R^1 \xrightarrow[\text{2. aq HCl, CH}_2\text{Cl}_2/\text{Et}_2\text{O}]{\substack{\text{1. } {}^t\text{BuN=S(Cl)Ph, DBU} \\ \text{CH}_2\text{Cl}_2, \text{-78°C, 1h}}} R\text{-CO-}R^1$$

73-98%

II.D. Amine Oxidations

II.D-1 Detomaso, A., Curci, R., *TL*, **42**, 755

Boc-N(H)-CH(R)-(CH$_2$)$_n$-CO$_2$Me + F$_3$C-C(Me)(O-O) $\xrightarrow{\text{5-8h, -20}\rightarrow\text{0°C}}$ Boc-N(OH)-CH(R)-(CH$_2$)$_n$-CO$_2$Me

57-82%

II.D-2 Katritzky, A.R. et al., *JOC*, **66**, 5585.

$$\text{benzotriazole-}N\text{-}R \xrightarrow[\text{CH}_2\text{Cl}_2, \text{rt}]{\text{DMD}} \text{benzotriazole } N\text{-oxide-}N\text{-}R$$

35-92%

II.D-3 Penkett, C.S., Simpson, I.D., *TL*, **42**, 3029.

$$\text{(MeO)}_2\text{C, NEt-Me aziridine} \xrightarrow[\text{MeCN, 0°C}]{\text{MCPBA, NaHCO}_3} \text{HO, (MeO)}_2\text{C, N}^+\text{(O}^-\text{)=CHEt, Me}$$

69%

II.D-4 Zolfigol, M.A. et al., *JCR(S)*, 390.

$$\text{HN-NH (Na) urazole-}R \xrightarrow[\substack{\text{or (NH}_2)_2\text{CO/H}_2\text{O}_2, \text{MCl}_x \\ M = \text{Al, Zr, W}}]{\text{C}_6\text{Cl}_5\text{-NCl}_2, \text{CH}_2\text{Cl}_2} \text{N=N triazolinedione-}R$$

75-99%

II.E. Sulfur Oxidations

II.E-1 Martin, S.E., Rossi, L.I., *TL*, **42**, 7147; Backvall, J.-E. et al., *ECJ*, **7**, 287; Noyori, R. et al., *T*, **57**, 2469; Patonay, T. et al., *JOC*, **66**, 2275; Koo, S. et al., *JOC*, **66**, 8192.

$$R\text{-}S\text{-}R^1 \xrightarrow[\text{MeCN, 25°C, air}]{\text{Fe(NO}_3)_3\text{, FeBr}_3\text{ or Cu(NO}_3)_2\text{, CuBr}_2} \overset{\overset{O}{\|}}{R\text{-}S\text{-}R^1}$$

82-92%

(similar oxidation to sulfoxides with riboflavin/H_2O_2;
30% H_2O_2, dioxirane or H_2O_2/LiNbMoO$_6$)

II.E-2 Skarzewski, J. et al., *JCR(S)*, 263.

≤90%
e.e. = 80%

≤92%
d.e. = ≤80%
e.e. = 95%

I.E-3 Saito, B., Katsuki, T., *TL*, **42**, 3873; Maguire, A.R. et al., *SL*, 41.

78-93%
e.e. = 92-99%

[similar enentioselective oxidation with Ti(O^{i}Pr)$_4$, DET, H_2O_2]

I.E-4 Matsugi, M. et al., *T*, **57**, 2739.

56-89%
e.e. = 31-77%

II.E-5 Scettri, A. et al., *TA*, **12**, 2775.

$$Ar-S-Me + \text{(furan)} \xrightarrow[\text{Ph-Me/H}_2\text{O, 0°C}]{\text{Ti(O}^i\text{Pr)}_4 \text{ (R)-BINOL}} Ar-S(=O)-Me$$

56-95%
e.e. = 49-93%

II.F. Oxidative Additions to C-C Multiple Bonds

II.F.1 Epoxidations

II.F.1-1 Jacobsen, E.N. et al., *JACS*, **123**, 7194; Ichihara, J., *TL*, **42**, 695; Lane, B.S., Burgess, K., *JACS*, **123**, 2933; Monfared, H.H., Ghorbani, M., *M*, **132**, 989.

$$\underset{R}{R^1}\!\!=\!\!R^2 \xrightarrow{\text{AcOH, H}_2\text{O}_2 \text{ MeCN, 4°C}} \underset{R}{R^1}\!\!-\!\!\overset{O}{\triangle}\!\!-\!\!R^2$$

$$\left[\text{Me-N} \underset{\text{N-N}}{\overset{\text{Me}}{\diagup}} \cdot \text{Fe(II)(MeCN)}_2 \right]^{2+} (\text{SbF}_6)_2$$

61-90%

[similar epoxidations with H_2WO_4/$(NH_2)_2CO \cdot H_2O_2$/fluoroapatite; $MnSO_4$, H_2O_2 or Fe^{2+}-SiO_2/H_2O_2]

II.F.1-2 Sheldon, R.A. et al., *SL*, 1305 and *JCS(P1)*, 224; Crousse, B. et al., *TL*, **42**, 4463.

$$\underset{R}{R^1}\!\!=\!\!R^2 \xrightarrow[\text{(CF}_3)_2\text{CHOH, rt}]{\text{(CF}_3)_2\text{CO, H}_2\text{O}_2} \underset{R}{R^1}\!\!-\!\!\overset{O}{\triangle}\!\!-\!\!R^2$$

55-95%

I.F.1-3 Katsuki, T. et al., *ECJ*, **7**, 3776; Che, C.-M. et al., *JOC*, **66**, 8145.

$$\underset{R \quad R^4}{R^1 \quad R^2}\!\!=\!\! \xrightarrow[\text{catalyst}]{\text{Ph-H or Et}_2\text{O, Cl-N}^+\text{(O}^-)\text{-Cl}} \underset{R \quad R^4}{R^1 \quad O \quad R^2}$$

26-70%
e.e. = ≤87%

II.F.1-4 Yudin, A.K. et al., *JOC, 66, 4713.*

70-96%

(other Re reagents were used)

II.F.1-5 Shi, Y. et al., *OL*, 3, 715, 1929 and *JOC*, **66**, 521, 1818.

67-99%
e.e. = 83-96%

II.F.1-6 Kureshy, R.I. et al., *TA*, **12**, 433.

68-99% conversion
e.e. = 23-100%

I.F.1-7 Page, P.C.B. et al., *JOC*, **66**, 6926.

34-73%
e.e. = ≤60%

II.F.1-8 Chen, R., Qian, C., deVries, J.G., *TL*, **42**, 6919; Shibasaki, M. et al., *JACS*, **123**, 9474; Jackson, R.F.W. et al., *CC*, 2712.

81-95%
e.e. = 38-95%

[similar chiral epoxidations wiyh La-(S)-BINOL, Ph₃As=O,
TBHP or TBHP, Mg Et₂Tartrate]

II.F.1-9 Adam, W. et al., *TA*, **12**, 121.
Asymmetric Weitz-Scheffer Epoxidation of Conformationally Flexible and Fixed Enones with Sterically Demanding Hydroperoxides Mediated by Optically Active Phase-Transfer Catalysts.

II.F.1-10 Spivey, A.C. et al., *AG(E)*, **40**, 769.

Zr(O^{i}Pr)$_4$•^{i}PrOH, ^{t}BuOOH, 4Å MS

(L)-(+)-DIPT, CH$_2$Cl$_2$, -20°C

59%
e.e. = 95%

I.F.1-11 Ikegami, S. et al., *OL*, **3**, 1837.

H$_2$O$_2$, rt

73-99%

II.F.1-12 Lygo, B., To, D.C.M., *TL*, **42**, 1343.

aq NaOCl, Ph-Me, rt, 24h

75-98%
d.e. = ≥95%
e.e. = 84-98%

II.F.1-13 Seki, M. et al., *TL*, **42**, 8201; Fogagnolo, M. et al., *TA*, **12**, 1113.

Oxone, NaHCO$_3$

0-95%
e.e. = 49-85%

[similarly with vinyl carboxylic acids, oxone and dehydrocholic acid]

II.F.2 Hydroxylations

II.F.1-1 Corey, E.J., Zhang, J., *OL*, **3**, 3211.
Highly Effective Transition Structure Designed Catalyst for the Enantio- and Position-Selective Dihydroxylation of Polyisoprenoids.

II.F.2-2 Krief, A., Castillo-Colaux, C., *SL*, 501.
Catalytic Asymmetric Dihydroxylation of C,C Double Bonds with Osmium Tetroxide Using Selenoxides as Co-Oxidents.

II.F.2-3 Choudary, B.M. et al., *JACS*, **123**, 9220; Hyeon, T., Kim, B.M. et al., *TA*, **12**, 1537; Song, C.E., Choi, J.H. et al., *TA*, **12**, 1533; Avenoza, A., Peregrina, J.M. et al., *TA*, **12**, 1383; Kobayashi, Y. et al., *JOC*, **66**, 7903.

$$R \diagup\!\!\diagup R^1 \xrightarrow[\text{'BuOH/H}_2\text{O}]{\text{LDH/OsO}_4,\ (\text{DHQD})_2\text{PHAL, NMO}} \quad \begin{array}{c} HO \quad * \quad R^1 \\ R \quad * \quad OH \end{array}$$

89-97%
e.e. = 77-99%

[Sharpless asymmetric dihydroxylation conditions used also with a silica-supported ligand; with high pressure; with a vinyl Weinreb's amide or with a vinyl phosphonate]

II.F.2-4 Backvall, J.-E. et al., *JACS*, **123**, 1365; Jacobs, P.A. et al., *AG(E)*, **40**, 586.

$$R \diagup\!\!\diagup \!\!\!\!^{R^1}_{R^2} \xrightarrow{\text{H}_2\text{O}_2,\ \text{OsO}_4\ \text{NMM, Et}_4\text{NOAc}} \quad \begin{array}{c} HO \quad R^2 \\ R^1 \diagup \\ R \quad OH \end{array}$$

72-95%

[similarly with HMM and a silica-supported Os catalyst]

II.F.2-5 Periasamy, M. et al., *AG(E)*, **40**, 512; Motorina, I., Crudden, C.M., *OL*, **3**, 2325.

$$\underset{Ph}{\overset{Ph}{\diagdown}}\!\!\diagup \xrightarrow[\text{'BuOH/H}_2\text{O}]{\text{K}_3[\text{Fe(CN)}_6],\ \text{K}_2\text{CO}_3,\ \text{K}_2\text{OsO}_2(\text{OH})_4} \quad \begin{array}{c} HO \quad Ph \\ Ph \quad OH \end{array}$$

(H₂Cinchonine)-O₂C / CO₂(H₂Cinchonine)

59-89%
e.e. = 20-42%

[similarly with an SBA-15 supported cinchona]

II.F.2-6 Que, L., Jr. et al., *JACS*, **123**, 6722.

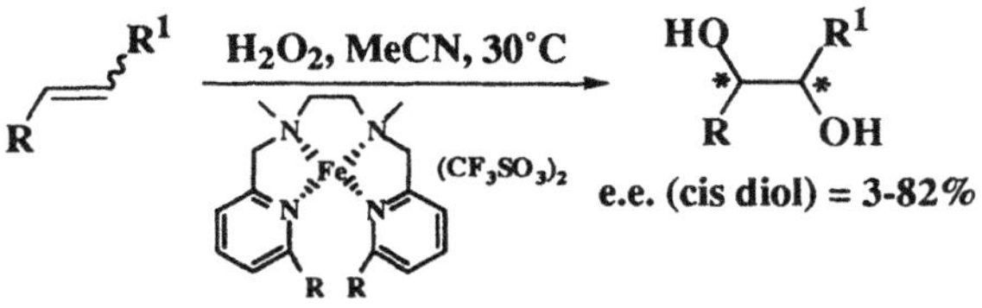

e.e. (cis diol) = 3-82%

II.F.2-7 Hazra, B.C. et al., *JCR(S)*, 500.

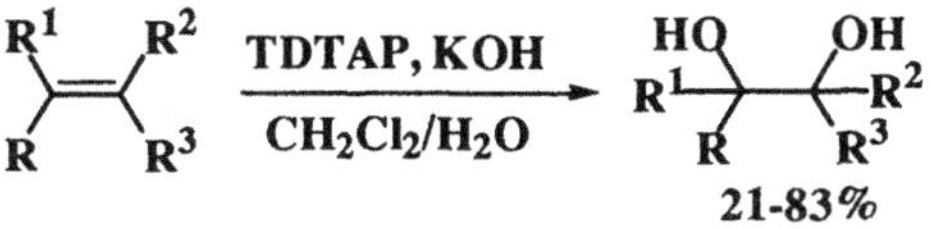

21-83%

II.F.2-8 Meyers, A.G. et al., *OL*, 3, 2923.

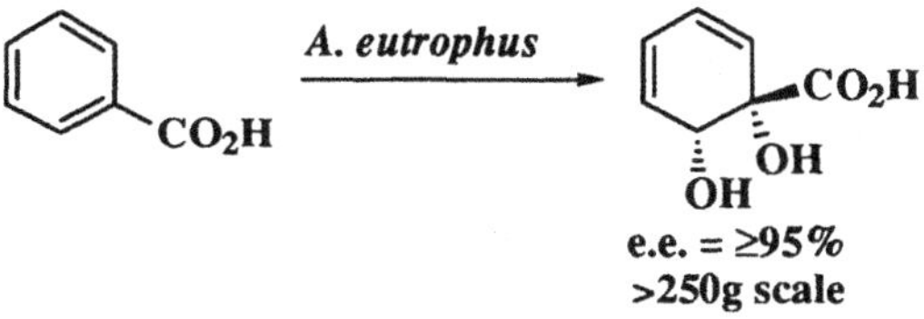

e.e. = ≥95%
>250g scale

II.F.3 Other Oxidative Additions to C-C Multiple Bonds

II.F.3-1 Lai, S., Lee, D.G., *S*, 1645.

$$ArCH=CHY \xrightarrow[\text{CH}_2\text{Cl}_2, \text{ rt}]{\text{KMnO}_4 \text{ solid support}} Ar\text{-}CHO$$

74-98%

II.F.3-2 Taber, D.F., Nakajima, K., *JOC*, **66**, 2515.

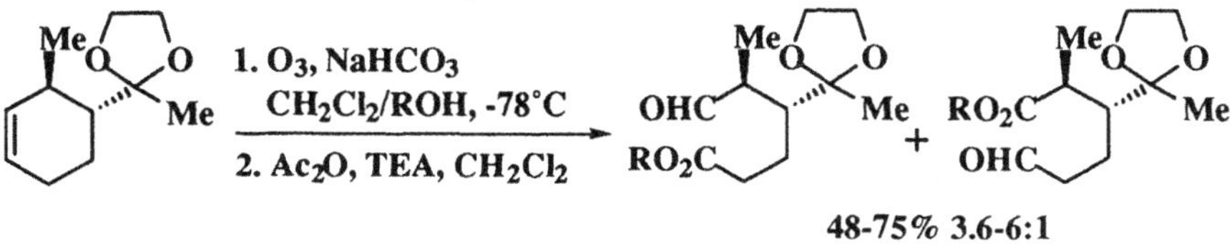

48-75% 3.6-6:1

II.F.3-3 Jun, C.-H. et al., *JACS*, **123**, 8600.

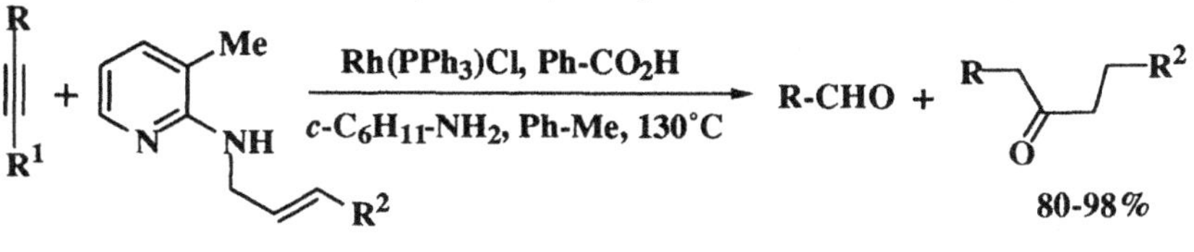

80-98%

II.F.3-4 Juge, S. et al., *JOC*, **66**, 4504; Yang, D., Zhang, C., *JOC*, **66**, 4814.

13-98%

[sililar oxidation cleavage with oxone, RuCl₃, NaHCO₃]

II.G. Phenol-Quinone Oxidations

II.G-1 Chi, D.Y. et al., *OL*, **3**, 445.

NBS, H₂SO₄, H₂O

THF, rt, 5min

98%

II.G-2 Yamazaki, S., *TL*, **42**, 3355.

CrO₃·H₅IO₆

44-79%

II.H. Dehydrogenations

II.H-1 Hauser, F.M. et al., *SC*, **31**, 77.

1. TMS-OTf, TEA

2. Pd(OAc)₂, MeCN

17-82%

II.H-2 Ishikawa, T., Saito, S. et al., *JOC*, **66**, 186.

PdCl₂(MeCN)₂, TEA

MeCN, rt, 2h

47-99%

II.H-3 Tilstam, U. et al., *TL*, **42**, 5385.

72-89%

II.H-4 Lu, J. et al., *SC*, **31**, 2625.

57-78%

II.H-5 Kibayashi, C. et al., *JOC*, **66**, 1494.

56-87%   e.e. = 24-82%
                (major enantiomer)

II.I Other Oxidations

II.I-1 Wong, M.-K., Yang, D. et al., *JOC*, **66**, 3606.

56-89%

II.I-2 Antoniotti, S., Dunach, E., *CC*, 2566.

48-83%

II.I-3 Ashford, S.W., Crega, K.C., *JOC*, **66**, 1523.

$R\text{-}CO_2H + R^1\text{-}CO_2H$

14-96%

II.I-4 Barluenga, J. et al., *AG(E)*, **40**, 3491.

$$\underset{R^1}{\overset{R}{\underset{(CH_2)_n}{\bigcirc}}}OH \xrightarrow[\text{CH}_2\text{Cl}_2]{\overset{h\nu}{\text{IPyr}_2\text{BF}_4,\ \text{CsCO}_3}} \underset{R^1}{\overset{R}{\underset{(CH_2)_n}{\bigcirc}}}\overset{O}{\underset{I}{}}$$

76-94%

II.I-5 Henry, P.M. et al., *JOC*, **66**, 180.

$$\underset{n}{\bigcirc}O \xrightarrow[\text{MeOH}]{\text{PdCl}_2,\ \text{CuCl}_2,\ \text{CO}} \begin{array}{c} \text{MeO}_2\text{C}(\text{CH}_2)_2(\text{CH}_2)_n\text{CH}_2\text{CO}_2\text{Me} \\ + \\ \text{MeO}_2\text{C}(\text{CH}_2)_2(\text{CH}_2)_n\text{CH}_2\text{Cl (minor)} \end{array}$$

30-83% 1.86-100:1

II.I-6 Shi, M. Feng, Y.-S., *JOC*, **66**, 3235.

$$R\underset{}{\bigcirc}X \xrightarrow[\text{R}_3\text{N}^+\text{Me HSO}_4^-,\ 90°\text{C}]{\text{H}_2\text{O}_2,\ \text{Na}_2\text{WO}_4\cdot2\text{H}_2\text{O},\ 4\text{Å MS}} R\underset{}{\bigcirc}\text{CO}_2\text{H}$$

30-91%

II.I-7 Kayser, M.M., Stewart, J.D. et al., *JOC*, **66**, 733; Sheldon, R.A. et al., *JOC*, **66**, 2429.

$$\underset{R}{\overset{R^1}{\bigcirc}}=O \xrightarrow{\text{engineered } E.\ coli} \underset{R}{\overset{R^1}{\bigcirc}}\overset{O}{\underset{O}{}}$$

54-91%
e.e. = 9-98%

[Baeyer-Villager reported also with H$_2$O$_2$, (ArSe)$_2$, CF$_3$CH$_2$OH]

III
REDUCTIONS

III.A. C=O Reductions

III.A-1 Bandgar, B.P., Kamble, V.T., *SC*, **31**, 3037.

$$R\text{—}C_6H_4\text{—CHO} \xrightarrow[\text{MeOH, 25°C}]{\text{SBER}} R\text{—}C_6H_4\text{—CH}_2\text{OH}$$

82-93%

III.A-2 Sengupta, S. et al., *SL*, 1464.

$$\xrightarrow[\text{Ph-H/THF, 0°C}]{\text{Zn(BH}_3)_4}$$

60-83%
d.e. = 80-90%

III.A-3 Cha, J.S. et al., *JOC*, **66**, 7514.

$$\xrightarrow[\text{reflux, 5days}]{\text{BH}_3\text{·THF}}$$

Me
99%
trans = 99%

III.A-4 Fu, I.-P., Uang, B.-J., *TA*, **12**, 45; Uang, B.-J. et al., *TA*, **12**, 3217.

$$\xrightarrow{\text{BH}_3\text{·SMe}_2\text{, chiral Al complex}}$$

91-97%
e.e. = 30-83%

III.A-5 Yoon, C.M. et al., *TL*, **42**, 2137.

$$\xrightarrow[\text{MeOH, 50°C}]{\text{B}_{10}\text{H}_{14}\text{, CeCl}_3\text{·7H}_2\text{O, pyrrolidine}}$$

92-98%

III.A-6 Yamada, T. et al., *OL*, **3**, 3421, 2543.

41-48%
anti = 94-99%
e.e. = 95-98%

III.A-7 Frejd, T. et al., *ECJ*, **7**, 2158.

≤99%
e.e. = ≤96%

III.A-8 Zhao, G. et al., *JOC*, **66**, 303 and *AG(E)*, **40**, 1109; **see also:** Cho, B.Y., Kim, D.J., *TA*, **12**, 2043; Xie, R.-G. et al., *TA*, **12**, 1907.

86-99%
e.e. = 52-95%

[other chiral amino alcohols used for similar reductions]

III.A-9 Bartoli, G. et al., *TL*, **42**, 8811; Dalpozzo, R. et al., *EJOC*, 2971.

87-99%
syn:anti = 45-99:1

III.A-10 Yamamoto, Y. et al., *JACS*, **123**, 6931; Nishiyama, Y., Sonoda, N. et al., *OL*, **3**, 3087.

90%
trans:cis = 1:7

[similarly with PhSe-TMS, Bu₃SnH, AIBN]

III.A-11 Kunieda, T. et al., *TL*, **42**, 8857.

$$\underset{Ph}{\overset{Et}{\diagdown}}C=O \xrightarrow[\text{\textit{i}PrOH, reflux}]{RuCl_2(PPh_3)_3,\ Yb(OTf)_3,\ NaOH} \underset{Ph}{\overset{Et}{\diagdown}}CH-OH$$

80%
35% without Yb(OTf)$_3$

III.A-12 Fuchikami, T. et al., *TL*, **42**, 2149.

$$R\text{-}CO_2R^1 \xrightarrow[\text{Ph-Me, 100°C, 10h}]{Et_3SiH,\ [RuCl_2(CO)_3]_2,\ Et\text{-}I,\ Et_2NH} R\underset{OEt}{\overset{OSiEt_3}{\diagdown}}$$

44-98%

III.A-13 Riant, O. et al., *OL*, **3**, 4111; Magnus, P., Fielding, M.R., *TL*, **42**, 6633.

$$\underset{R}{\overset{R^1}{\diagdown}}C=O \xrightarrow[\text{Ph-Me, rt, air}]{PhSiH_3,\ CuF_2,\ (S)\text{-}BINAP} \underset{R}{\overset{R^1}{\diagdown}}CH\text{-}^{\prime\prime\prime\prime}OH$$

79-99%
e.e. = 20-92%

[similarly with Mn(dpm)$_3$]

III.A-14 Lipshutz, B.H. et al., *JACS*, **123**, 12917.

$$\underset{Ar}{\overset{R}{\diagdown}}C=O \xrightarrow[\text{Ph-Me, -50°C}]{PMHS,\ CuCl,\ (R)\text{-}(Ph\text{-}Me_2)\text{-}MeO\text{-}BIPHEP} \underset{Ar}{\overset{R}{\diagdown}}CH\text{-}^{\prime\prime}OH$$

87-99%
e.e. = 78-97%

III.A-15 Talukdar, S., Fang, J.-M., *JOC*, **66**, 330.

$$\underset{R}{\overset{R^1}{\diagdown}}C=O \xrightarrow[\text{THF}]{Sm,\ aq\ HCl} \underset{R}{\overset{R^1}{\diagdown}}CH\text{-}OH$$

85-99%

III.A-16 Yasohara, Y., *TA*, **12**, 1713; Ema, T., Sakai, T. et al., *JOC*, **66**, 8682; Bornscheuer, U.T. et al., *TA*, **12**, 1207.

$$\underset{R}{\overset{RO_2C}{\diagdown}}C=O \xrightarrow{\text{carbonyl reductase S1}} \underset{R}{\overset{RO_2C}{\diagdown}}CH\text{-}OH$$

e.e. = ≤99%

III.A-17 Deng, J.-G. et al., *CC*. 1488.
Dendritic Catalysts for Asymmetric Transfer Hydrogen.

III.A-18 Kawamoto, A.M., Wills, M., *JCS(P1)*, 1916; Wills, M. et al., *TA*, **12**, 1801; Hage, A. et al., *TA*, **12**, 1025.

[similarly with a Rh catalyst or a chiral amino alcohol ligand]

III.A-19 Ikariya, T. et al., *OM*, **20**, 379; Abdur-Rashid, K., Lough, A.J., Morris, R.M., *OM*, **20**, 1047.

70-99% conversion
e.e. = 64-95%

[similarly with other Ru complexes and chiral diamine]

III.A-20 Leadbetter, N.E., *JOC*, **66**, 2168; Faller, J.W., Lavoie, A.R., *OM*, **20**, 5245; Nolan, S.P. et al., *OM*, **20**, 4246; Nguyen, S.T. et al., *OL*, **3**, 2391.

40-83%

[similar reductions, including enantioselective examples,
using Ru, Rh, Ir or Al complexes]

III.A-21 Gefflaut, T. et al., *JOC*, **66**, 2296; Rebolledo, F. et al., *TA*, **12**, 513; Yadav, J.S. et al., *TA*, **12**, 63; Wei, Z.-L., Li, Z.-Y., Lin, G.-O., *TA*, **12**, 229; Maffei, M. et al., *T*, **57**, 537; Moran, P.J.S. et al., *TA*, **12**, 847; Rodriquez, S., Kayser, M.M., Stewart, J.D., *JACS*, **123**, 1547.

e.e. = 96-98%   e.e. = 97-98%
67-98 1:1%

[similarly with *Curvularia Lunata*, Baker's yeast and genetically
engineered Baker's yeast]

III.B. C-N Multiple Bond Reductions

III.B.1. Imine Reductions

IIII.B.1-1 Blackburn, L., Taylor, R.J.K., *OL*, **3**, 1637; Apodaca, R., Xiao, W., *OL*, **3**, 1745.

$$R-\text{C}_6\text{H}_4-\text{CH}_2\text{OH} + \text{HNRR}^1 \xrightarrow[\text{CH}_2\text{Cl}_2, \text{ refluz}]{\text{MnO}_2, \; \bullet\text{-CNBH}_4, \text{ 4Å MS}} R-\text{C}_6\text{H}_4-\text{CH}_2\text{NRR}^1$$

82-95%

[similar reductive amination reported with PhSiH$_3$, Bu$_2$SnCl$_2$]

III.B.1-2 Toth, M., Somsak, L., *TL*, **42**, 2723.

$$\text{R-CN} + \text{TsNHNH}_2 \xrightarrow[\text{AcOH/H}_2\text{O/pyr, rt}]{\text{RaNi, NaH}_2\text{PO}_2} \text{R-CH=NNHTs}$$

20-96%

III.B.1-3 Moghaddam, F.M. et al., *JCR(S)*, 525; Shimizu, M. et al., *CL*, 792.

$$\text{Ar-CH=N-Ar}^1 \xrightarrow[\text{HCO}_2\text{H}]{\mu\text{W, HCO}_2\text{HNEt}_3} \text{Ar-CH}_2\text{-NH-Ar}^1$$

63-90%

[similar imine reduction with TiI$_4$]

III.B.1-4 Fontaine, E. et al., *TA*, **12**, 2185; Demir, A.S. et al., *TA*, **12**, 2309.

$$\xrightarrow[\text{THF, rt}]{}$$

42-96%
e.e. = 96-99%

[similarly with a different chiral amino alcohol]

III.B.1-5 Maeda, K. et al., *SL*, 1808.

$$\xrightarrow[\text{2. Boc}_2\text{O}]{\text{1. H}_2, \text{Pt/C}}$$

89%

III.B.1-6 Williams, D.R. et al., *TL*, **42**, 8597.

70%
anti:syn = 2:1

III.B.1-7 Matsumura, Y. et al., *TL*, **42**, 2525; Xiao, D., Zhang, X., *AG(E)*, **40**, 3424; Abe, H., Amil, H., Uneyama, K., *OL*, **3**, 313.

52-99%
e.e. = 49-66%

[similar chiral imine reductions with an Ir-f-binaphthane complex
or with H_2/Pd(OCOCF$_3$)$_2$, (R)-BINAP]

III.B.2. Reduction of Heterocycles

III.B.2-1 Zacharie, B. et al., *JOC*, **66**, 5264.

43-99%

III.C. Reduction of Sulfur Compounds

III.C-1 Yadav, J.S. et al., *SL*, 854

75-92%

III.D. N-O Reduction

III.D-1 Toyokuni, T. et al., *TL*, **42**, 5601; Varma, R.S. et al., *TL*, **42**, 5347; Couturier, M. et al., *TL*, **42**, 2285; Desai, D.G. et al., *SC*, **31**, 1249.

55-95%

[similarly with $N_2H_4 \cdot H_2O$, $FeCl_3 \cdot 6H_2O$, mW; $Me_2N \cdot BH_3$,
Pd(OH)$_2$/C or FeS/NH$_4$Cl]

III.D-2 Bode, J.W., Carreira, E.M., *OL*, **3**, 1587.

$$\underset{R}{\overset{R^1}{\diagdown}}\overset{N-O}{\diagup}R^2 \quad \xrightarrow[\text{2. B(OH)}_3, \text{H}_2\text{O}]{\text{1. SmI}_2, \text{THF, 0°C}} \quad R^1\diagdown\diagup\overset{O}{\diagup}\diagdown\diagup\overset{OH}{\diagup}R^2 \quad 56\text{-}84\%$$

III.D-3 Yus, M. et al., *S*, 914.

$$\underset{R}{\overset{O}{\diagup}}\underset{OR^2}{\overset{R^1}{N}} \quad \xrightarrow[\text{THF, }\Delta]{\text{Li, DTBB}} \quad \underset{R}{\overset{O}{\diagup}}\overset{R^1}{\underset{NH}{}} \quad 62\text{-}88\%$$

III.D-4 Cho, Y.S. et al., *S*, 81.

$$\text{R-NO}_2 \quad \xrightarrow[\text{aq THF}]{\text{In, HCl}} \quad \text{R-NH}_2 \quad 73\text{-}98\%$$

III.D-5 Pratap, T.V., Baskaran, S., *TL*, **42**, 1983.

$$\text{Ar-NO}_2 \quad \xrightarrow[\text{MeCN, reflux}]{\text{HCO}_2\text{NH}_4, \text{Pd/C}} \quad \text{Ar-NHCHO} \quad 40\text{-}94\%$$

III.D-6 Yus, M. et al., *S*, 427.

$$\underset{R}{\overset{R^1}{\diagup}}\overset{+}{N}\diagdown R^2 \quad \xrightarrow[\text{THF}]{\text{Li, NiCl}_2\cdot 2\text{H}_2\text{O, (}^t\text{BuPh)}_2} \quad \underset{R}{\overset{R^1}{\diagup}}\overset{R^2}{\underset{NH}{}} \quad 41\text{-}89\%$$

III.D-7 Hu, L. et al., *JOC*, **66**, 919.

$$\text{R}\diagdown\text{NO}_2 \quad \xrightarrow[\text{C}_8\text{H}_{17}\text{-N}^+\diagup\diagdown\text{N}^+\cdot\text{C}_8\text{H}_{17} \ 2\text{Br}^-]{\text{Sm, MeOH, rt}} \quad \text{R}\diagdown\text{NH}_2 \quad 78\text{-}99\%$$

III.E. C-C Multiple Bond Reductions

III.E.1. C≡C Reductions

III.E.1-1 Bianchini, C., Vizza, F. et al., *OM*, **20**, 2660.
Synthesis of Polymer-Supported Rhodium(I)-1,3-bis(diphenylphosphino)-propane Moieties and their Use in the Heterogeneous Hydrogenation of Quinoline and Benzylideneacetone.

III.E.1-2 Maki, S. et al., *SL*, 1590.
Hydrogenolysis-Free Hydrogenation by Pd Black Powder Catalyst.

III.E.1-3 Jessop, P.G. et al., *JACS*, **123**, 1254.
Asymmetric Hydrogenation and Catalyst Recyling Using Ionic Liquid and Supercritical Carbon Dioxide.

III.E.1-4 Crooks, R.M. et al., *JOC*, **66**, 6840.
Size-Selective Hydrogenation of Olefins by Dendrimer-Encapsulated Palladium Nanoparticles.

III.E.1-5 van Leeuwen, P.W.N.M. et al., *JACS*, **123**, 8468.
A Silica-Supported, Switchable and Recyclable Hydroformylation-Hydrogenation Catalyst.

III.E.1-6 Crudden, C.M. et al., *CC*, 1154.
Rhodium Bis-phosphine Catalysts on Mesoporous Silica Supports: New Highly Efficient Catalysts for the Hydrogenation of Alkenes.

III.E.1-7 Moglioni, A.G., Ortuno, R.M. et al., *TA*, **12**, 25.
On the Stereoselective Hydrogenation of Chiral Cyclobutyl Dehydro-Amino Acid Derivatives: Influence of the Catalyst in the π-Facial Diastereoselection.

III.E.1-8 Chen, W., Xiao, J., *TL*, **42**, 8737.
Asymmetric Activation of Conformationally Flexible Monodentate Phosphites for Enantiselective Hydrogenation.

III.E.1-9 Maki, S. et al., *TL*, **42**, 8323.
Effect of Solvent and Hydrogen during Selective Hydrogenation.

III.E.1-10 Robinson, A.J., Lim, C.Y., *JOC*, **66**, 4141, 4148; Heller, D., Borner, A. et al., *JOC*, **66**, 6816; **see also:** Bruneau, C. et al., *TA*, **12**, 863.

0-99%
e.e. = 77-99%

III.E.1-11 Comins, D.L. et al., *JOC*, **66**, 2181.

$$R^1 \quad \overset{O}{\underset{RO_2C}{\bigg|}} \xrightarrow{\text{Zn, AcOH}} R^1 \quad \overset{O}{\underset{RO_2C}{\bigg|}}$$

91-95%

III.E.1-12 Fan, Q.-H., Chan, A.S.C. et al., *TA*, **12**, 1241.

MeO — CO_2H $\xrightarrow[\text{MeOH}]{\text{H}_2,\ \text{Ru(BINAP)}}$ MeO — $\overset{Me}{\underset{}{}}CO_2H$

37-99% conversion
e.e. = 80-96%

III.E.1-13 Masuno, M.N., Molinski, T.F., *TL*, **42**, 8263.

$$Ar \diagup \overset{H}{\underset{N}{}} CO_2Et \xrightarrow[\text{TFA, -10°C}]{\text{Et}_3\text{SiH}} Ar \diagdown \overset{H}{\underset{N}{}} CO_2Et$$

92-99%

III.E.1-14 Rajanbabu, T.V. et al., OL, *3*, 2053; Yan, Y.-Y., Rajanbabu, T.V., *JOC*, **66**, 3277; Claver, C., van Leeuwen, P.W.N.M., *JOC*, **66**, 7826; Ruiz, A. et al., *JOC*, **66**, 8364; Knochel, P. et al., *TA*, **12**, 909; Shieh, W.-C. et al., *TA*, **12**, 2421; Zanotti-Gerosa, A. et al., *OL*, 3, 3687; Adamczyk, M. et al., *OL*, 3, 3157.

MeO — $\overset{CO_2Me}{\underset{NO_2 \ NHAc}{}}$ $\xrightarrow{\text{H}_2,\ \text{Rh}^+[\text{L}](\text{cod}),\ \text{THF}}$ MeO — $\overset{CO_2Me}{\underset{NO_2 \ NHAc}{}}$

$$L = \overset{R}{\underset{Ar_2P \ \ Ar_2P}{\bigg|}} OPh$$

98%
e.e. = 98%

[similar reductions reported with various chiral Rh complex]

III.E.1-15 Ishii, Y. et al., *JOC*, **42**, 4710.

$$\overset{R^1}{\underset{R}{\bigg|}} \overset{O}{\underset{R^2}{\bigg|}} \xrightarrow[{}^i\text{PrOH, Ph-Me, 80°C}]{[\text{Ir(cod)Cl}]_2\ \text{dppp, Cs}_2\text{CO}_3} \overset{R^1}{\underset{R}{\bigg|}} \overset{O}{\underset{R^2}{\bigg|}}$$

70-98% conversion

III.E.1-16 Kawai, Y. et al., *TA*, **12**, 309.

$$\overset{R^1}{\underset{R}{\bigg|}} \overset{NO_2}{\underset{R^3}{\bigg|}} \xrightarrow{\text{Baker' yeast}} \overset{R^1}{\underset{R}{\bigg|}} \overset{NO_2}{\underset{R^3}{\bigg|}}$$

16-81%
e.e. = 0-45%

III.E.1-17 Nolan, S.P. et al., *OM*, **20**, 1255; Burgess, K. et al., *JACS*, **123**, 8878; Yi, C.S. et al., *OM*, **20**, 794.

≤99%

[similarly with a different Ir complex or with a Ru complex]

III.E. C≡C Reductions

III.E.2-1 Ranu, B.C. et al., *JOC*, **66**, 5624.

18-95%

III.E.2-2 van Koten, G. et al., *JOC*, **66**, 1647; Campos, K.R. et al., *JOC*, **66**, 3634.

28-99%
cis:trans = 10-100:1

[high Z selectivity also reported with H₂, (NH₂CH₂)₂, Lindlar's catalyst]

III.F. Hetero Bond Reductions

III.F.1. C-O → C-H

III.F.1-1 Gordon, P.E., Fry, A.J., *TL*, **42**, 831; Cain, G.A., Holler, E.R. et al., *CC*, 1168.

95-99%

[similarly with NaI, TMS-Cl]

III.F.1-2 Pashkovsky, F.S. et al., *SL*, 1391.

75-96%

III.F.1-3 Jang, D.O. et al., *TL*, **42**, 1073.

$$CF_3CO_2R \xrightarrow[\text{140°C, 15h}]{\text{Ph}_2\text{SiH}_2 \; (^t\text{BuO})_2} \text{R-H}$$
$$65\text{-}85\%$$

III.F.1-4 Kunishima, M., Tani, S. et al., *CPB*, **49**, 97.

$$\xrightarrow[\text{MeCN, rt}]{\text{SmI}_2, \text{AlCl}_3}$$
0-88%

III.F.1-5 Nyerges, M. et al., *S*, 1479.

$$\xrightarrow[\text{EtOH, rt}]{\text{NaBH}_4}$$
87-95%

III.F.1-6 Homma, K. et al., *T*, **57**, 5353.

$$R\text{-CO}_2R^1 \xrightarrow{\text{Et}_3\text{SiH, TiCl}_4, \text{TMS-OTf}} R\text{---}OR^1$$
0-89%

III.F.1-7 Meyers, A.I. et al., *JOC*, **66**, 1413.

$$\xrightarrow{^i\text{Bu}_2\text{AlH}}$$
53-95%
d.r. = 1-19:1

(other hydrides were also used)

III.F.1-8 Dechoux, L. et al., *TL*, **42**, 8629; Igarashi, M., Fuchikami, T., *TL*, **42**, 1945.

$$\xrightarrow[\text{2. aq NH}_4\text{Cl}]{\text{1. BH}_3}$$
61-87%

[amide reduction also reported with HSiMe₂Ph, RuCl₂(CO)₂(PPh₃)₂, EtI]

III.F.1-9 Baba, A. et al., *JOC*, **66**, 7741.

$$\text{R-OH} + \text{Ph}_2\text{SiHCl} \xrightarrow{\text{InCl}_3, \text{CH}_2\text{Cl}_2 \text{ or DCE}} \text{R-H} \quad 0\text{-}99\%$$

III.F.1-10 Gevorgyan, V., Yamamoto, Y. et al., *JOC*, **66**, 1672.

$$\underset{X}{\overset{O}{\underset{\|}{R-C}}} \xrightarrow[\text{CH}_2\text{Cl}_2, \text{ rt}]{\text{Et}_3\text{SiH, B(C}_6\text{F}_5)_3} \underset{91\text{-}99\%}{\text{R-Me}}$$

III.F.2. C-Hal → C-H

III.F.2-1 Ranu, B.C. et al., *JOC*, **66**, 4102.

$$\underset{\text{Br}}{\overset{R}{\diagup}}\text{Br} \xrightarrow[\text{EtOH/H}_2\text{O, 95-100°C}]{\text{In, NH}_4\text{Cl}} \underset{\substack{70\text{-}95\% \\ \text{E:Z} = 1\text{-}19:1}}{\overset{R}{\diagup}\text{Br}}$$

III.F.2-2 Baba, A. et al., *TL*, **42**, 4661.

$$\text{R-X} \xrightarrow[\text{THF, rt}]{\text{Bu}_3\text{SnH, InCl}_3} \underset{61\text{-}91\%}{\text{R-H}}$$

(also intramolecular radical cyclization to tetrahydrofurans)

III.F.2-3 Oshima, K. et al., *BCJ*, **74**, 747.

$$\text{R-X} + \left[\underset{O}{\overset{}{\diagdown}} \right]_3 \text{GeH} \xrightarrow[\text{THF or H}_2\text{O}]{\text{Et}_3\text{B}} \underset{73\text{-}99\%}{\text{R-H}}$$

(other examples with appended olefin to give cyclized product)

III.F.2-4 Davies, I.W. et al., *JOC*, **66**, 251.

$$\underset{RR^1N}{\overset{Cl \quad R^2}{\diagup\!\!\!\diagup N^+_{R^3}}} \xrightarrow[\text{dioxane, 25°C, 14h}]{\text{HI}} \underset{49\text{-}85\%}{\overset{R^2}{RR^1N\diagup\!\!\!\diagup N^+_{R^3}}}$$

III.F.2-5 Barrero, A.F., Alvarez-Manzaneda, E.J. et al., *SL*, 485.
Raney Nickel: An Effective Reagent for Reductive Dehalogenation of Organic Halides

III.F.2-6 Deleure, H. et al., *JCS(P1)*, 366.
The Preparation and Use of PolyHIPE-Grafted Reactants to Reduce Alkyl Halides Under Free-Radical Conditions.

III.F.2-7 Knetttle, B.W., Flowers, R.A., II, *OL*, **3**, 2321.

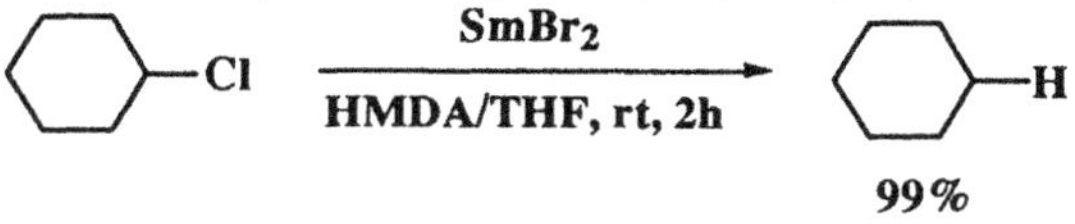

Cl
SmBr2
HMDA/THF, rt, 2h
H
99%

III.F.2-8 Kiyooka, S. et al., *TL*, **42**, 7299.

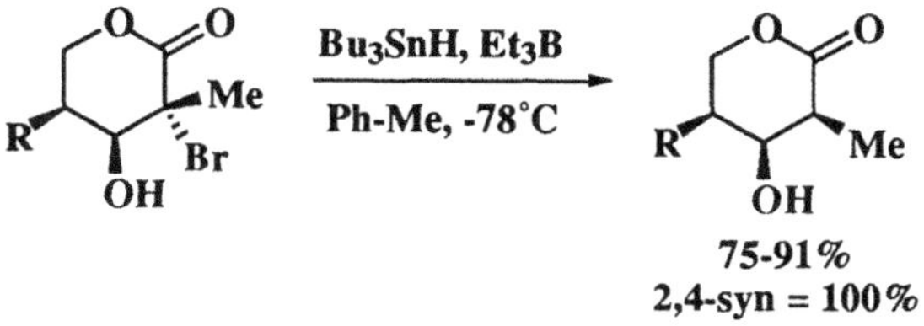

O
O
Me
R
Br
OH
Bu3SnH, Et3B
Ph-Me, -78°C
O
O
R
Me
OH
75-91%
2,4-syn = 100%

III.F.2-9 Lipshutz, B.H. et al., *TL*, **42**, 7737 and *SL*, 970; Sako, M. et al., *JOC*, **66**, 3610; Nolan, S.P. et al., *OM*, **20**, 3607.

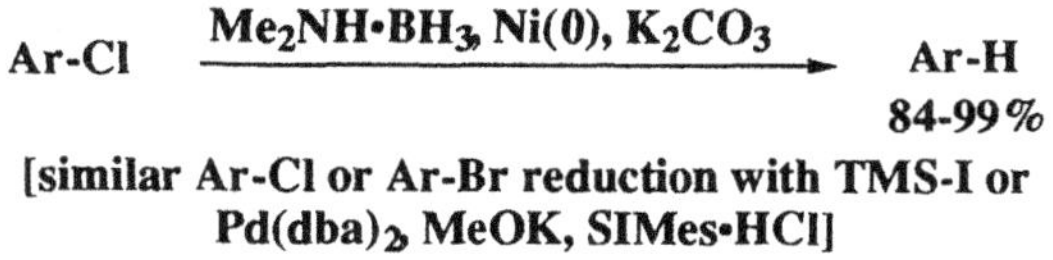

Ar-Cl
Me2NH•BH3, Ni(0), K2CO3
Ar-H
84-99%
[similar Ar-Cl or Ar-Br reduction with TMS-I or
Pd(dba)2, MeOK, SIMes•HCl]

III.F.3. C-S → C - H

III.F.3-1 Kunishima, M. et al., *TL*, **42**, 415.

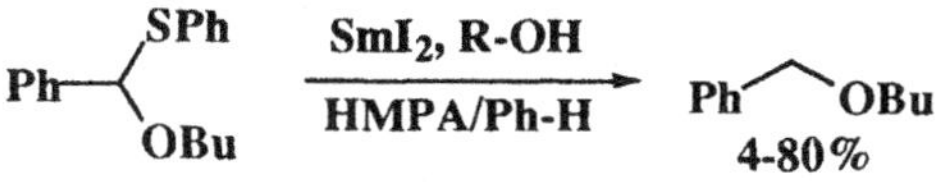

SPh
Ph
OBu
SmI2, R-OH
HMPA/Ph-H
Ph
OBu
4-80%

III.F.3-2 Liu, Y., Zhang, Y., *OPP*, **33**, 372.

SO2Ar1
Ar
NHR
O
Sm
AcOH/EtOH, rt
NHR
Ar
O
76-86%

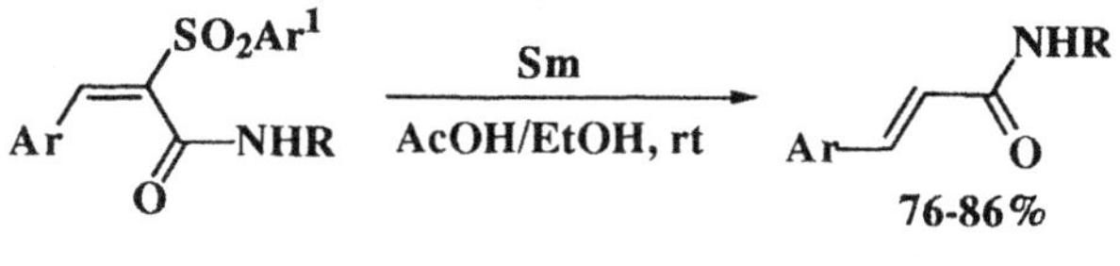

III.G. Reductive Cleavages

III.G.1. Oxiranes

III.G.1-1 Doris, E. et al., *JOC*, **66**, 1046; Chakraborty, T.K., Tapadar, S. et al., *TL*, **42**, 1375.

$$59\text{-}90\%$$

III.G.1-2 Couturier, M. et al., *TL*, **42**, 2763.

81%

(other functional groups reduced are olefins, acetylenes, debenzylation, dehalogenation)

III.G.1-3 Archelas, A. et al., *JOC*, **66**, 538.

6-28%	34-55%
e.e. = 96-99%	e.e. = 7-56%

III.G.1-4 Sato, F. et al., *OL*, **3**, 2205.

1. HCO$_2$H, Pd$_2$(dba)$_3$•CHCl$_3$
   TEA, PBu$_3$, THF
2. TBS-Cl, imidazole, DMF

90%

III.G.2. Other Reductive Cleavages

III.G.2-1 Gauthier, D.R. et al., *TL*, **42**, 7011.

DIBAL
Ph-Me, 0°C

89% 17:1

III.G.2-2 Deshong, P. et al., *JOC*, **66**, 4352.

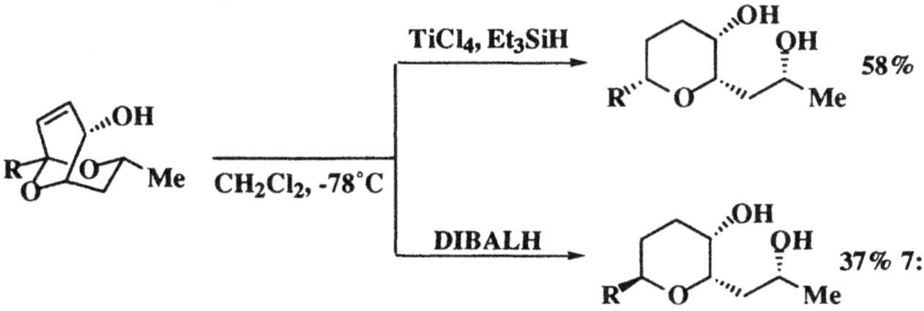

III.H. Reduction of Azides

III.H-1 Suzuki, K. et al., *SL*, 1003.

$$\text{R-N}_3 \xrightarrow[\text{2. MeI, Aq NaOH, 100°C}]{\text{1. PMe}_3\text{, CH}_2\text{Cl}_2\text{, rt}} \text{R-NHMe}$$

68-86%

III.I. Other Reductions

III.I-1 Torisawa, Y. et al., *BMCL*, **11**, 829.

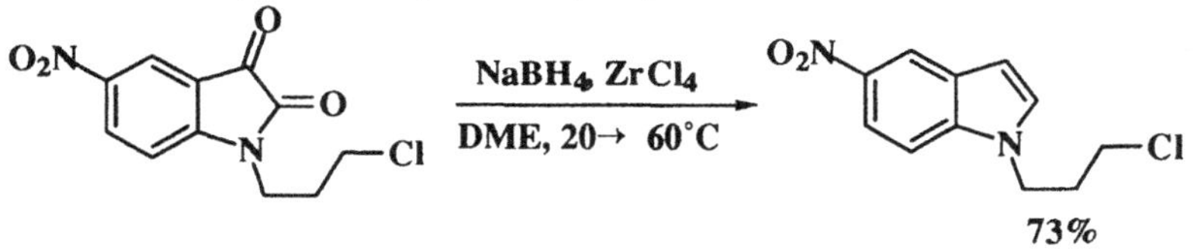

73%

III.I-2 Eguchi, T., Hoshino, Y., *BCJ*, **74**, 967.

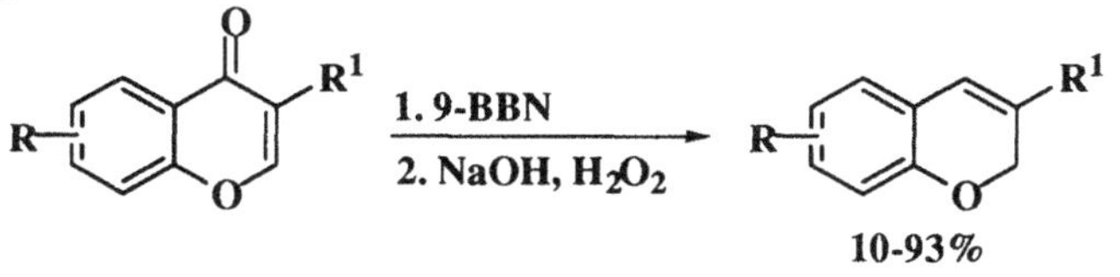

10-93%

III.I-3 Meier, G.P. et al., *TL*, **42**, 5367; Wang, Y., Guziec, F.S., *JOC*, **66**, 8293.

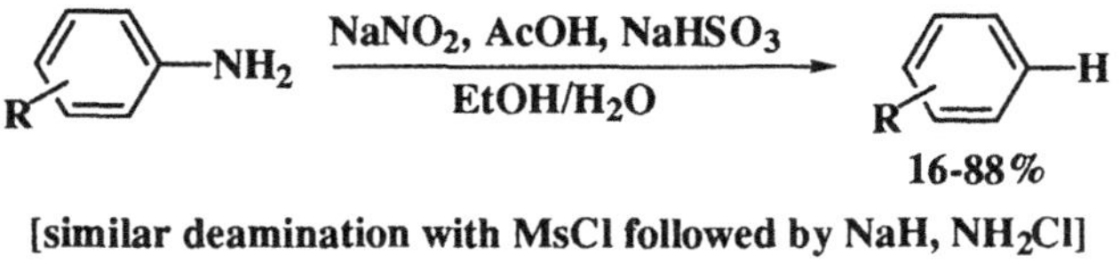

16-88%

[similar deamination with MsCl followed by NaH, NH$_2$Cl]

III.I- Roman, M.V.G., Light, M.E. et al., *TA*, **12**, 1673.

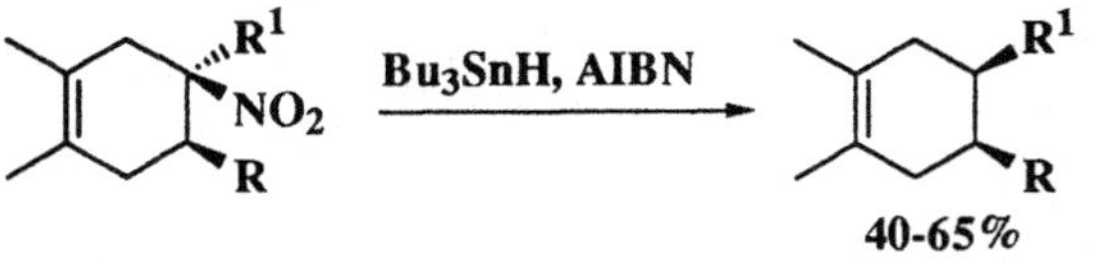

R1
NO2
R
Bu3SnH, AIBN
R1
R
40-65%

III.I-5 Chavan, S.P. et al., *SL*, 857.

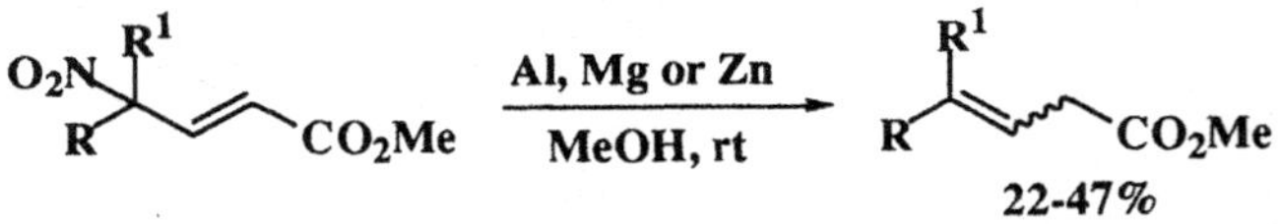

R1
O2N
R
CO2Me
Al, Mg or Zn
MeOH, rt
R1
R
CO2Me
22-47%

III.I-6 Fujiwara, S., Sonoda, N. et al., *JOC*, **66**, 2183.

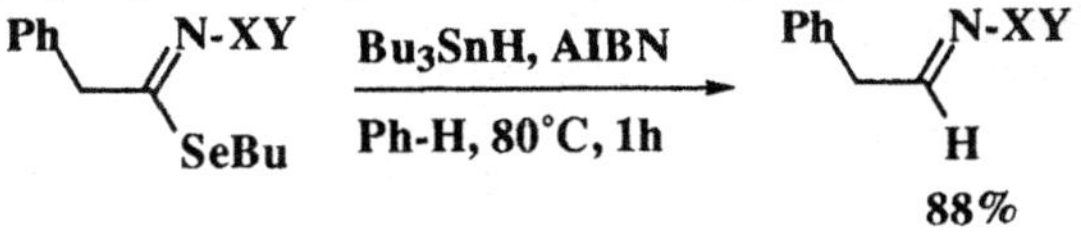

Ph
N-XY
SeBu
Bu3SnH, AIBN
Ph-H, 80°C, 1h
Ph
N-XY
H
88%

IV
SYNTHESIS OF HETEROCYCLES

IV.A. Oxiranes, Aziridines and Thiiranes

IV.A-1 Reetz, M.T., Lee, W.K., *OL*, **3**, 3119; Saito; T. et al., *TL*, **42**, 5451.

$$R\overset{CHO}{\underset{NBn_2}{|}} + \underset{H_2N}{\bigcirc}\text{—OMe} \xrightarrow[\text{2. BuLi, TMS-I, THF}]{\text{1. 4Å MS, CH}_2\text{Cl}_2} R\underset{NBn_2}{|}\overset{H}{\triangle}\text{N-PMP}$$

85-91%
cis:trans = 4-7:1

IV.A-2 Tanaka, T. et al., *OL*, **3**, 2269.

$$\underset{Ts}{\overset{R}{\underset{NH}{|}}}\diagup\diagdown_{Br} \xrightarrow[\substack{76\text{-}93\% \\ \text{cis:trans} = 3.4\text{-}89:1}]{\text{NaH, DMF}} R\triangle NTs \xleftarrow[\substack{50\text{-}99\% \\ \text{cis:trans} = 10.1\text{-}99:1}]{\text{NaH, DMF}} \underset{Ts}{\overset{R}{\underset{NH}{|}}}\diagup\diagdown_{Br}$$

IV.A-3 Aggarwal, V.K. et al., *TL*, **42**, 1587; **see also:** Crousse, B., Bonnet-Delpon, D. et al., *SL*, 679.

$$\underset{R}{\overset{R^1}{N=}} \xrightarrow[\underset{S}{\bigcirc}]{\text{Et}_2\text{Zn, ClCH}_2\text{I, CH}_2\text{Cl}_2} \underset{R^1\quad R}{\overset{N}{\triangle}}$$

3-79%

IV.A-4 Handy, S.T., Czopp, *OL*, **3**, 1423; Hutchings, G.J., *JCS(P2)*, 1714.

$$R\diagup\diagdown + \text{PhI=NTs} \xrightarrow[\substack{\text{N—B·H Na}^+ \\ \text{N}}]{\text{CuCl}_2, \text{MeCN}} R\triangle\text{NTs} \quad 18\text{-}89\%$$

IV.A-5 Dodd, R.H. et al., *JACS*, **123**, 7707.

$$\underset{R^2}{\overset{R\quad R^1}{\diagup\diagdown}} \xrightarrow[\text{3Å MS, MeCN}]{\text{R}^3\text{-SO}_2\text{NH}_2, \text{PhI=O, Cu(MeCN)}_4\text{PF}_6} \underset{R^2\quad R^1}{\overset{R}{\triangle}}\text{N—SO}_2R^3$$

40-78%

IV.A-6 Dauban, P., Dodd, R.H., *TL*, **42**, 1037; Chanda, B.H., Bedekar, A.V. et al., *JOC*, **66**, 30.

$$\underset{\substack{n}}{\text{ArCH}-\text{SO}_2\text{NH}_2} \quad \xrightarrow[\text{2. PTAB, MeCN, rt}]{\text{1. }^t\text{BuOCl, NaOH, H}_2\text{O}} \quad \text{product}$$

0-70%

IV.A-7 Aggarwal, V.K. et al., *AG(E)*, **40**, 1433, 1430.

$$\underset{R^1 \quad R}{\overset{NR^2}{\diagup\diagdown}} \;+\; \text{Ph}-\text{N}=\text{N}-\text{Ts} \quad \xrightarrow[\text{dioxane, 40°C}]{\text{Rh}_2(\text{OAc})_4 \; \text{Et}_3\text{N}^+\text{Bn Cl}^-} \quad \underset{R}{\overset{R^1 \quad NR^2}{\diagup\diagdown}}\text{Ph}$$

50-82%
trans:cis = 2-8:1
e.e. = 73-98%

IV.A-8 Penkett, C.S., Simpson, I.D., *TL*, **42**, 1179.

$$\xrightarrow[\text{MeOH}]{h\nu, \text{NaOH}}$$

65%

IV.A-9 Fioravanti, S. et al., *S*, 1975; Loreto, M.A. et al., *TL*, **42**, 2185.

$$\xrightarrow[\text{CH}_2\text{Cl}_2, \text{rt}]{\text{NsONHCO}_2\text{R}^2, \text{CaO}}$$

64-91%

IV.A-10 Cardillo, G. et al., *JOC*, **66**, 8657.

$$\xrightarrow[\text{CH}_2\text{Cl}_2/\text{THF, rt}]{^t\text{BuOK}}$$

54-78%

IV.A-11 Concellon, J.M. et al., *T*, **57**, 8983.

$$\underset{R^1}{\overset{R}{>}}\!\!=\!\text{O} \;+\; \text{CH}_2\text{I}_2 \quad \xrightarrow[\text{THF, 0°C}\rightarrow\text{rt}]{\text{MeLi}} \quad \underset{R^1}{\overset{R}{\diagup}}\!\!\overset{}{\underset{O}{\triangle}}$$

35-95%

IV.A-12 Doyle, M.P. et al., *OL*, **3**, 933; Davies, H.M.L., DeMeese, J., *TL*, **42**, 6803.

0-37%

0-64%

IV.A-13 Yang, K.-S., Chen, K., *JOC*, **66**, 1676.

86-95%
d.r. = 19:1-1:19

IV.A-14 Ishikawa, T. et al., *JACS*, **123**, 7705.

76-96%
cis:trans = 1,4:1-1:11.5
e.e.(trans) = 72-97%

IV.A-15 Florio, S. et al., *S*, 2299 and *JOC*, **66**, 3049 and *T*, **57**, 6775.

38-79%

IV.A-16 Metzner, P. et al., *JOC*, **66**, 5620.

37-97%
d.e. = 75-95%
e.e. = 64-93%

IV.A-17 Chelma, F. et al., *EJOC*, 3295.

67-93%

IV.A-18 Hamaguchi, M. et al., *JOC*, **66**, 5395.

IV.A-19 Katsuki, T. et al., *CL*, 984.

76%
e.e. = 98%

IV.B. Oxetanes, Azetidines and Thietanes

IV.B-1 Bach, T., Schroder, J., *S*, 1117.

Ph-CHO +

46-71%

IV.B-2 Padwa, A. et al., *OL*, **3**, 1781.

68%

IV.B-3 Rousseau, G. et al., *TL*, **42**, 2477, 2481.

15-87%

IV.B-4 Warren, S. et al., *JCS(P1)*, 2983.

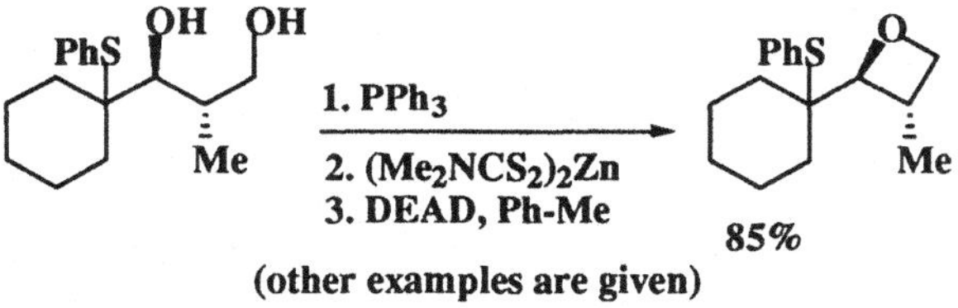

PhS
OH
OH
Me
1. PPh_3
2. $(Me_2NCS_2)_2Zn$
3. DEAD, Ph-Me
PhS
O
Me
85%
(other examples are given)

IV.B-5 Iesce, M.R. et al., *JOC*, **66**, 4732.

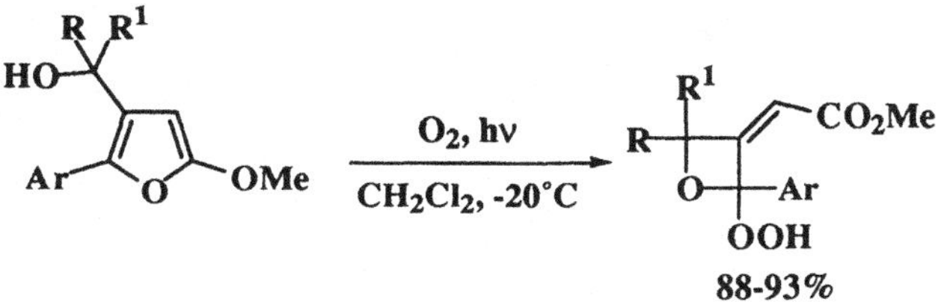

R
R1
HO
Ar
O
OMe
O_2, hν
CH_2Cl_2, -20°C
R1
R
CO_2Me
O
Ar
OOH
88-93%

IV.B-6 Tanaka, T. et al., *JOC*, **66**, 4904.

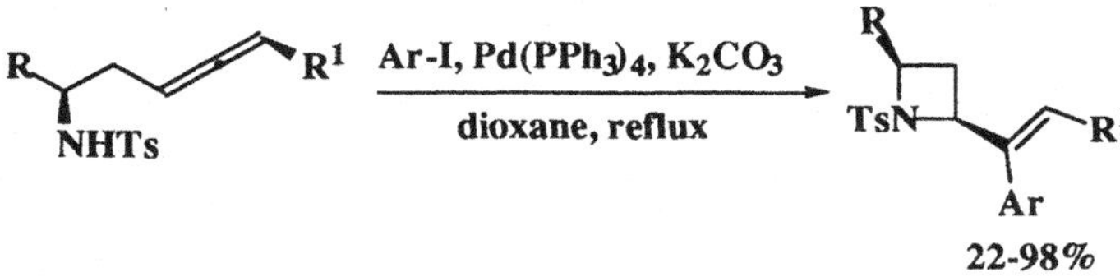

R
R1
NHTs
Ar-I, $Pd(PPh_3)_4$, K_2CO_3
dioxane, reflux
R
TsN
R1
Ar
22-98%

IV.C. Lactams

IV.C-1 Lee, J.-C., Kang, S.H. et al., *TL*, **42**, 4519.

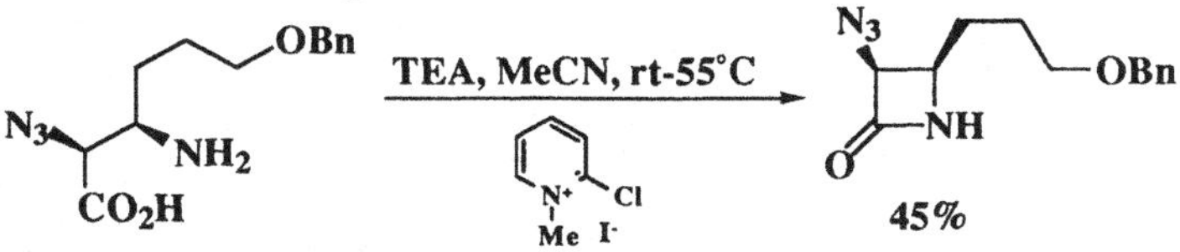

OBn
N_3
NH_2
CO_2H
TEA, MeCN, rt-55°C
N+
Cl
Me
I-
N_3
OBn
NH
O
45%

IV.C-2 Lin, X., Weinreb, S.M., *TL*, **42**, 2631; Alcaide, B. et al., *TL*, **42**, 1503; Sharma, S.D. et al., *JCR(S)*, 321.

NBu
O
N
H
+ $ClCH_2COCl$
TEA, CH_2Cl_2
O
Cl
NBu
O
N
H
87%

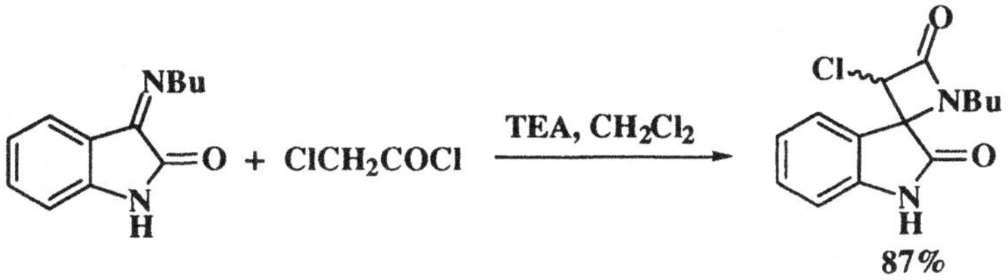

IV.C-3 Parsons, A.F. et al., *TL*, **42**, 2901.

IV.C-4 Langer, P., Doring, M., *SL*, 1437.

27-60%

IV.C-5 Sano, Y. et al., *CPB*, **49**, 1132.

IV.C-6 Novikov, M.S., Khlebnikov, A.F. et al., *TL*, **42**, 533.

56%

IV.C-7 Pays, C., Mangency, P., *TL*, **42**, 589; Poli, G., Malacria, M. et al., *TL*, **42**, 6287.

60%
d.e. = 95%

IV.C-8 Murai, S. et al., *JOC*, **66**, 169.

$$TMS\text{---}\!\!\equiv\!\!\text{---}OLi \;+\; \underset{R}{\overset{R^1}{\diagup}}\!\!\diagdown NTs \xrightarrow{\text{THF, -78}\rightarrow 20°C} $$

36-77%

IV.C-9 Dominguez, E. et al., *T*, **57**, 5403.

$$\xrightarrow[\text{CH}_2\text{Cl}_2, \text{-20°C}]{\textbf{PIFA}}$$

83%

IV.C-10 Clark, A.J. et al., *TL*, **42**, 1999, 2003.

$$\xrightarrow[\text{CH}_2\text{Cl}_2, \text{rt, 24h}]{\textbf{CuCl}}$$

48%

IV.C-11 Ikeda, M. et al., *H*, **54**, 1021; Murphy, J.A. et al., *CC*, 2732.

$$\xrightarrow[\text{Ph-Me, reflux}]{\textbf{Bu}_3\textbf{SnH, AIBN}}$$

43%   34%

IV.C-12 Poli, G. et al., *TL*, **42**, 5179; Hirama, M. et al., *OL*, **3**, 2863.

$$\xrightarrow[\text{DMA, 140°C}]{\textbf{NaH, Herrmann's cat., NaOAc}}$$

38-60%

IV.C-13 Poisson, J.-F., Normant, J.F., *OL*, **3**, 1889.

86%
d.r. = 40:1
e.e. = 95%

IV.C-14 Weinreb, S.M. et al., *T*, **57**, 8779.

^{t}BuSOCl or (EtO)$_2$PCl

(PhSe)$_2$ or TEMPO
CH$_2$Cl$_2$, -50°C→rt

56-84%

IV.C-15 Reissig, H.-U. et al., *S*, 1649.

^{t}Bu-N≡C

MeOH, reflux

35-79%

IV.C-16 Iyer, S. et al., *SL*, 1241.

PdCl$_2$(PPh$_3$)$_2$, ZnCl$_2$, Na$_2$CO$_3$

NMP, 90-95°C

20-70%

IV.C-17 Beak, P. et al., *JACS*, **123**, 1004.

1. HCl, CHCl$_3$
2. NaClO$_2$, NaH$_2$PO$_4$
3. HCl, MeOH
4. H$_2$, Raney Ni

57-71%
d.r. = 99:1

IV.C-18 Alper, H. et al., *JACS*, **123**, 10214; Gabriele, B., Salerno, G. et al., *EJOC*, 4607.

73-82%

IV.C-19 Nishiwaki, N., Ariga, M. et al., *H*, **55**, 1581.

18-49%

IV.C-20 Chmielewski, M. et al., *TA*, **12**, 979.

82%
d.r. = 100:1

IV.C-21 Honda, T. et al., *OL*, **3**, 631.

81%

IV.C-22 Ila, H., Junjappa, H. et al., *OL*, **3**, 229.

60-80%

IV.C-23 Du, Y., Wiemer, D.F., *TL*, **42**, 6069; Reitz, A.B. et al., *OL*, **3**, 893; Guibe, F. et al., *SL*, 37; Iqbal, J. et al., *TL*, **42**, 339.

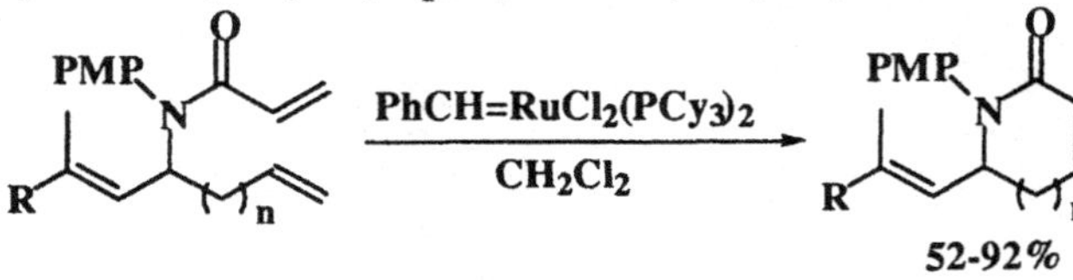

$$\xrightarrow[\text{CH}_2\text{Cl}_2]{\text{PhCH=RuCl}_2(\text{PCy}_3)_2}$$

52-92%

IV.C-24 Kang, S.-K. et al., *OL*, **3**, 2851.

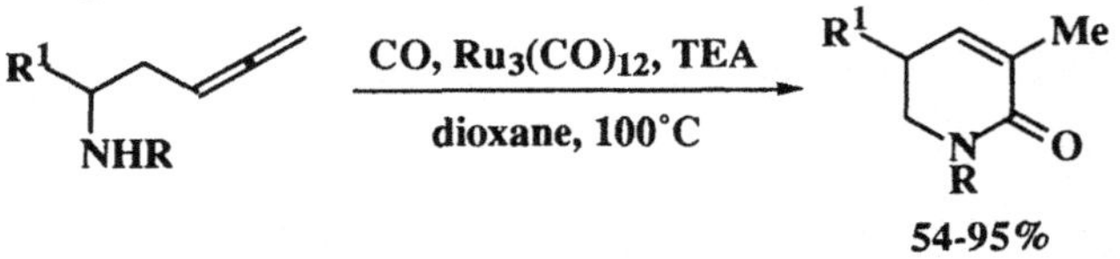

$$\xrightarrow[\text{dioxane, 100°C}]{\text{CO, Ru}_3(\text{CO})_{12}, \text{TEA}}$$

54-95%

IV.C-25 Merour, J.-Y. et al., *SL*, 848.

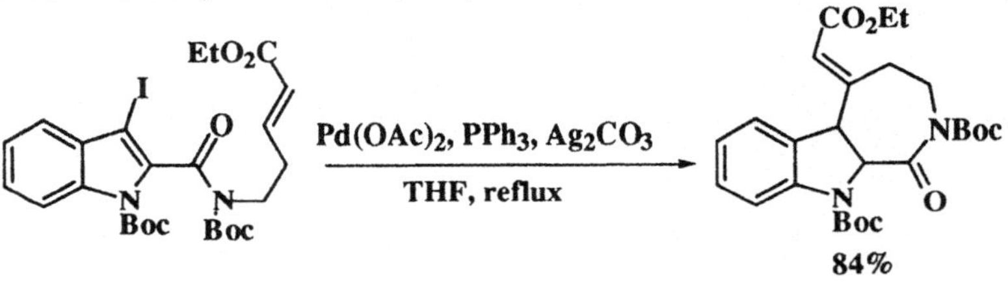

$$\xrightarrow[\text{THF, reflux}]{\text{Pd(OAc)}_2, \text{PPh}_3, \text{Ag}_2\text{CO}_3}$$

84%

IV.C-26 Alcaide, B. et al., *TL*, **42**, 3081.

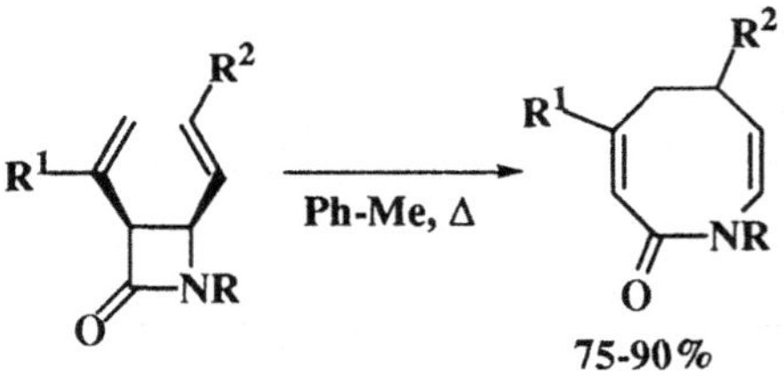

$$\xrightarrow{\text{Ph-Me, }\Delta}$$

75-90%

IV.D. Lactones

IV.D-1 Valentin, E. et al., *TA*, **12**, 1039.

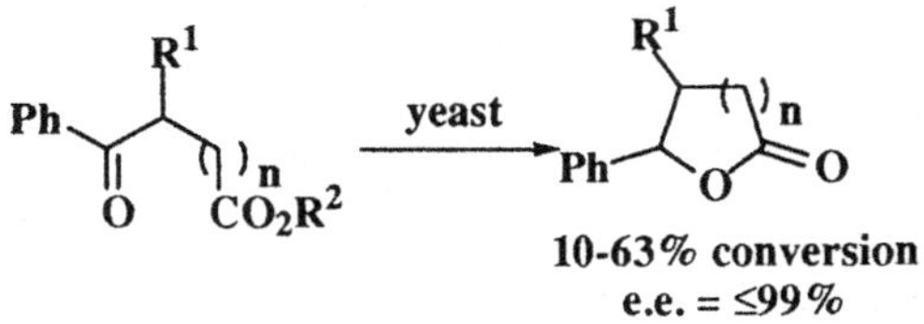

$$\xrightarrow{\text{yeast}}$$

10-63% conversion
e.e. = ≤99%

IV.D-2 Romo, D. et al., *S*, 1731.

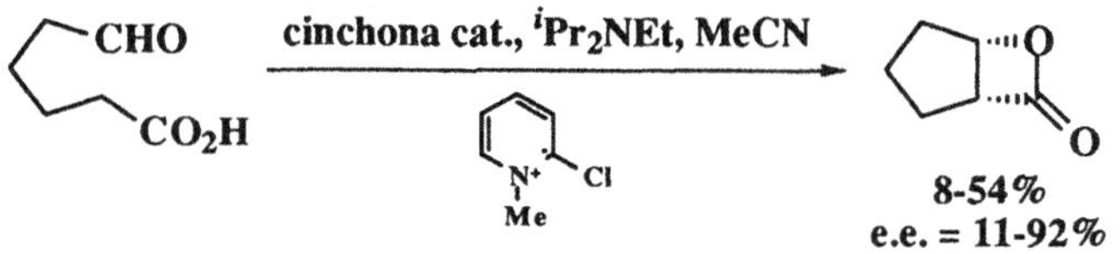

CHO
CO2H
cinchona cat., iPr2NEt, MeCN
Cl
Me
O
O
8-54%
e.e. = 11-92%

IV.D-3 Cossy, J. et al., *JOC*, **66**, 7195.

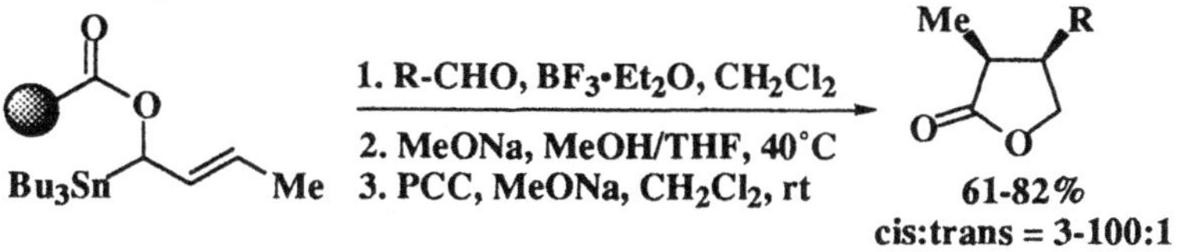

O
O
Bu3Sn
Me
1. R-CHO, BF3•Et2O, CH2Cl2
2. MeONa, MeOH/THF, 40°C
3. PCC, MeONa, CH2Cl2, rt
Me
R
O
O
61-82%
cis:trans = 3-100:1

IV.D-4 Langer, P. et al., *JOC*, **66**, 2222.

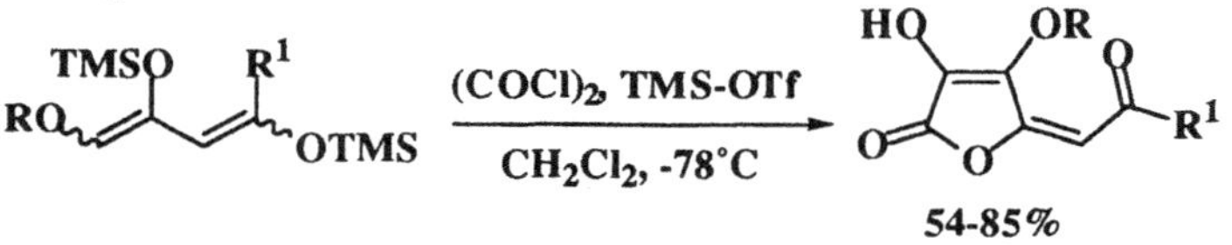

TMSO
R1
RO
OTMS
(COCl)2, TMS-OTf
CH2Cl2, -78°C
HO
OR
O
R1
O
O
54-85%

IV.D-5 Hu, Y. et al., *JCS(P1)*, 66.

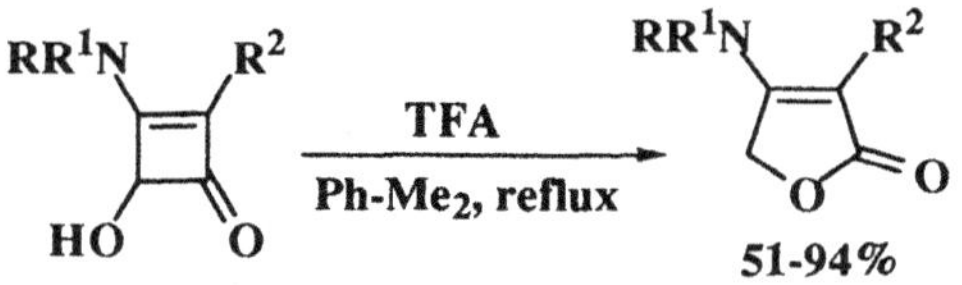

RR1N
R2
HO
O
TFA
Ph-Me2, reflux
RR1N
R2
O
O
51-94%

IV.D-6 Bolm, C. et al., *SL*, 1461; Uchida, T., Katsuki, T., *TL*, **42**, 6911.

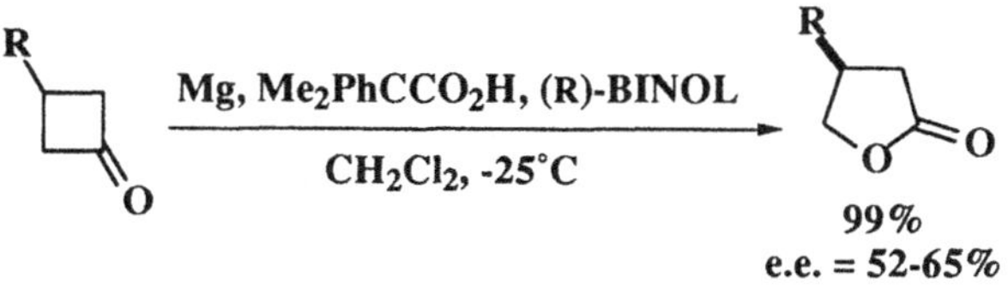

R
O
Mg, Me2PhCCO2H, (R)-BINOL
CH2Cl2, -25°C
R
O
O
99%
e.e. = 52-65%

IV.D-7 Ma, S., Wu, S., *TL*, **42**, 4075; Ma, S. et al., *T*, **57**, 1585.

R1
R2
R
CO2Et
CuBr2
EtOH/H2O, 80-85°C
Br
R2
R1
R
O
O
54-97%

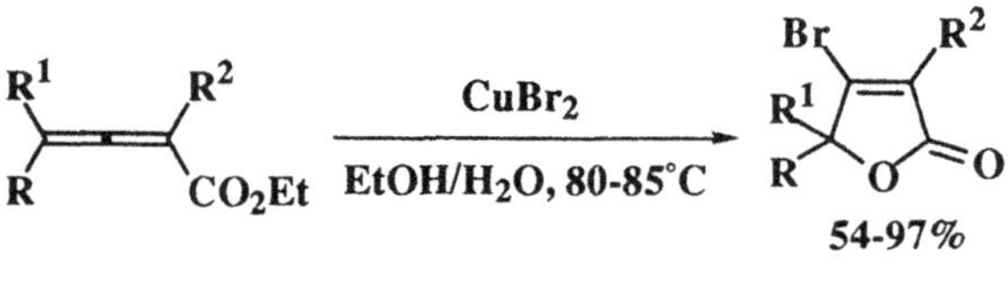

IV.D-8 Huang, X. et al., *JCR(S)*, 480; Hsu, J.-L., Fang, J.-M., *JOC*, **66**, 8573; **see also:** Chen, J., Zhang, Y., *JCR(S)*, 394.

55-73%

IV.D-9 Matsuda, I. et al., *TL*, **42**, 1301; Jiang, Z.-X., Qing, F.-L., *TL*, **42**, 9051.

67-99%

IV.D-10 Zhang, W., Pugh, G., *TL*, **42**, 5617.

78% 1:2

IV.D-11 Furstner, A. et al., *OL*, **3**, 449; Lee, C.W., Grubbs, R.H., *JOC*, **66**, 7155; **see also:** Choi, T.-L., Grubbs, R.H., *CC*, 2688.

37%

IV.D-12 Cox, C., Danishefsky, S.J., *OL*, **3**, 2899.

58%

IV.D-13 Shimizu, I. et al., *BCJ*, **74**, 1437.

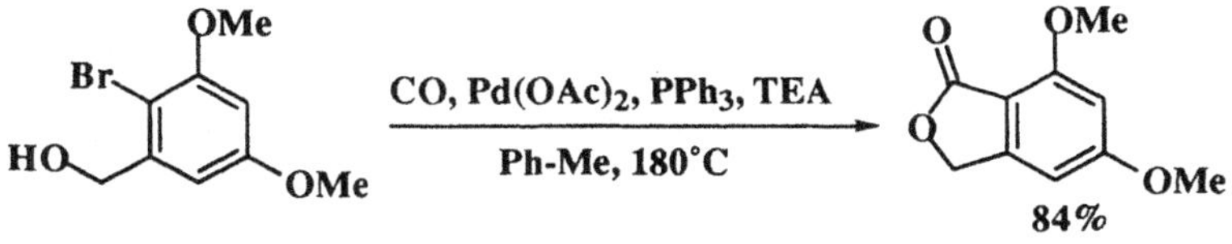

OMe
Br
HO
OMe
CO, Pd(OAc)2, PPh3, TEA
Ph-Me, 180°C
O
OMe
O
OMe
84%

IV.D-14 Takahashi, S. et al., *TL*, **42**, 5459.

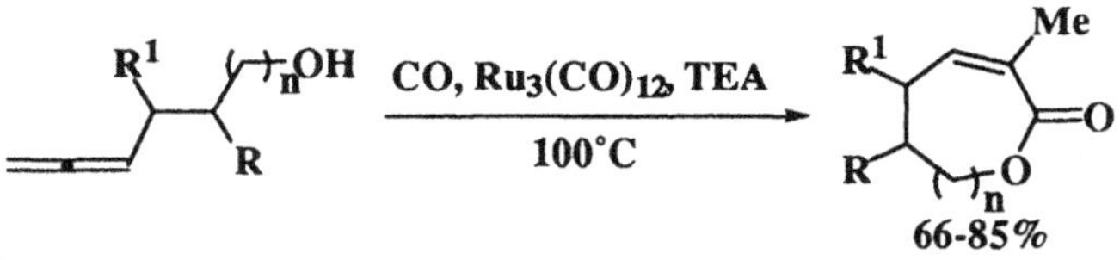

R1
OH
n
R
CO, Ru3(CO)12, TEA
100°C
Me
R1
O
R
O
n
66-85%

IV.D-15 Metz, P. et al., *TL*, **42**, 5377.

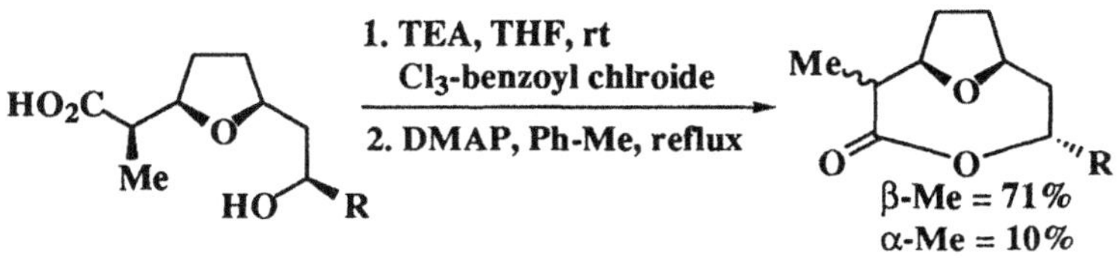

HO2C
O
Me
HO
R
1. TEA, THF, rt
Cl3-benzoyl chlroide
2. DMAP, Ph-Me, reflux
Me
O
O
O
R
β-Me = 71%
α-Me = 10%

IV.D-16 Ballini, R. et al., *S*, 1519.

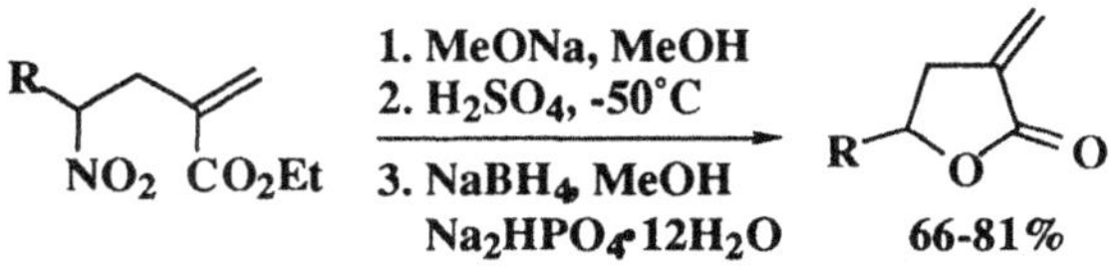

R
NO2
CO2Et
1. MeONa, MeOH
2. H2SO4, -50°C
3. NaBH4, MeOH
Na2HPO4•12H2O
R
O
O
66-81%

IV.D-17 Domling, A. et al., *OL*, **3**, 2875.

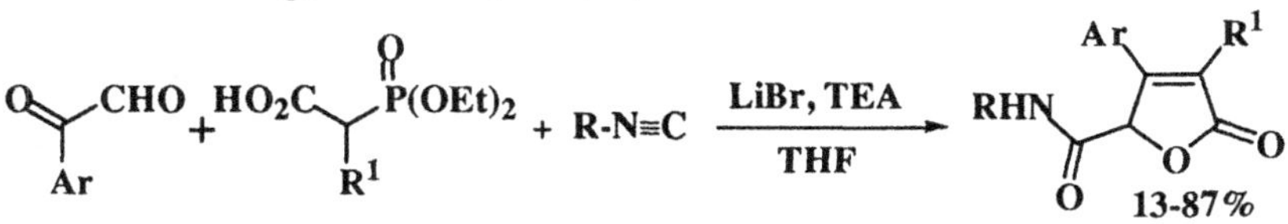

O
CHO
Ar
+
HO2C
O
P(OEt)2
R1
+ R-N≡C
LiBr, TEA
THF
Ar
R1
RHN
O
O
O
13-87%

IV.D-18 Campagne, J.-M. et al., *OL*, **3**, 3807; Bach, T., Kirsch, S., *SL*, 1974.

R-CHO +
Me
OTMS
OMe
CuF, (S)-tolBINAP
Me
O
O
R
30-76%
anti:syn = ≤49:1
e.e. = 82-91%

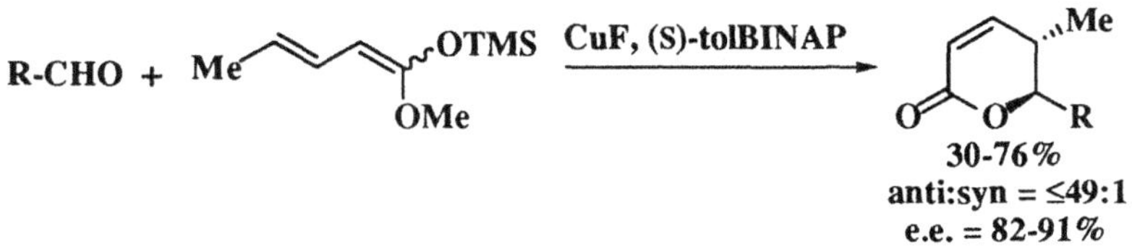

IV.D-19 Bouyssi, D., Balme, G., *SL*, 1191.

$$\xrightarrow[\text{DMSO. rt}]{\text{Cs}_2\text{CO}_3, \text{PPh}_3}$$

64%

IV.D-20 Bellina, F., Rossi, R. et al., *T*, **57**, 2857.

$$\xrightarrow[\text{MeCN}]{\text{I}_2, \text{NaHCO}_3}$$

59-72%   +   32-22%

IV.D-21 Zhang, Y., Herndon, J.W., *TL*, **42**, 777.

72%

IV.D-22 Hinterding, K. et al., *TL*, **42**, 8463; **see also:** Armstrong, A. et al., *TL*, **42**, 4585.

$$\xrightarrow{\text{LHDMS, -78°C}}$$

95%

IV.D-23 Sugino, T., Tanaka, K., *CL*, 110.

$$\xrightarrow[\text{neat, rt, 5min}]{\text{piperdine}}$$

73-99%

IV.D-24 Malacria, M. et al., *SL*, 138.

70-90%

IV.E. Furans and Thiophenes

IV.E-1 Aurrecoechea, J.M., Perez, E., *TL*, **42**, 3839 and *JOC*, **66**, 564; Knight, D.W. et al., *TL*, **42**, 5945.

1. SmI_2, THF, -5°C

2. Ar-X, $Pd(PPh_3)_4$, TEA
   DMF/H_2O, 60-80°C

40-73%

IV.E-2 Burke, S.D., Jiang, L., *OL*, **3**, 1953; Hara, O. et al., *H*, **54**, 419; Yamamoto, Y. et al., *JOC*, **66**, 7142; Balme, G. et al., *JOC*, **66**, 175, 4069; Ling, Y.-C. et al., *JOC*, **66**, 6014.

$Pd_2(dba)_3$, THF

97%

IV.E-3 Yamamoto, Y. et al., *AG(E)*, **40**, 1298; Sugita, Y. et al., *H*. **55**, 135; Yadav, V.K., Balamurugan, R., *OL*, **3**, 2717; Evans, D.A. et al., *JACS*, **123**, 12095.

+ R-CHO

$Pd(PPh_3)_4$, PBu_3

120°C

38-86%

IV.E-4 Kabalka, G.W. et al., *T*, **57**, 8017; Chaplin, J.H., Flynn, B.L., *CC*, 1594.

μM
Pd, CuI, PPh_3

KF/Al_2O_3

33-54%

IV.E-5 Macsari, I., Szabo, K.J., *ECJ*, **7**, 4097.

40-86%

IV.E-6 Jiang, H., Chen, M. et al., *TL*, **42**, 6923.

$$R \!-\!\!\equiv \quad \xrightarrow[\text{H}_2\text{O/dioxane}]{\text{PdCl}_2,\ \text{CO, CuCl}_2}$$

84-98%

IV.E-7 Langer, P. et al., *JOC*, **66**, 6057.

60-80%
E:Z = 10-49:1

IV.E-8 Kitching, W. et al., *JOC*, **66**, 7487.

e.e. = 80-98%

IV.E-9 Tse, B., Jones, A.B., *TL*, **42**, 6429.

60-72%

IV.E-10 Hoffmann-Roder, A., Krause, N., *OL*, **3**, 2537; Mukai, C. et al., *OL*, **3**, 3385; Sato, F. et al., *TL*, **42**, 5501.

65-99%

IV.E-11 Langer, P., Krummel, T., *ECJ*, **7**, 1720.

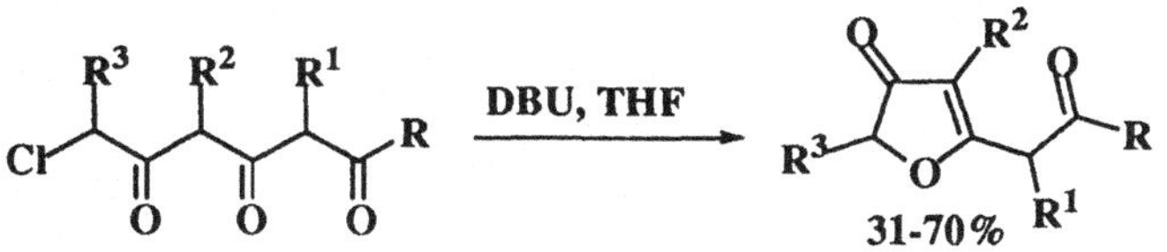

R3
R2
R1
Cl
R
O
O
O
DBU, THF
O
R2
O
R3
O
R
31-70%
R1

IV.E-12 Benbow, J.W., Katoch-Rouse, R., *JOC*, **66**, 4965.

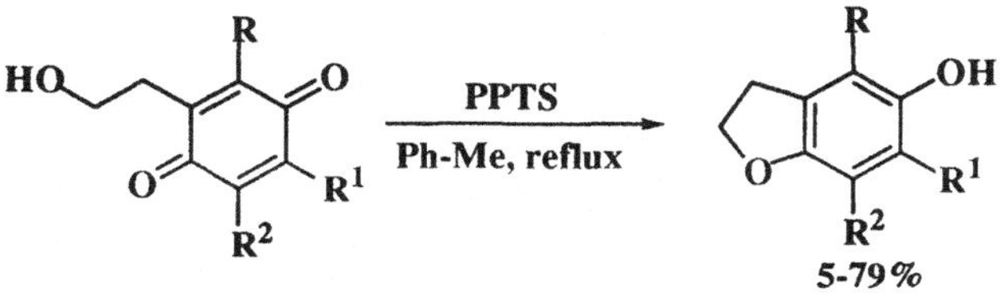

R
HO
O
O
R1
R2
PPTS
Ph-Me, reflux
R
OH
O
R1
R2
5-79%

IV.E-13 Pearson, A.J., Mesaros, E.F., *OL*, **3**, 2665.

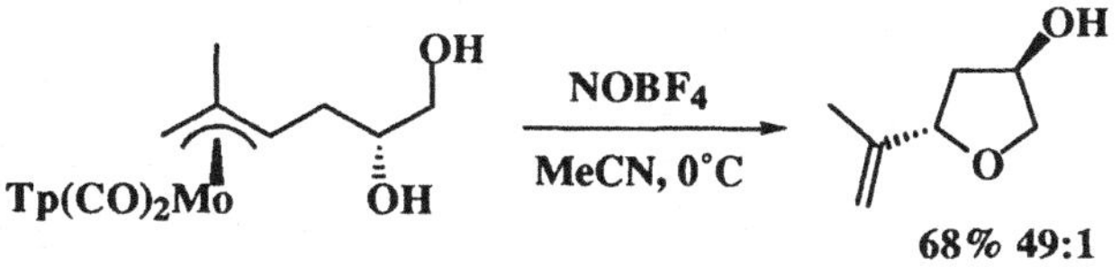

OH
Tp(CO)2Mo
OH
NOBF4
MeCN, 0°C
OH
O
68% 49:1

IV.E-14 Nair, V. et al., *CC*, 1682.

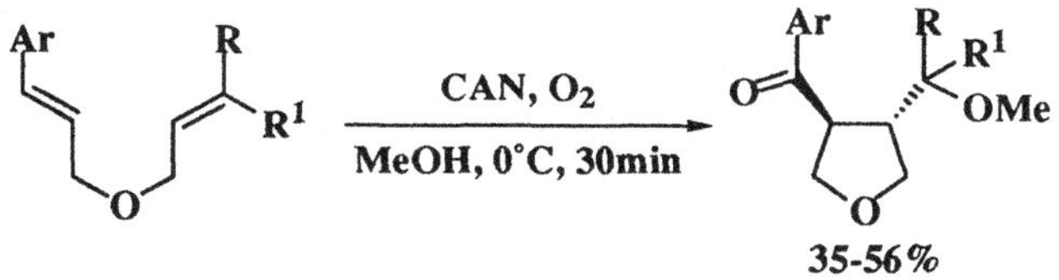

Ar
R
R1
O
CAN, O2
MeOH, 0°C, 30min
Ar
R
R1
O
OMe
O
35-56%

IV.E-15 Langer, P., Freifeld, I., *ECJ*, **7**, 565.

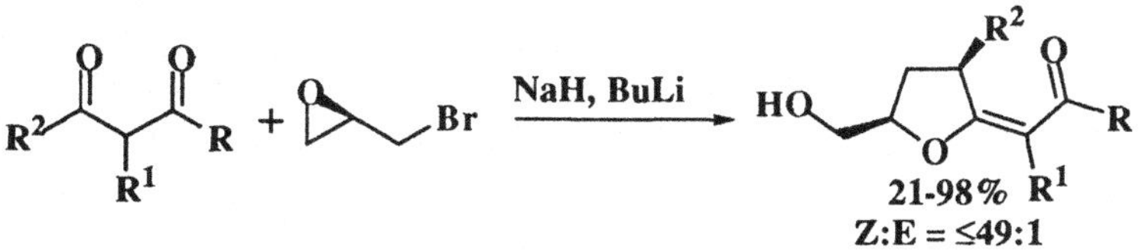

O
O
R2
R
R1
+
O
Br
NaH, BuLi
R2
O
HO
O
R
21-98% R1
Z:E = ≤49:1

IV.E-16 Kabalka, G.W. et al., *TL*, **42**, 6049.

R
OH
µW
(CH2O)n, HNRR1
CuI, Al2O3
R
O
NRR1
38-70%

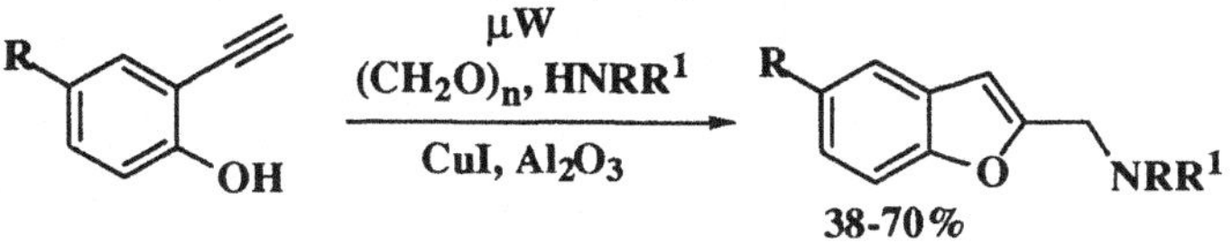

IV.E-17 Verkade, J.G. et al., *SL*, 670; **see also:** Ballini, R. et al., *S*, 2003.

80-99%

IV.E-18 Mann, A. et al., *TL*, **42**, 6499; **see also:** Maruoka, K. et al., *T*, **57**, 135.

79-90%

IV.E-19 Parsons, A.F. et al., *T*, **57**, 4719; Micalizio, C.C., Roush, W.R., *OL*, **3**, 1949.

31-65%

IV.E-20 Davies, H.M.L. et al., *OL*, **3**, 1475; **see also:** Angle, S.R., Chann, K., *TL*, **42**, 1819; Lo, M,M.-C., Fu, G.C., *T*, **57**, 2621.

98%
e.e. = 94%

IV.E-21 Grubbs, R.H. et al., *OL*, **3**, 3225; Wallace, D.J., Kennedy, D.J. et al., *SL*, 357.

82%
e.e. = 90%

IV.E-22 Johnson, T. et al., *SL*, 646; Suga, H. et al., *BCJ*, **74**, 1115; Muthusamy, S. et al., *TL*, **42**, 523.

IV.E-23 Greedy, B., Gouverneur, V., *CC*, 233.

IV.E-24 Katritzky, A.R. et al., *JOC*, **66**, 5613.

IV.E-25 Larock, R.C., Yue, D., *TL*, **42**, 6011.

(also using I_2, NBS or Ph-SeCl as electrophile)

IV.E-26 Basavaiah, D. et al., *TL*, **42**, 1147; **see also** Zhang, Y. et al., *S*, 1004; Takai, K. et al., *SL*, 1614.

IV.E-27 Fevig, T.L. et al., *JOC*, **66**, 2493; Shevelev, S.A. et al., *TL*, **42**, 8539.

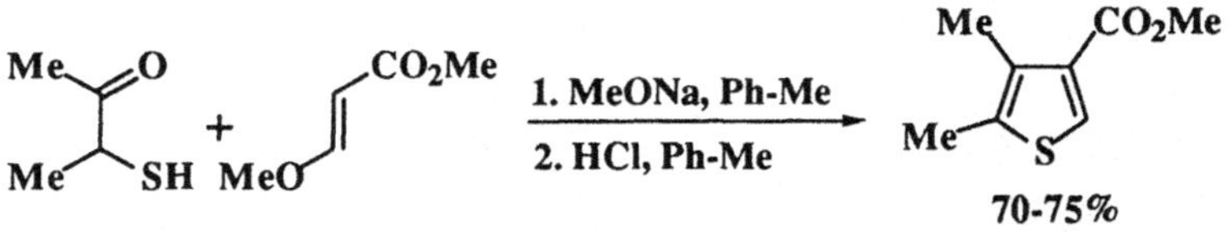

Me
O
Me
SH
+
MeO
CO_2Me
1. MeONa, Ph-Me
2. HCl, Ph-Me
Me
CO_2Me
Me
S
70-75%

IV.E-28 Armengol, M., Joule, J.A., *JCS(P1)*, 154.

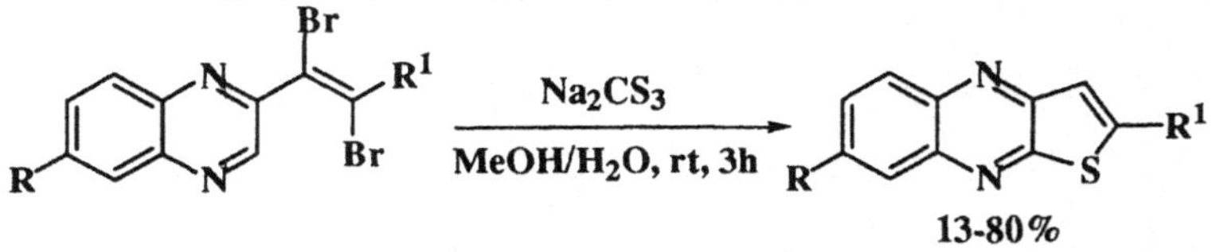

Br
N
R[1]
R
N
Br
Na_2CS_3
$MeOH/H_2O$, rt, 3h
N
R[1]
R
N
S
13-80%

IV.E-29 Loh, T.-P. et al., *JACS*, **123**, 2450.

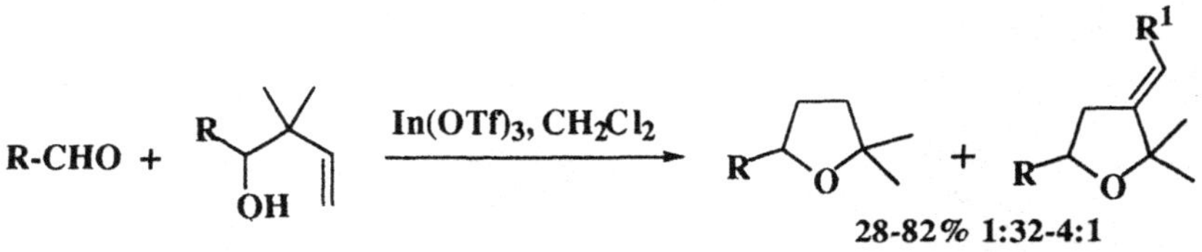

R-CHO +
R
OH
$In(OTf)_3$, CH_2Cl_2
R
O
+
R[1]
R
O
28-82% 1:32-4:1

IV.E-30 Ajamian, A., Gleason, J.L., *OL*, **3**, 4161.

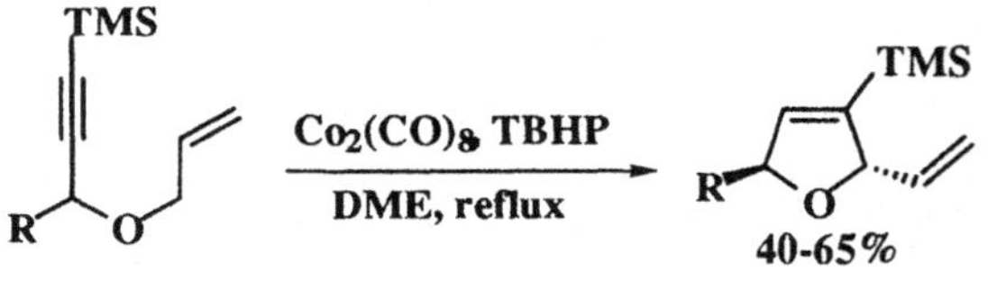

TMS
R
O
$Co_2(CO)_8$, TBHP
DME, reflux
TMS
R
O
40-65%

IV.E-31 Kiryanov, A.A., Sampson, P., Seed, A.J., *JOC*, **66**, 7925; Seed, A.J. et al., *JOC*, **66**, 7283.

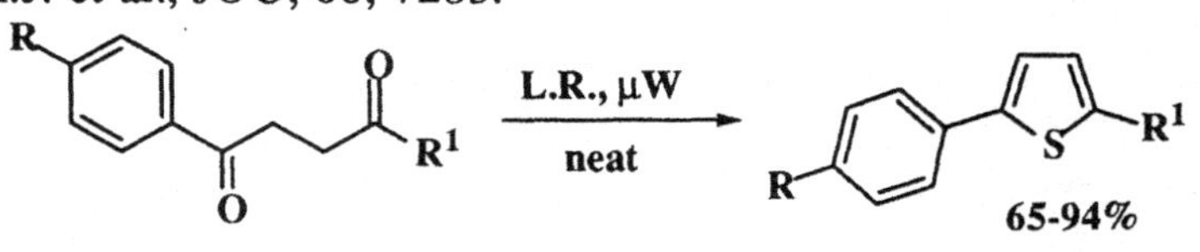

R
O
R[1]
O
L.R., µW
neat
S
R[1]
R
65-94%

IV.E-32 Katritzky, A.R. et al., *JOC*, **66**, 2850.

Bt
R[1]
+ R-N=C=S
1. BuLi
2. $ZnBr_2$, CH_2Cl_2
R[1]
S
NHR
25-80%

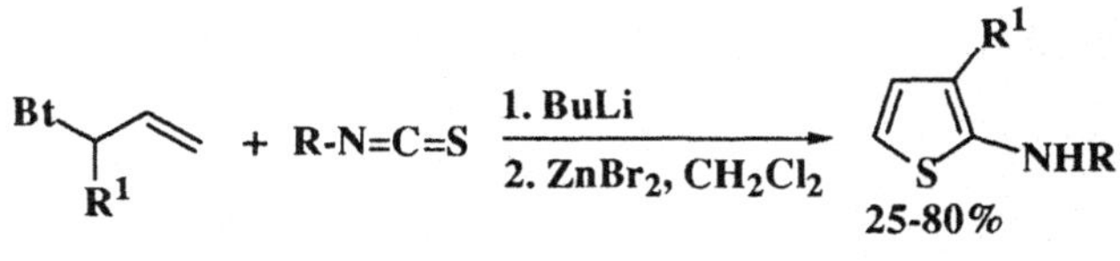

IV.E-33 Guidon, Y. et al., *JOC*, **66**, 8992.

20-86%

IV.F Pyrroles, Indoles, etc.

IV.F-1 Dieter, R.K., Yu, H., *OL*, **3**, 3855; Bates, R.W., Satcharoen, V., *SL*, 532.

60-90%

IV.F-2 Oshima, K. et al., *OL*, **3**, 2709; Somfai, P. et al., *JCS(P1)*, 891; Lebreton, J. et al., *TA*, **12**, 1121; Livinghouse, T. et al., *TL*, **42**, 2933.

28-96%

IV.F-3 Kitamura, T., Mori, M., *OL*, **3**, 1161.

0-75%

IV.F-4 Rai, K.M.L. et al., *T*, **57**, 6993; O'Neil, I.A. et al., *TL*, **42**, 8243; Yadav, J.S. et al., *TL*, **42**, 9089.

40-58%

IV.F-5 Barluenga, J. et al., *ECJ*, **7**, 2896; Hayes, C.J. et al., *OL*, **3**, 3377.

56-91%

IV.F-6 Livinghouse, T. et al., *OL*, **3**, 2961; Duncan, D., Livinghouse, T., *JOC*, **66**,m 5237.

60%

IV.F-7 Ogasawara, K. et al., *TL*, **42**, 4523; **see also:** Sackus, A. et al., *JCR(S)*, 540.

71%

IV.F-8 Johnson, J.N. et al., *OL*, **3**, 1009.

30-90%

IV.F-9 Guindon, Y. et al., *OL*, **3**, 2293.

84% 1:16

(also reaction with aminyl radicals)

IV.F-10 Blechart, S. et al., *JOC*, **66**, 6896.

34-54%
cis:trans = 1:2-99

IV.F-11 Farcas, S., Namy, J.-L., *T*, **57**, 4881,

62-96%

IV.F-12 Lovely, C.J. et al., *T*, **57**, 4095; **see also:** Butin, A.V. et al., *TL*, **42**, 2031.

60-88%

IV.F-13 Coldham, I. et al., *JCS(P1)*, 1758; Subramaniyan, G., Raghunathan, R., *T*, **57**, 2909; Gan, L. et al., *JOC*, **66**, 6369.

55%

IV.F-14 Dechoux, L. et al., *SL*, 1440.

5-96%

IV.F-15 Pearson, W.H. et al., *TL*, **42**, 7361.

46%
e.e. = 98%

IV.F-16 Suarez, E. et al., *JOC*, **66**, 1861.

15-78%

IV.F-17 Narasaka, K. et al., *SL*, 974.

64-82%

IV.F-18 Roesch, K.R., Larock, R.C., *JOC*, **66**, 412; **see also:** Ujjainwalla, F., Walsh, T.F., *TL*, **42**, 6441.

46-94%

IV.F-19 Muller, T.J.J. et al., *OL*, **3**, 3297.

49-59%

IV.F-20 Kobayashi, K. et al., *H*, **55**, 1561.

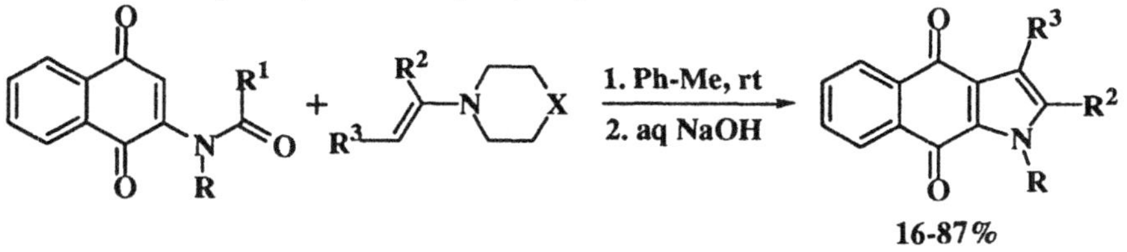

1. Ph-Me, rt
2. aq NaOH

16-87%

IV.F-21 Ryu, J.-S., Marks, T.J., McDonald, F.E., *OL*, **3**, 3091.

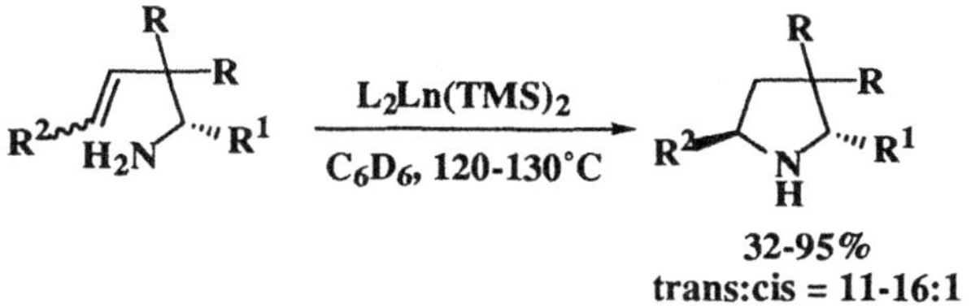

$L_2Ln(TMS)_2$

C_6D_6, 120-130°C

32-95%
trans:cis = 11-16:1

IV.F-22 Dobbs, A., *JOC*, **66**, 642.

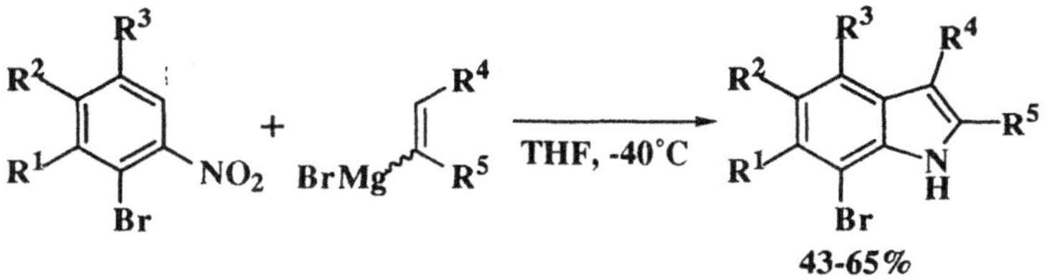

THF, -40°C

43-65%

IV.F-23 Gabriele, B., Salerno, G. et al., *TL*, **42**, 1339.

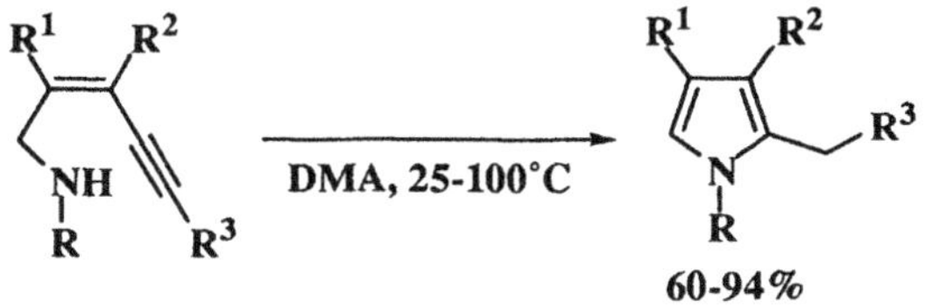

DMA, 25-100°C

60-94%

IV.F-24 Grigg, R. et al., *T*, **57**, 10335.

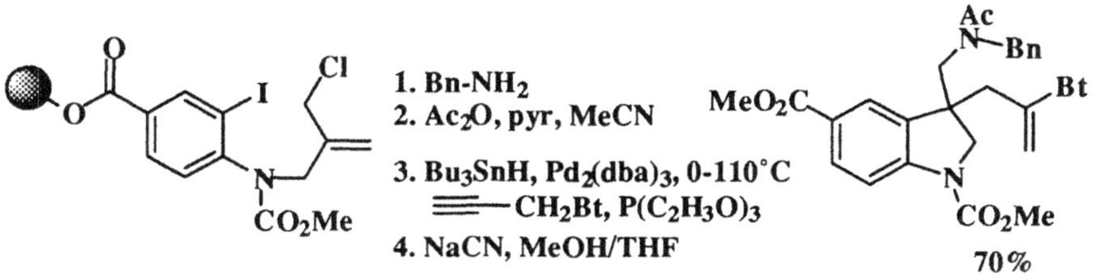

1. Bn-NH$_2$
2. Ac$_2$O, pyr, MeCN
3. Bu$_3$SnH, Pd$_2$(dba)$_3$, 0-110°C
   ≡—CH$_2$Bt, P(C$_2$H$_3$O)$_3$
4. NaCN, MeOH/THF

70%

IV.F-25 Nishiwaki, N., Ariga, M. et al., *JOC*, **66**, 7535.

33-74%

IV.F-26 Bonnet-Delpon, D. et al., *JOC*, **66**, 2098.

65-98% 3.76-24:1

IV.F-27 Paulmier, C. et al., *JCS(P1)*, 37.

68-85%

IV.F-28 Jung, K.W. et al., *OL*, **3**, 3539.

93-95%

IV.F-29 Cacchi, C. et al., *SL*, 1605; **see also:** Back, T.G. et al., *JOC*, **66**, 8599; Grigg, R. et al., *TL*, **42**, 8677.

22-97%

IV.F-29 Gevorgyon, V. et al., *JACS*, **123**, 2074.

$$R^1 \text{—}\!\!\!\equiv\!\!\!\text{—} \overset{R^2}{\underset{NR}{\diagup}} \quad \xrightarrow[\text{DMA, 110°C}]{\text{CuI, TEA}} \quad R^1 \text{———} R^2$$

50-93%

IV.F-30 Nair, V. et al., *JOC*, **66**, 4427; Murahashi, S. et al., *OL*, **3**, 421.

$$\underset{R^1}{\overset{\|}{N}} \;+\; \underset{CO_2Me}{\overset{CO_2Me}{\|\|}} \;+\; \underset{Ts}{\overset{R}{N\!\!=\!\!\diagdown}} \quad \xrightarrow[\text{Ph-H, rt}]{} \quad R^1HN \overset{MeO_2C \quad CO_2Me}{\underset{Ts}{\diagup\!\!\diagdown}} R$$

79-94%

IV.F-31 Khadilkar, B.M., Rebeiro, G.L., *S*, 370.

$$R\text{—}\langle\!\!\rangle\text{—NHNH}_2 \;+\; \underset{O}{\overset{R^1}{\diagdown\!\!\diagup}} R^2 \quad \xrightarrow{\text{AlCl}_3,\ \text{pyr}^+\text{-Bu Cl}^-} \quad R\text{—}\langle\!\!\rangle \overset{R^1}{\underset{H}{\diagup\!\!\diagdown}} R^2$$

41-92%

IV.F-33 Tokunaga, M., Wakatsuki, Y. et al., *TL*, **42**, 3865.

$$R\text{—}\langle\!\!\rangle\text{—NH}_2 \;+\; \underset{\|\|}{\overset{HO\diagdown R^1}{\diagdown}} \quad \xrightarrow[\text{air, neat}]{\text{Ru}_3(CO)_{12},\ \text{lig.}} \quad R\text{—}\langle\!\!\rangle \overset{R^1}{\underset{H}{\diagup\!\!\diagdown}}\text{Me}$$

5-97%%

IV.F-34 Grigg, R. et al., *T*. **57**, 1347, 1361.

$$\underset{SO_2Ph}{\overset{I}{\langle\!\!\rangle\text{—N}}}\diagdown \quad \xrightarrow[\text{Pd(OAc)}_2,\ \text{PPh}_3,\ \text{Ph-Me}]{\text{CO, Ph}_2\text{MeSiH, Et}_4\text{NCl}} \quad \underset{SO_2Ph}{\langle\!\!\rangle}\diagdown\text{CHO}$$

61%

IV.F-35 Ogura, K. et al., *H*, **55**, 231.

$$\underset{R}{\overset{Ts}{\diagdown}}\!\!\equiv\!\!\equiv\!\!\overset{R}{\underset{R}{\diagdown}}\text{Ts} \quad \xrightarrow[\text{DMF, 90°C, 8h}]{\text{Ar-NH}_2,\ \text{CuCl, CuCl}_2} \quad R\overset{}{\underset{Ar}{\langle\!\!\rangle}}\text{—}\!\!\equiv\!\!\overset{R}{\underset{Ts}{\diagdown}}$$

47-63%

IV.F-36 Noguchi, Y., Kobayashi, S. et al., *TL*, **42**, 5253.

74% 9:1

IV.G. Pyridines, Quinolines, etc.

IV.G-1 Furstner, A. et al., *JACS*, **123**, 11863.

47-87%

IV.G-2 Itoh, K. et al., *CC*, 1102.

50-87% 1:6.7-100:1

IV.G-3 Zhang, H., Larock, R.C., *OL*, **3**, 3083.

54-96%

IV.G-4 Kamatani, A., Overman, L.E., *OL*, **3**, 1229.

0-91%
e.e. = 0-90%

IV.G-5 Gottlich, R., Noack, M., *TL*, **42**, 7771.

$$BuN \xrightarrow[\text{Ph-H, 50°C}]{\text{SmI}_2} BuN$$

54%

IV.G-6 Wallace, D.J., Kennedy, D.J. et al., *OL*, **3**, 671; Martin, S.F. et al., *TL*, **42**, 1635; Couty, F. et al., *T*, **57**, 5393; Sabat, M., Johnson, C.R., *TL*, **42**, 1209; Kumareswaran, R., Hassner, A., *TA*, **12**, 2269; Takahata, H. et al., *TA*, **12**, 817; Rutjes, F.P.J.T. et al., *OL*, **3**, 2045; Grigg, R. et al., *TL*, **42**, 8673.

$$\xrightarrow{\text{PhCH=RuCl}_2(\text{PCy}_3)_2}$$

86%
d.s. = 70%

IV.G-7 Prevost, N., Shipman, M., *OL*, **3**, 2383; Harrowven, D.C. et al., *TL*, **42**, 9061.

$$\xrightarrow[\text{Ph-H, reflux}]{\text{Bu}_3\text{SnH, AIBN}}$$

68%

IV.G-8 Harrity, J.P.A. et al., *SL*, 1596.

$$R-\text{NSO}_2\text{Ar} + \text{AcO} \diagup\!\!\!\diagup \text{TMS} \xrightarrow[\text{THF, 65°C}]{\text{Pd(OAc)}_2, \text{P}(^i\text{Pr})_3}$$

44-82%

IV.G-9 Tiecco, M. et al., *TA*, **12**, 3297.

$$\xrightarrow[\text{CH}_2\text{Cl}_2, \text{-78°C}]{}$$

51-96%
d.r. = 9-24:1

IV.G-10 Maruoka, K. et al., *S*, 1716.

56-64%
e.e. = 84-88%

IV.G-11 Lee, J. et al., *TL*, **42**, 6223; Cossy, J., Pardo, D.G. et al., *TL*, **42**, 5705.

85%
e.e. = 99%

IV.G-12 Zhu, J. et al., *TL*, **42**, 4503; Silveira, C.C. et al., *TL*, **42**, 8947; Bonin, M. et al., *TL*, **42**, 2111; Katritzky, A.R. et al., *TA*, **12**, 2427 and *JOC*, **66**, 148; Loucher, C., *SC*, **31**, 2895.

96%

IV.G-13 Fuchs, J.R., Funk, R.L., *OL*, **3**, 3349.

61%

(other examples given)

IV.G-14 Kawecki, R., *S*, 828.

38-95%

IV.G-15 Wenkert, D., McPhail, A.T. et al., *OL*, **3**, 2301; Ohba, M., Izuta, R., *H*, **55**, 823.

30%

IV.G-16 Hundsdorf, T., Neunhoeffer, H., *S*, 1800; Rykowski, A. et al., *OPP*, **33**, 501; **see also:** Diaz-Ortiz, A., de la Hoz, A. et al., *SL*, 236.

81-89%

IV.G-17 Collin, J. et al., *TL*, **42**, 7405; Chou, S.-S.P., Hung, C.-C., *SC*, **31**, 1097.

61-87%

IV.G-18 Kotsuki, H. et al., *SL*, 1323; Katsumura, S. et al., *JOC*, **66**, 3099.

30-74%

IV.G-19 Yu, C., Hu, L., *TL*, **42**, 5167.

84-99%

IV.G-20 Nagarajan, R., Perumal, R.T., *SC*, **31**, 1733; Ramalingham, T. et al., *SC*, **31**, 1075.

49-67%

IV.G-21 Troin, Y. et al., *TL*, **42**, 4815.

53-74%

IV.G-22 Davies, I.W., Marcoux, J.-F., Reider, P.J., *OL*, **3**, 209; Marcoux, J.-F. et al., *JOC*, **66**, 4194; **see also:** Koike, T., *CPB*, **49**, 558.

48-85%

IV.G-23 Bagley, M.C. et al., *SL*, 1149, 1523.

65-95%

IV.G-24 Rodriguez, J., *OL*, **3**, 2145.

0-85%

IV.G-25 Ungureanu, I., Mann, A. et al., *TL*, **42**, 6087.

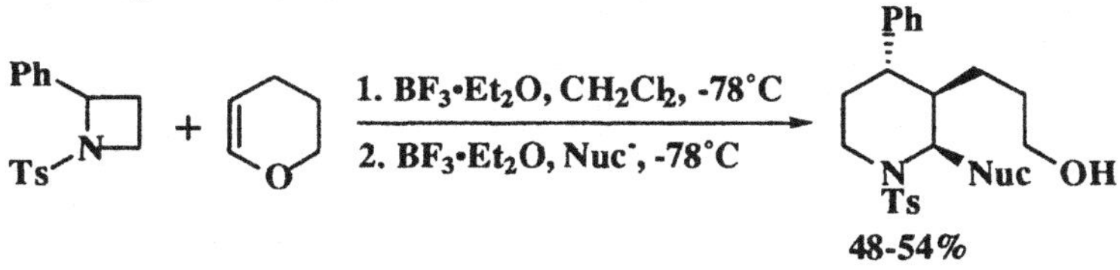

Ph
Ts
N
O
1. BF3•Et2O, CH2Cl2, -78°C
2. BF3•Et2O, Nuc⁻, -78°C
Ph
N
Ts
Nuc
OH
48-54%

IV.G-26 Haddad, M., Larcheveque, M., *TL*, **42**, 5223.

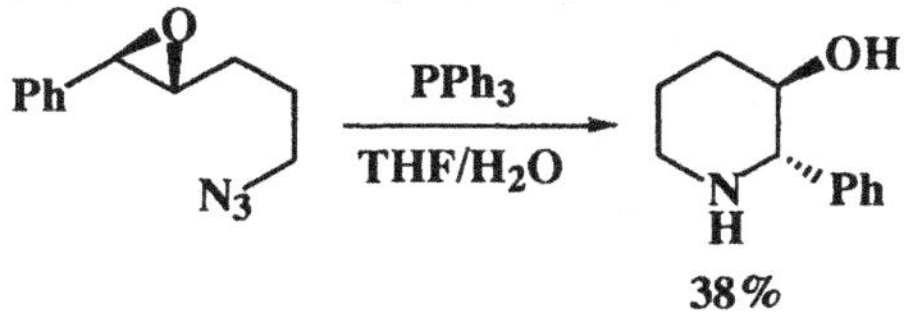

O
Ph
N3
PPh3
THF/H2O
OH
N
H
Ph
38%

IV.G-27 Dow, R.L., Schneider, S.R., *JHC*, **38**, 535.

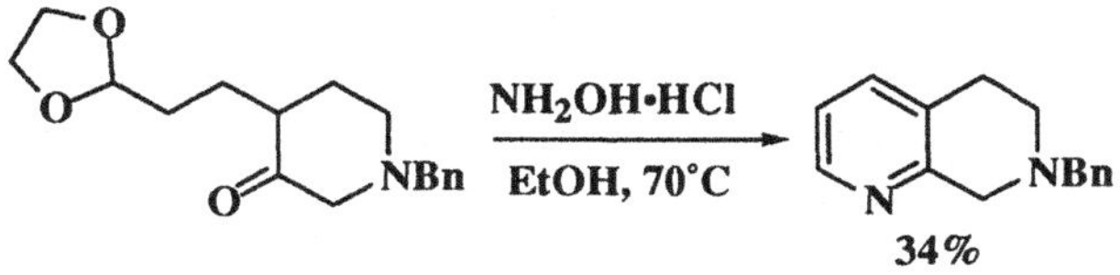

O
O
O
NBn
NH2OH•HCl
EtOH, 70°C
N
NBn
34%

IV.G-28 Cho, C.S., Shim, S.C. et al., *CC*, 2576; Cladiali, S., Thummel, R.P. et al., *JOC*, **66**, 400.

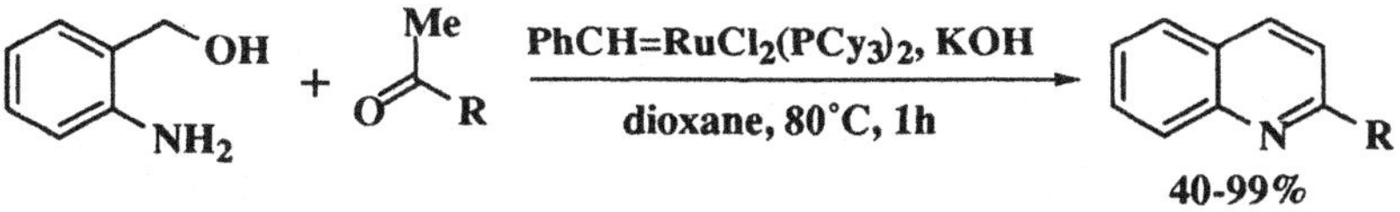

OH
NH2
Me
O
R
PhCH=RuCl2(PCy3)2, KOH
dioxane, 80°C, 1h
N
R
40-99%

IV.G-29 Kim, J.N. et al., *TL*, **42**, 3737; Kim, J.N. et al., *TL*, **42**, 8341.

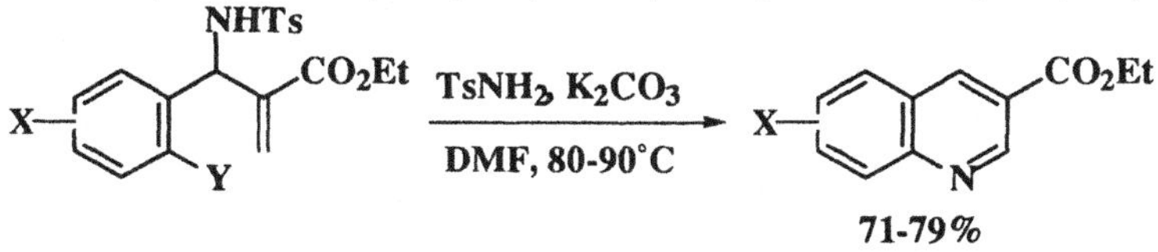

NHTs
CO2Et
X
Y
TsNH2, K2CO3
DMF, 80-90°C
X
CO2Et
N
71-79%

IV.G-30 Rajanna, K.C., *SL*, 251.

R
NHAc
POCl3, DMF, SDS or CTAB
MeCN, reflux
R
CHO
N
Cl
45-90%

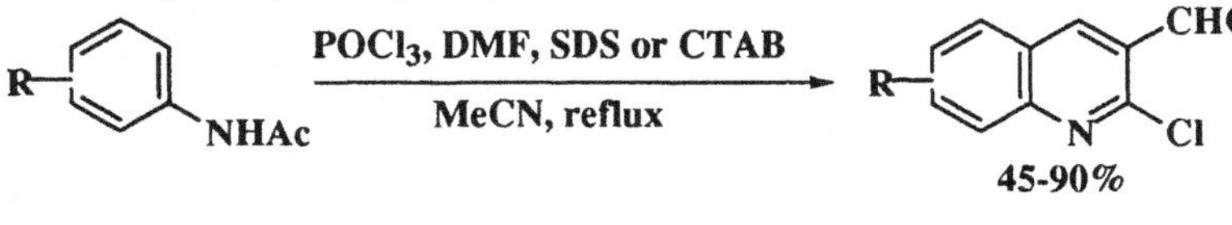

IV.G-31 Dai, G., Larock, R.C., *OL*, **3**, 4035; Larock, R.C. et al., *OL*, **3**, 2973 and *JOC*, **66**, 8042.

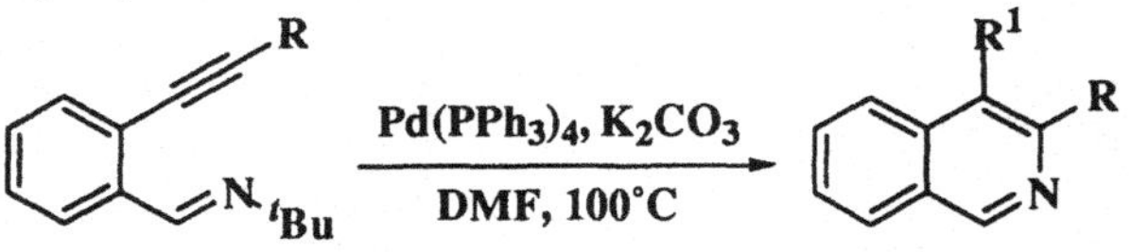

R
N
*t*Bu
Pd(PPh3)4, K2CO3
DMF, 100°C
R1
R
N

23-80%

IV.G-32 Rossi, E. et al., *TL*, **42**, 3705.

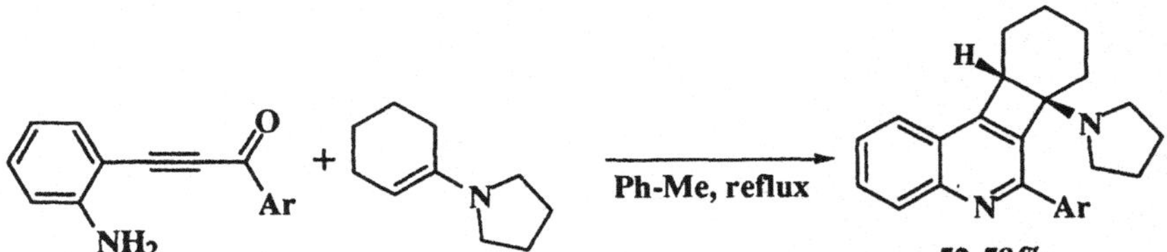

O
Ar
NH2
+
N
Ph-Me, reflux
H
N
N
Ar

52-58%

IV.G-33 Sole, D., Bonjoch, J. et al., *JOC*, **66**, 5266.

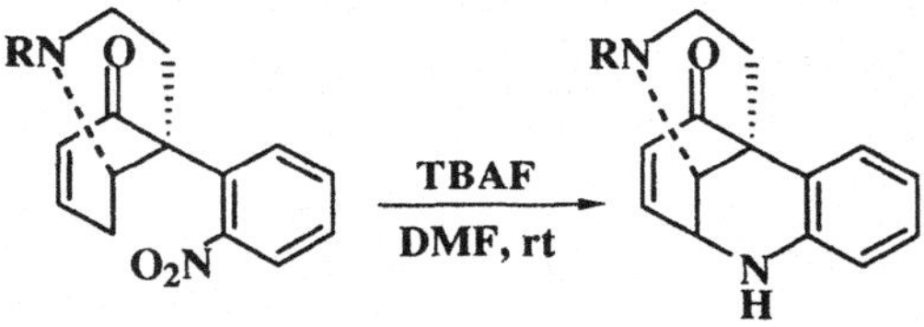

RN
O
O2N
TBAF
DMF, rt
RN
O
N
H

37%

IV.G-34 Bunce, R.A. et al., *JOC*, **66**, 2822.

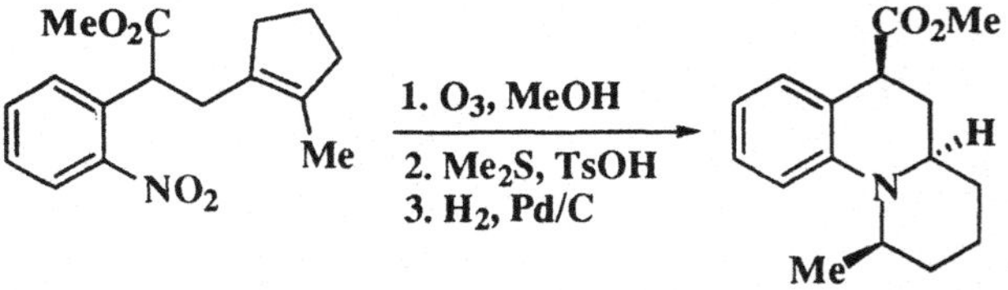

MeO2C
Me
NO2
1. O3, MeOH
2. Me2S, TsOH
3. H2, Pd/C
CO2Me
H
N
Me

65%

IV.G-35 Toda, T. et al., *H*, **55**, 1249.

NH2
+
Br
Ph
O
Yb(OTf)3, CH2Cl2
Ph
OH
N
H

43%

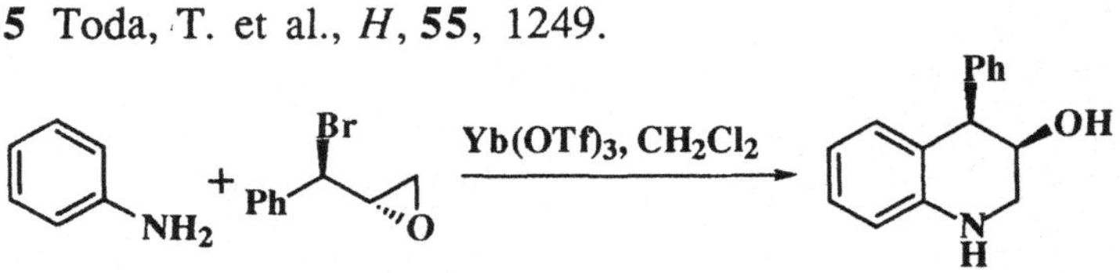

IV.G-36 Kobayashi, K. et al., *CL*, 602.

0-87%

IV.H Pyrans, Pyrones and Sulfur Analogues

IV.H-1 Heck, M.P., Mioskowski, C. et al., *OL*, **3**, 1989; Nadolski, G.T., Davidson, B.S., *TL*, **42**, 797; Messinger, B.T., Davidson, B.S., *TL*, **42**, 801; Rainier, J.D. et al., *TL*, **42**, 179; Schmidt, B. et al., *JOC*, **66**, 7658; Burke, S.D., Voight, E.A., *OL*, **3**, 237.

59-75%

IV.H-2 Hoveyda, A.H. et al., *JACS*, **123**, 3139.

90-93%
e.e. = 74-98%

IV.H-3 Smith, A.B., III et al., *OL*, **3**, 3979.

90%

IV.H-4 Li, J., Li, C.-J., *TL*, **42**, 793.

73-96%

IV.H-5 Willis, M. et al., *JOC*, **66**, 3284.

≤99%

IV.H-6 Alami, M. et al., *TL*, **42**, 2657; Yamaguchi, S. et al., *TL*, **42**, 1091.

79%

IV.H-7 Parker, K.A., Mindt, T.L., *OL*, **3**, 3875; Snieckus, V. et al., *S*, 140.

59%

IV.H-8 Liu, R.-S. et al., *JOC*, **66**, 8106; Mukai, C., Hanaoka, M. et al., *CPB*, **49**, 613.

51-74%

IV.H-9 Feng, J. et al., **31**, *SC*, 2663.

61-90%

IV.H-10 Hashimoto, S. et al., *OL*, **3**, 4075.

1. aq HCl, THF, 0°C
2. MeOLi, THF/MeOH

72%

IV.H-111 Zaveri, T., *OL*, **3**, 843.

NaBH₄

THF/EtOH

55%

IV.H-12 Hsung, R.P. et al., *TL*, **42**, 609.

EtOAc, 85°C

Ac₂O,

55-78% 1-2.3:1

IV.H-13 Li, C.-J. et al., *JOC*, **66**, 739; Yadav, J.S. et al., *TL*, **42**, 89 and *S*, 885; Vakalopoulos, A., Hoffmann, H.M.R., *OL*, **3**, 2185.

$+ R^3\text{-CHO}$ InCl₃, CH₂Cl₂

Y = O 23-69%
Y = S 62-82%

IV.H-14 Huang, H., Panek, J.S., *OL*, **3**, 1693.

TMS-OTf

CH₂Cl₂, 20°C

60-70%

V.H-15 Nicolaou, K.C. et al., *AG(E)*, **40**, 1262.

EtO2C-CHO + [diene structure with OMe]

IV.H-16 Kalesse, M. et al., *TL*, **42**, 1263; Yamada, T. et al., *BCJ*, **74**, 1333; **see also:** Bogdanowicz-Szwed, K., Palasz, A., *ZN*, **56B**, 416; Bear, B.R., Shea, K.J., *OL*, **3**, 723.

$$\text{EtO}_2\text{C-CHO} + \quad \xrightarrow[\text{CH}_2\text{Cl}_2, \text{reflux}]{\text{Ti(O}^i\text{Pr)}_4 \text{ (R)-BINOL}}$$

65%
e.e. = 98%

IV.H-17 Roush, W.R., Dilley, G.J., *SL*, 955; Leroy, B., Marko, I.E., *TL*, **42**, 8685; Suginome, M., Ohmori, Y., Ito, Y., *JACS*, **123**, 4601; Suginome, M., Ito, Y. et al., *SL*, 1042; Oriyama, T. et al., *BCJ*, **74**, 569.

$$\xrightarrow[\text{CH}_2\text{Cl}_2, -78°\text{C}]{\text{TMS-OTf, 4Å MS}}$$

82%
d.s. = 15.7:1

IV.H-18 Ishino, Y. et al., *SC*, **31**, 439.

$$\xrightarrow[\text{DCE, rt}]{\text{TsOH}}$$

48-90%

IV.H-19 Nair, V. et al., *T*, **57**, 5807.

$$\xrightarrow[\text{dioxane, reflux}]{\text{(CHO)}_n}$$

58%

IV.H-20 Dunach, E. et al., *CC*, 2284.

29-76%

IV.H-21 Ishino, Y. et al., *SL*, 1317.

39-83%

IV.H-22 Hanson, P.R. et al., *SL*, 605.

19-99%

IV.H-23 Katritzky, A.R., Button, M.A.C., *JOC*, **66**, 5595.

40-99%
d.r. = ≤23:1

IV.H-24 Abe, H., Harayama, T. et al., *CPB*, **49**, 1223.

0-86%

IV.I. Other Heterocycles with One Heteroatom

IV.I-1 Rosini, C. et al., *TA*, **12**, 1225.

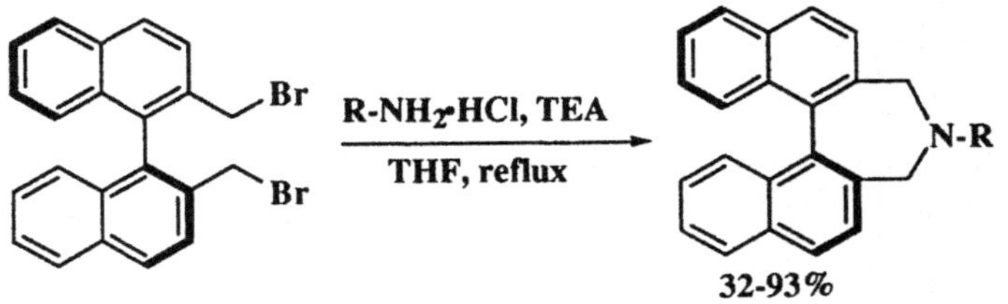

Br
Br
R-NH2•HCl, TEA
THF, reflux
N-R
32-93%

IV.I-2 Brechbiel, M.W. et al., *JOC*, **66**, 7745.

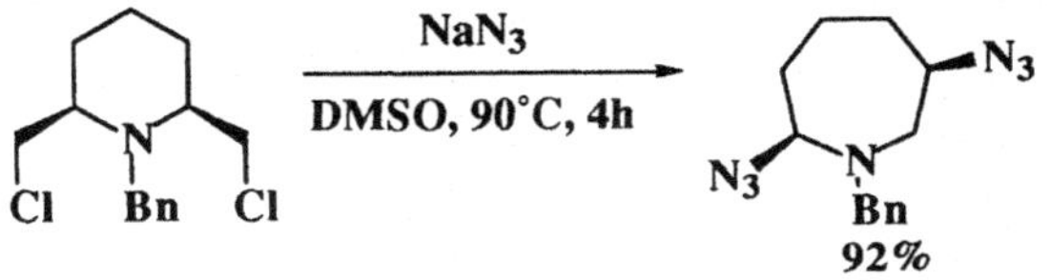

NaN3
DMSO, 90°C, 4h
N
Cl
Bn
Cl
N3
N3
N
Bn
92%

IV.I-3 Perez-Castells, J. et al., *CC*, 2602; **see also:** Pearson, W.H., Aponick, A., *OL*, **3**, 1327.

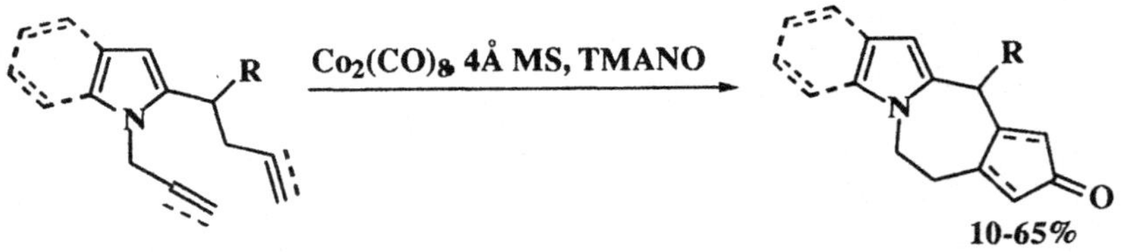

R
N
Co2(CO)8, 4Å MS, TMANO
R
N
O
10-65%

IV.I-4 Dominguez, D. et al., *TL*, **42**, 665.

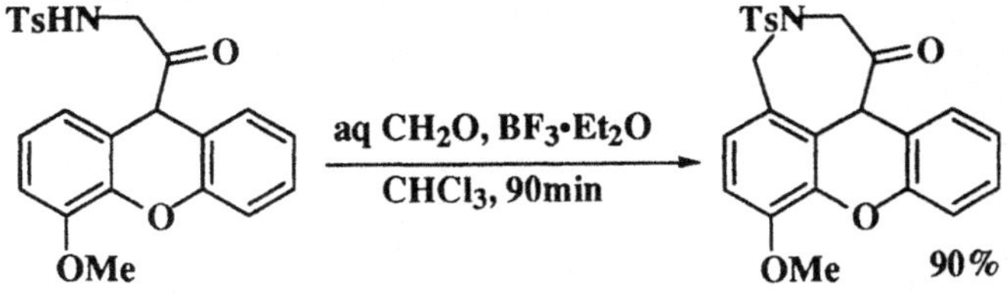

TsHN
O
O
OMe
aq CH2O, BF3•Et2O
CHCl3, 90min
TsN
O
O
OMe
90%

IV.I-5 Lee, J.Y., Lee, Y.S. et al., *H*, **55**, 1519,

Ph
OH
N
O
HCO2H, reflux
N
O
88%

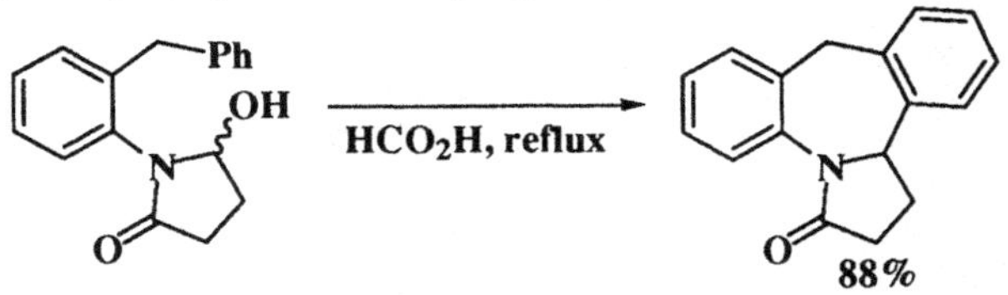

IV.I-6 Node, M. et al., *AG(E)*, **40**, 3060.

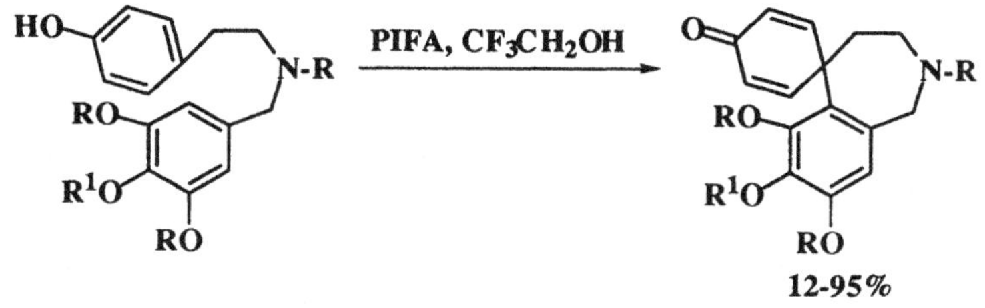

12-95%

IV.I-7 Lohse, O. et al., *TL*, **42**, 385.

96%

IV.I-8 Sugita, Y. et al., *H*, **55**, 855.

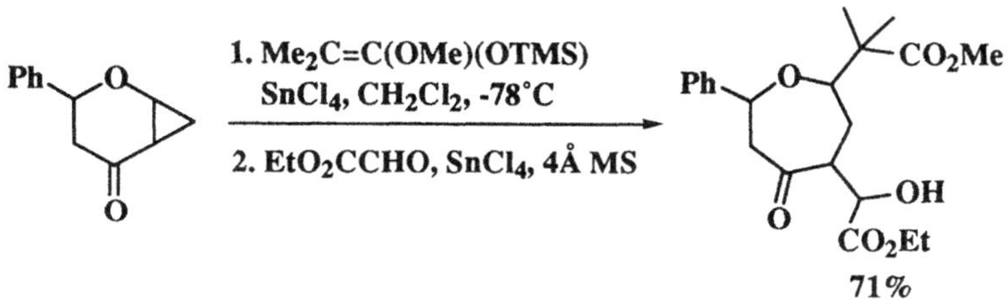

1. Me$_2$C=C(OMe)(OTMS)
   SnCl$_4$, CH$_2$Cl$_2$, -78°C

2. EtO$_2$CCHO, SnCl$_4$, 4Å MS

71%

IV.I-9 Chiu, P. et al., *OL*, **3**, 1721.

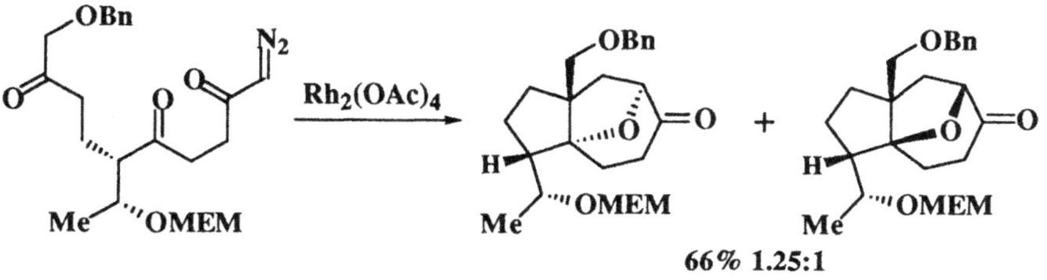

Rh$_2$(OAc)$_4$

66% 1.25:1

IV.I-10 Kira, K., Isobe, M., *TL*, **42**, 2821.

IV.J. Heterocycles with a Bridgehead Heteroatom

IV.J-1 Genet, J.-P. et al., *TL*, **42**, 2461.

IV.J-2 Alcaide, B. et al., *JOC*, **66**, 1351.

IV.J-3 Katritzky, A.R. et al., *JCS(P1)*, 1767.

IV.J-4 Kalaus, G., Szantay, C. et al., *H*, **55**, 873.

IV.J-5 Paquette, L.A. et al., *JOC*, **66**, 3564.

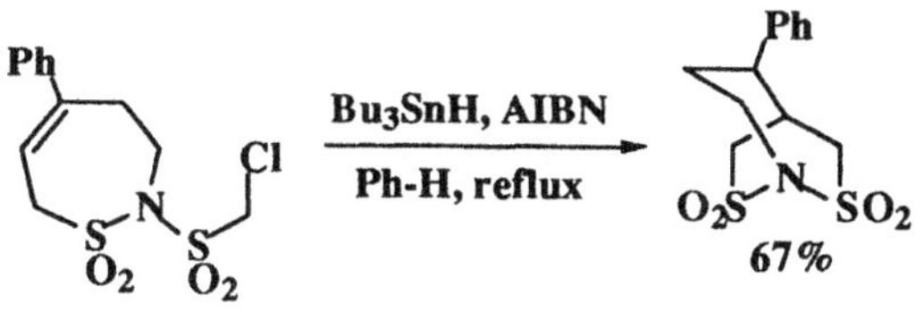

Bu₃SnH, AIBN
Ph-H, reflux
67%

IV.J-6 Hodgson, D.M. et al., *JCS(P1)*, 2161.

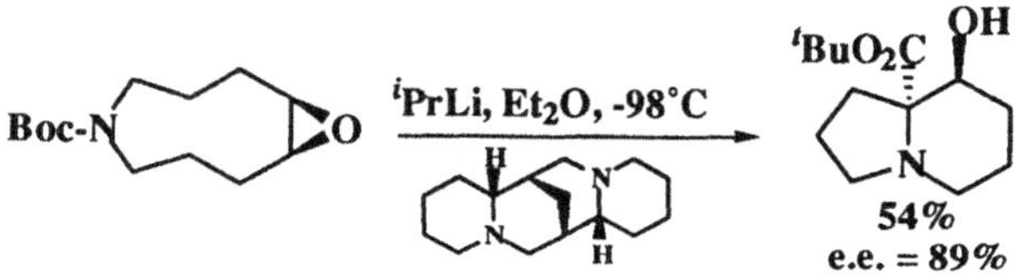

ᶦPrLi, Et₂O, -98°C
54%
e.e. = 89%

IV.J-7 Brummond, K.M., Lu, J., *OL*, **3**, 1347.

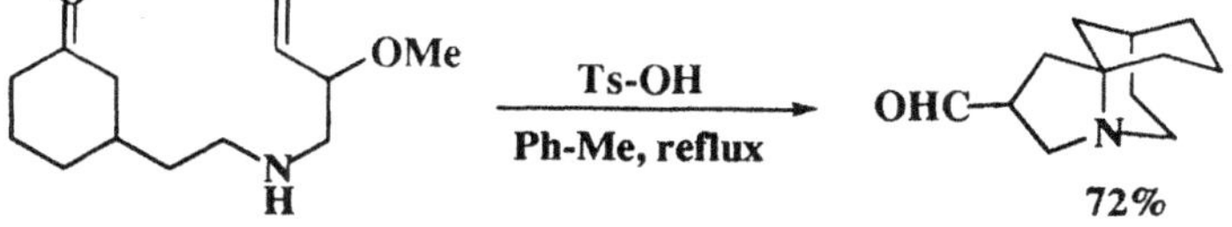

Ts-OH
Ph-Me, reflux
72%

IV.J-8 Svete, J., Stanovnik, B. et al., *JHC*, **38**, 1307.

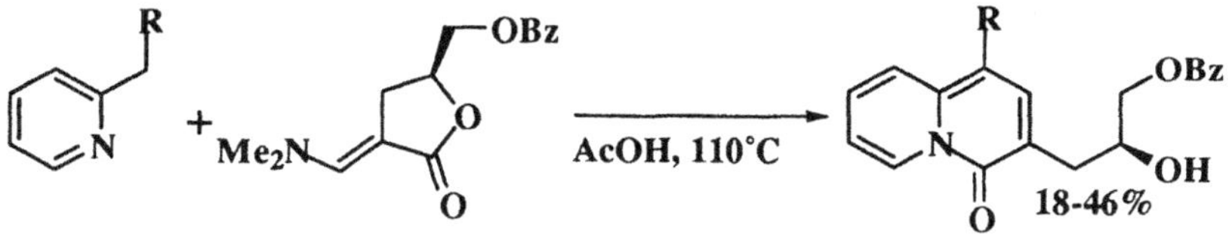

AcOH, 110°C
18-46%

IV.J-9 Yoda, H. et al., *TL*, **42**, 2509.

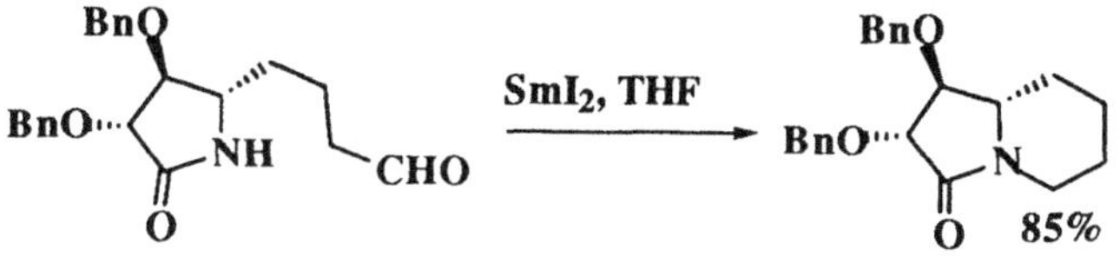

SmI₂, THF
85%

IV.J-10 Hu, Y. et al., *JHC*, **38**, 853.

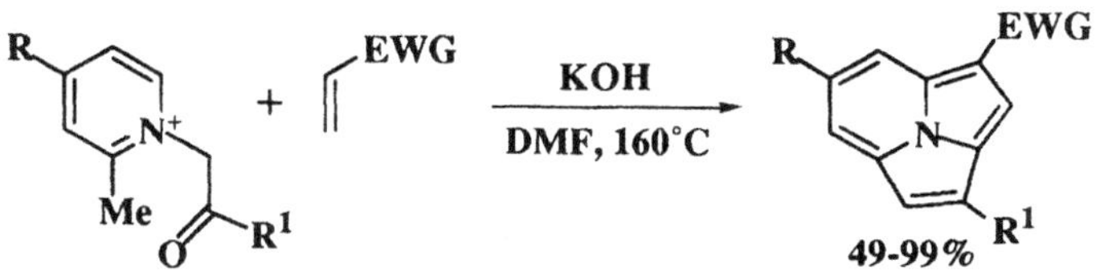

KOH
DMF, 160°C
49-99%

IV.J-11 Mootoo, D.R. et al., *JOC*, **66**, 1761; Pearson, W.H. et al., *TL*, **42**, 8267, 8273; Paolucci, C., Mattioli, L., *JOC*, **66**, 4787.

$$\text{NaCNBH}_3, \text{NH}_4\text{HCO}_3 \quad \text{MeOH}$$

78%

IV.J-12 Wu, P.-L. et al., *JOC*, **66**, 6585.

300°C

33-44%

IV.J-13 Oku, A. et al., *JOC*, **66**, 1638, 2618.

$$\text{Me}_2\text{CO}, 55°\text{C}$$

50%

IV.J-14 Najera, C. et al., *EJOC*, 3133; Aldous, D.J., Harwood, L.M. et al., *SL*, 1836; 1341.

1. $(\text{CH}_2\text{O})_n$, Ph-Me, 80°C

2. $R^1 \equiv\!\!\equiv R$

22-75%

IV.J-15 Almquist, F. et al., *JOC*, **66**, 6756.

HCl

DCE, 0→ 64°C

63-86%
e.e. = 75-97%

IV.J-16 Evans, P.A., Manangan, T., *TL*, **42**, 6637; Teulade, J.C. et al., *JOC*, **66**, 6576.

(TMS)₂SiH, Et₃B
Ph-H, rt

70-81%

IV.J-17 Allin, S.M. et al., *TL*, **42**, 3943 and *JCS(P1)*, 3029.

TiCl₄
CH₂Cl₂, -10°C

87%
α:β = 2:1

IV.K. Heterocycles with Two or More Heteroatoms

IV.K.1a. 5-Membered Heterocycles with 2 N's

IV.K.1a-1 Yoshida, M., Iyoda, M. et al., *TL*, **42**, 33; **see also:** Chandrasekhar, S. et al., *TL*, **42**, 6599.

1. (Bu₃Sn)₂, O₂, hv, Ph-H
2. N₂H₄, AcOH/EtOH

40-48%

(formation of other perfluoroalkylated heterocycles given)

IV.K.1a-2 Katritzky, A.R. et al., *JOC*, **66**, 6787; Gonzalez-Ortega, A. et al., *H*, **55**, 331; **see also:** Sandhu, J.S. et al., *JHC*, **38**, 491.

R-NHNH₂, NaOEt
EtOH, reflux

51-80%

IV.K.1a-3 Muller, D., Beckert, R. et al., *S*, 601.

+ Me₂N⁺=CHCl X⁻

58-80%

IV.K.1a-4 Song, J.J. Yee, N.K., *TL*, **42**, 2937.

81-90%

IV.K.1a-5 Kawase, M. et al., *CPB*, **49**, 461.

94%

IV.K.1a-6 Li, G. et al., *AG(E)* **40**, 4277.

45-82%
anti:syn = 20-50:1

IV.K.1a-7 Li, Z., Zhang, Y., *OPP*, **33**, 185.

60-75%

IV.K.1a-8 Zhang, P.-F., Chen, Z.-C., *S*, 2075.

42-67%

IV.K.1a-9 Soufiaoui, M. et al., *T*, **57**, 163.

84-95%

IV.K.1a-10 Mignani, S. et al., *SL*, 135.

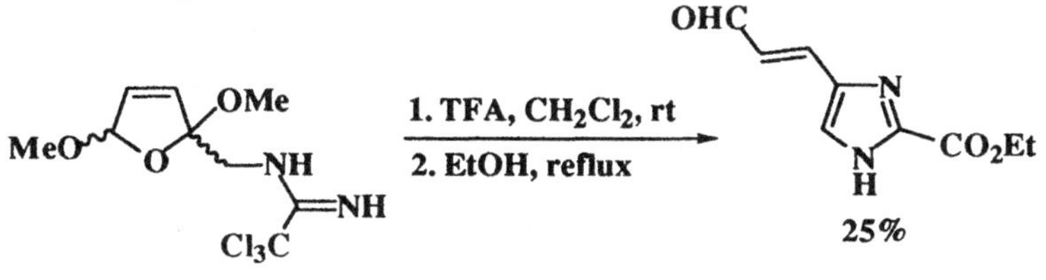

1. TFA, CH₂Cl₂, rt
2. EtOH, reflux

25%

IV.K.1a-11 Polanc, S. et al., *SL*, 1237.

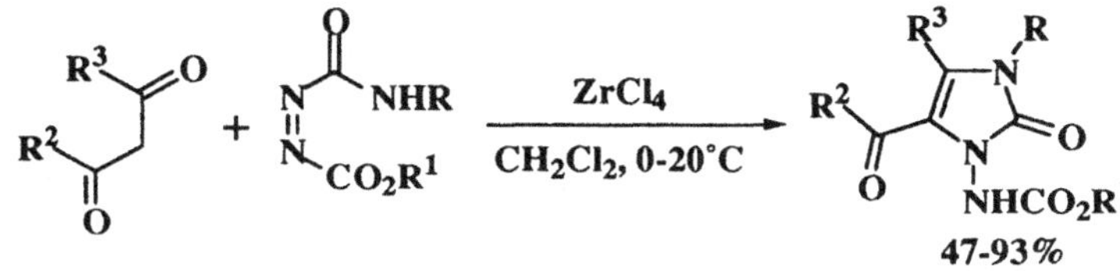

ZrCl₄

CH₂Cl₂, 0-20°C

47-93%

IV.K.1a-12 Katritzky, A.R. et al., *JOC*, **66**, 2862.

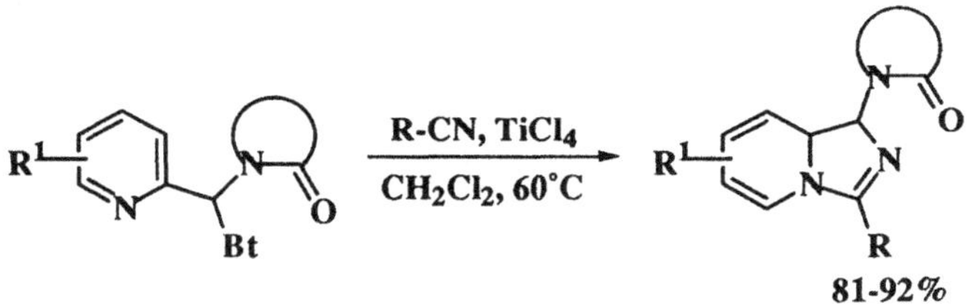

R-CN, TiCl₄

CH₂Cl₂, 60°C

81-92%

IV.K.1a-13 Ohberg, L., Westman, J., *SL*, 1893.

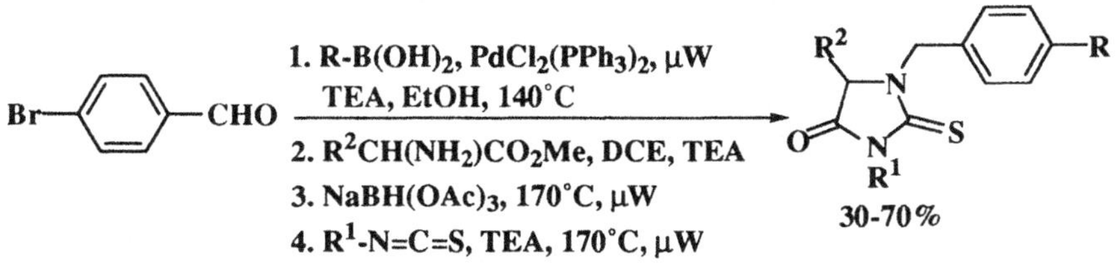

1. R-B(OH)₂, PdCl₂(PPh₃)₂, μW
   TEA, EtOH, 140°C
2. R²CH(NH₂)CO₂Me, DCE, TEA
3. NaBH(OAc)₃, 170°C, μW
4. R¹-N=C=S, TEA, 170°C, μW

30-70%

IV.K.1a-14 Chen, P. et al., *TL*, **42**, 4293, 4297.

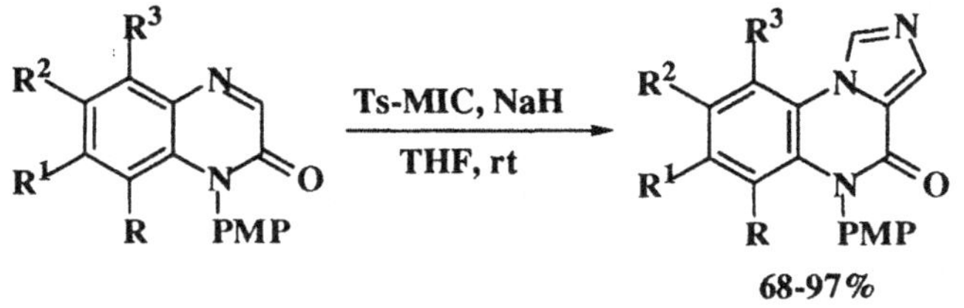

Ts-MIC, NaH

THF, rt

68-97%

IV.K.1a-15 Gutschow, M., Powers, J.C., *JOC*, **66**, 4723.

36-91%

IV.K.1a-16 Yamauchi, M., Yajima, M., *CPB*, **49**, 1638.

88-98%

IV.K.1a-17 Prakash, O., Moriarty, R.M. et al., *S*, 541,

61-82%

IV.K.1a-18 Katritzky, A.R. et al., *JOC*, **66**, 2858.

70-82%

IV.K.1a-19 Sarodnick, G., Linker, T., *JHC*, **38**, 829

36-81%

IV.K.1a-20 Weixing, Q., Yongzhou, H., *JCR(S)*, 320.

25-76%

IV.K.1b 6 Membered Heterocycles with 2 N's

IV.K.1b-1 Snyder, J.K. et al., *T*, **57**, 5497; **see also:** Sohar, P. et al., *JCS(P1)*, 558.

87%

IV.K.1b-2 Curini, M. et al., *H*, **55**, 1599.

89-99%

IV.K.1b-3 Shiori, T., Aoyama, T. et al., *H*, **55**, 2283.

80-96%   3-7%

IV.K.1b-4 Yavari, I., Adib, M., *JCR(S)*, 543.

78-95%

IV.K.1b-5 Marcaccini, S. et al., *S*, 85.

59-81%

IV.K.1b-6 Agamy, S.M., *JCR(S)*, 349.

IV.K.1b-7 Yadav, J.S. et al., *S*, 1341; Kaimal, T.N.B. et al., *SL*, 863; Reddy, C.S. et al., *TL*, **42**, 7873.

75-90%

IV.K.1b-8 Kundu, N.G., Chaudhuri, G., *T*, **57** 6833.

31-69%

IV.K.1b-9 Sandhu, J.S. et al., *SL*, 1299.

70-98%

IV.K.1b-10 Proenca, M.F. et al., *JOC*, **66**, 8436.

50-90%

IV.K.1b-11 Connolly, D.J., Guiry, P.J., *SL*, 1707.

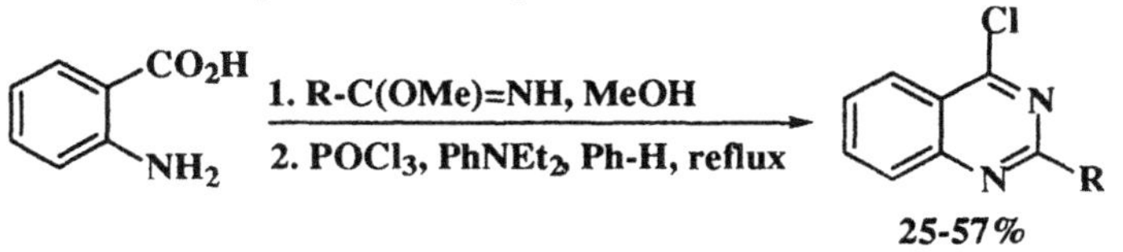

1. R-C(OMe)=NH, MeOH
2. POCl$_3$, PhNEt$_2$, Ph-H, reflux

25-57%

IV.K.1b-12 Alper, H. et al., *SL*, 914,

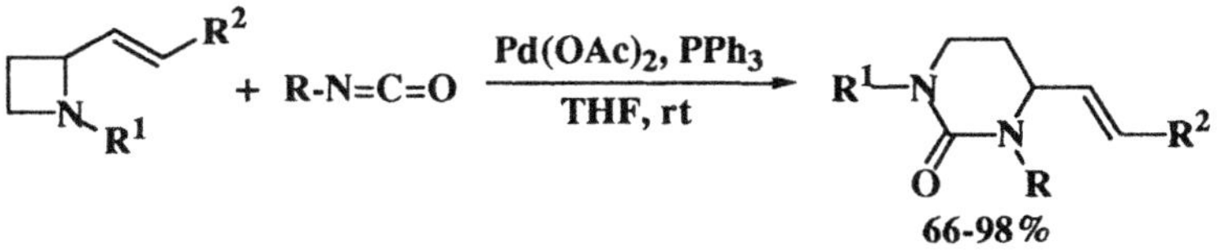

+ R-N=C=O

Pd(OAc)$_2$, PPh$_3$
THF, rt

66-98%

IV.K.1b-13 Fomun, Z.T. et al., *JCS(P1)*, 457.

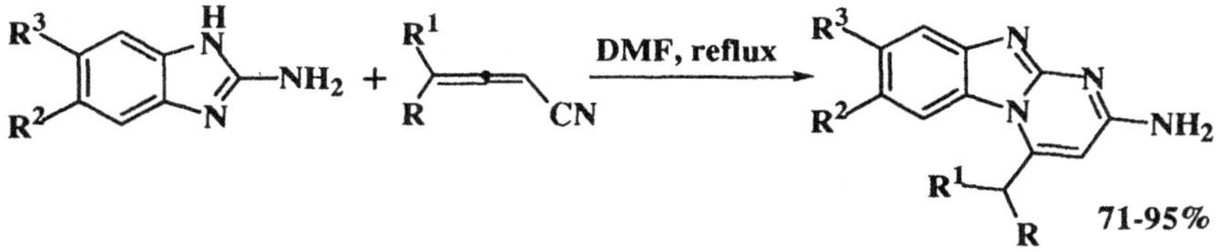

DMF, reflux

71-95%

IV.K.1b-14 Smith, B.D. et al., *TL*, **42**, 1851.

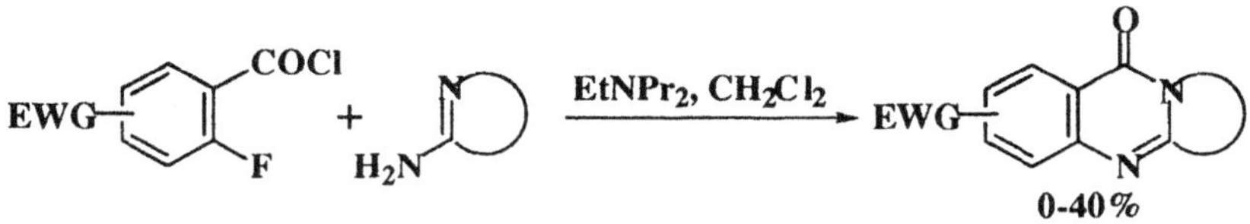

EtNPr$_2$, CH$_2$Cl$_2$

0-40%

IV.K.1b-15 Kobayashi, K. et al., *BCJ*, **74**, 1109.

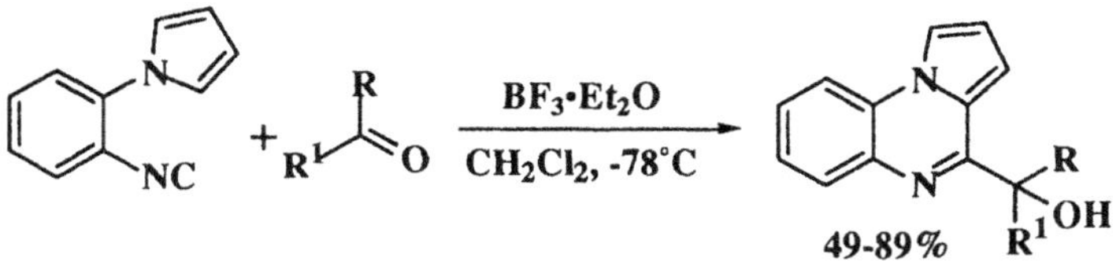

BF$_3$•Et$_2$O
CH$_2$Cl$_2$, -78°C

49-89%

IV.K.1b-16 Mukhopadhyay, R., Kundu, N.G., *SL*, 1143.

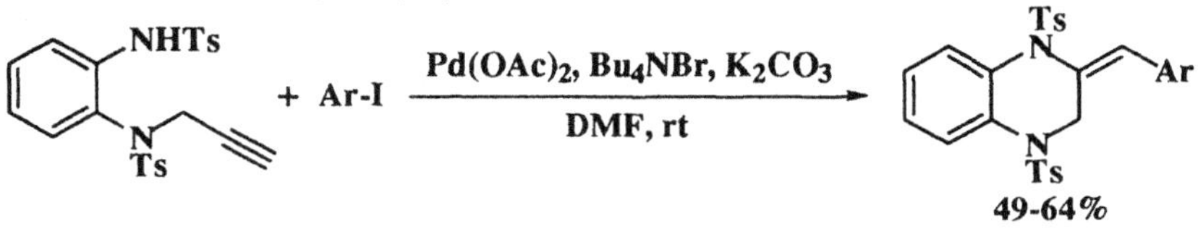

+ Ar-I

Pd(OAc)$_2$, Bu$_4$NBr, K$_2$CO$_3$
DMF, rt

49-64%

IV.K.1c 7-Membered Heterocycles with 2 N's

IV.K.1c-1 Kaboudin, B., Navaee, K., *H*, **55**, 1443; Curini, M. et al., *TL*, **42**, 3193.

73-85%

IV.K.1c-2 Zhang, Y. et al., *TL*, **42**, 73.

0-89%

IV.K.1c-3 Garanti, L. et al., *TA*, **12**, 1201,

75-89%

78-91%

IV.K.1c-4 Kamal, A. et al., *BMCL*, **10**, 2311.

70-75%

IV.K.1c-5 Griesbeck, A.G. et al., *AG(E)*, **40**, 577.

54-83%

IV.K.1c-6 Ferraccioli, R. et al., *SL*, 803.

57-85%
e.e. = 84-99%

IV.K.2. Heterocycles with 2 O's or 2 S's

IV.K.2-1 Sinou, D. et al., *JOC*, **66**, 6634.

47-98%

IV.K.2-2 Sinou, D. et al., *S*, 1456.

22-99%

IV.K.2-3 Babin, P., Bennetau, B., *TL*, **42**, 5231.

27-67%

IV.K.2-4 Mohammadpoor-Baltork, I. et al., *SC*, **31**, 3411.

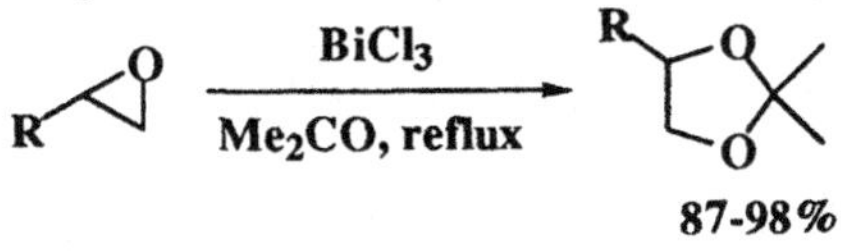

$$R \xrightarrow[\text{Me}_2\text{CO, reflux}]{\text{BiCl}_3} R$$

87-98%

IV.K.2-5 Paddock, R.L., Nguyen, S.T., *JACS*, **123**, 11498.

$$R \xrightarrow[]{\text{CO}_2, \text{Cr(III)salen, DMAP}} R$$

94-99%

IV.K.2-6 Krupadanam, G.L.D. et al., *BCJ*, **74**, 2397.

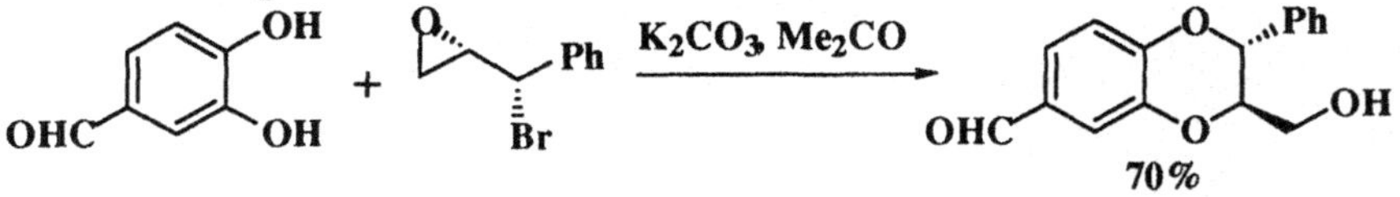

$$\text{OHC} \cdots \text{OH, OH} + \text{Ph, Br} \xrightarrow{\text{K}_2\text{CO}_3, \text{Me}_2\text{CO}} \text{OHC} \cdots \text{Ph, OH}$$

70%

IV.K.2-7 Shimizu, H. et al., *H*, **54**, 139.

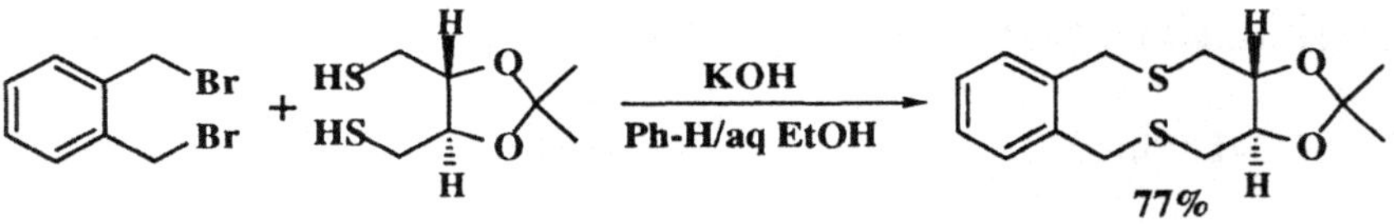

$$\text{Br, Br} + \text{HS, HS} \xrightarrow[\text{Ph-H/aq EtOH}]{\text{KOH}} $$

77%

IV.K.3. Heterocycles with 1 N and 1 O

IV.K.3-1 Lin, Y.-M., Miller, M.J., *JOC*, **66**, 8282.

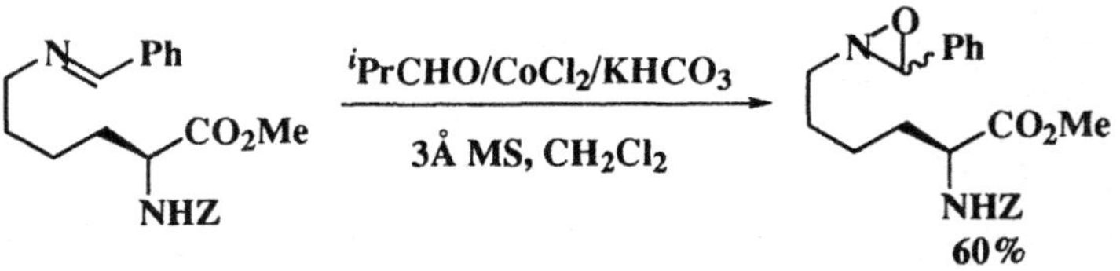

$$\text{N, Ph, CO}_2\text{Me, NHZ} \xrightarrow[\text{3Å MS, CH}_2\text{Cl}_2]{^i\text{PrCHO/CoCl}_2/\text{KHCO}_3} \text{N, O, Ph, CO}_2\text{Me, NHZ}$$

60%

IV.K.3-2 White, J.D. et al., *OL*, **3**, 413.

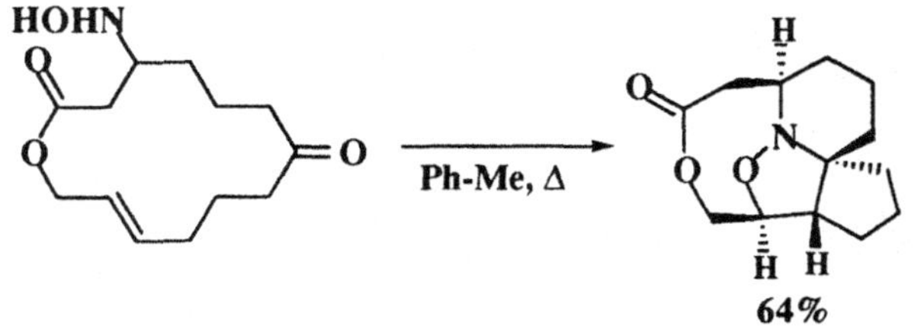

$$\text{HOHN, O, O, O} \xrightarrow{\text{Ph-Me, }\Delta} $$

64%

IV.K.3-3 Broggini, G. et al., *H*, **55**, 1987; Noguchi, M. et al., *H*, **55**, 223 and *BCJ*, **74**, 917; Kim, K. et al., *JOC*, **66**, 7334.

24%

IV.K.3-4 Grigg, R. et al., *T*, **57**, 7951.

36-69%

IV.K.3-5 Fukuyama, T. et al., *OL*, **3**, 2575.

57%

IV.K.3-6 Ku, Y.-Y. et al., *OL*, **3**, 4185.

75%

IV.K.3-7 Kidwai, M., Sapra, P., *OPP*, **33**, 381.

$R-NO_2$ + Ar-CHO

82-96%

IV.K.3-8 Rodriguez-Franco, M.I. et al., *S*, 1711.

25-35%

IV.K.3-9 Khripach, V.A. et al., *ST*, **66**, 569.

99%

IV.K.3-10 Righi, P. et al., *OL*, **3**, 727; Denmark, S.E., Gomez, L., *OL*, **3**, 2907; Fringuelli, F. et al., *JOC*, **66**, 4661.

62-97%

IV.K.3-11 Sibi, M.P., Liu, M., *OL*, **3**, 4181; Iwasa, S. et al., *TL*, **42**, 5897; Cheng, Q. et al., *JCS(P1)*, 452.

79-91%
e.e. = 57-88%

(reversal of enantioselectivity with Zn(OTf)₂)

IV.K.3-12 Radspieler, A., Liebscher, J., *S*, 745.

36-77%

IV.K.3-13 Coskun, N. et al., *TA*, **12**, 1463; deMarch, P. et al., *TA*, **12**, 1747; Iwasa, S. et al., *TL*, **42**, 6715; Tamura, O., Sakamoto, M. et al., *JOC*, **66**, 2602; Tranmer, G.K., Tam, W., *JOC*, **66**, 5113; Tam, W. et al., *JOC*, **66**, 276.

24-85%

IV.K.3-14 Melnyk, O. et al., *TL*, **42**, 1875.

52%

IV.K.3-15 Kim, T.H. et al., *T*, **57**, 7137.

64-94%

IV.K.3-16 Komatsu, M.et al., *TL*, **42**, 9019.

75-85%
e.e. = 88-92%

IV.K.3-17 Cook, G.R., Shanker, P.S., *JOC*, **66**, 6818.

93-97%

IV.K.3-18 Cacchi, S. et al., *OL*, **3**, 2501.

11-83%

IV.K.3-19 Miller, B.L. et al., *JOC*, **66**, 991.

35-81%

IV.K.3-20 Smith, A.B., III, et al., *SL*, 1739.

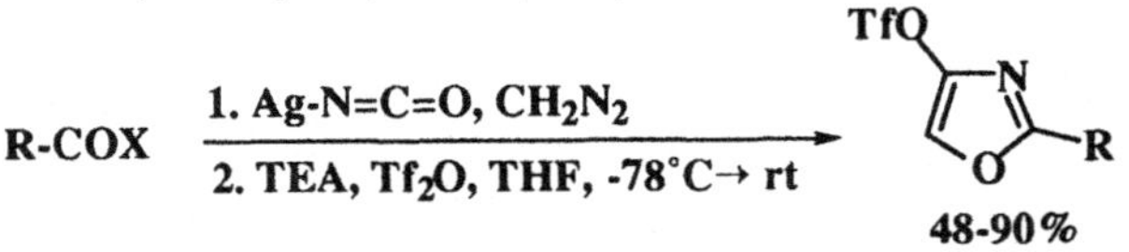

R-COX
1. Ag-N=C=O, CH_2N_2
2. TEA, Tf_2O, THF, -78°C→ rt
TfO
N
O
R
48-90%

IV.K.3-21 Janda, K.D. et al., *OL*, **3**, 2173.

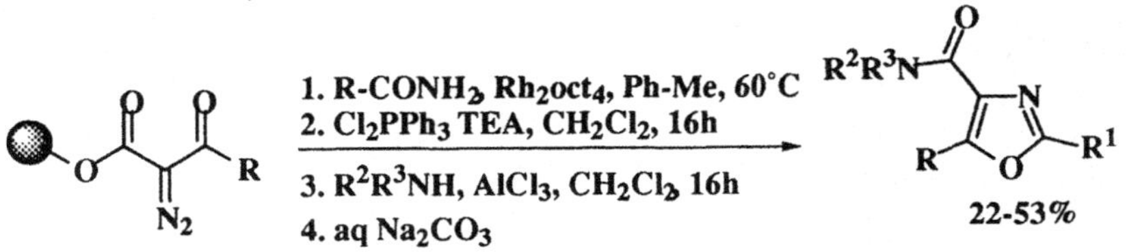

O
O
O
R
N_2
1. R-$CONH_2$, Rh_2oct_4, Ph-Me, 60°C
2. Cl_2PPh_3 TEA, CH_2Cl_2, 16h
3. R^2R^3NH, $AlCl_3$, CH_2Cl_2, 16h
4. aq Na_2CO_3
R^2R^3N
N
R
O
R^1
22-53%

IV.K.3-22 Saitz, C. et al., *SC*, **31**, 135.

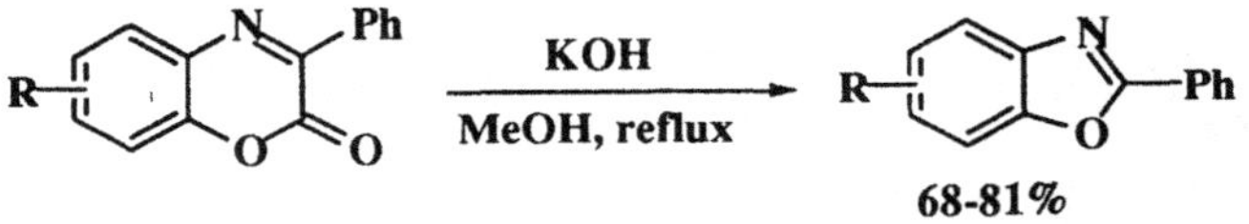

R
N
Ph
O
O
KOH
MeOH, reflux
N
Ph
O
68-81%

IV.K.3-23 Karmakar, S., Mohapatra, D.K., *SL*, 1326.

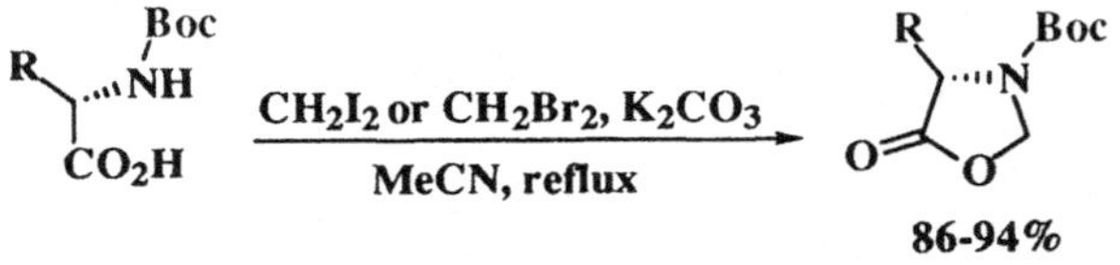

Boc
R
NH
CO_2H
CH_2I_2 or CH_2Br_2, K_2CO_3
MeCN, reflux
R
Boc
N
O
O
86-94%

IV.K.3-24 Roers, R., Verdine, G.L. et al., *TL*, **42**, 3563.

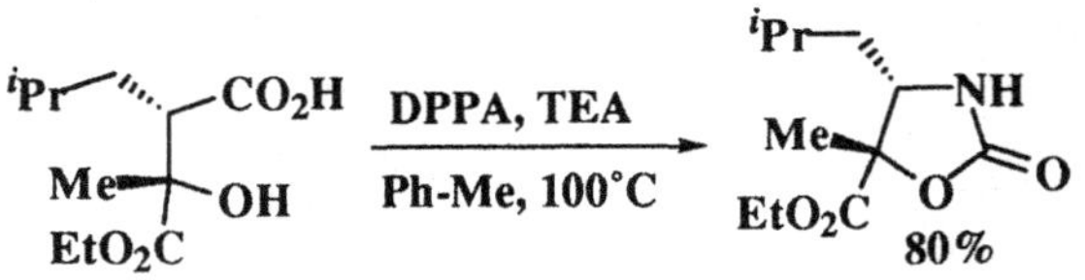

^iPr
CO_2H
Me
OH
EtO_2C
DPPA, TEA
Ph-Me, 100°C
NH
O
80%

IV.K.3-25 Trost, B.M. et al., *SL*, 907.

O
O
O
O
NHTs
TsHN
$PdCl/_2$
PPh_2
PPh_2
H
N
HN
Ts
N
O
69-89%
e.e. = 73-96%

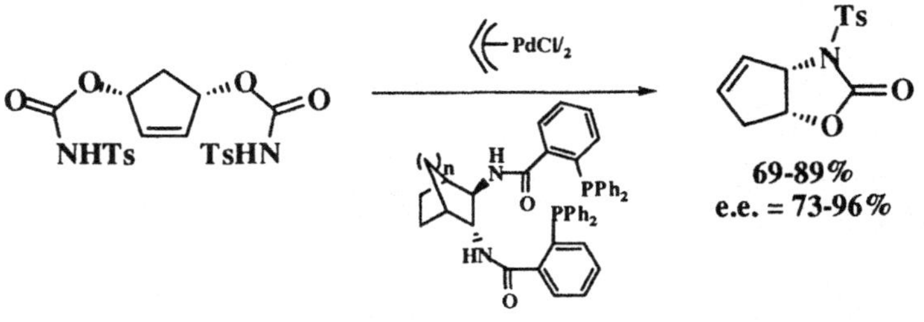

IV.K.3-26 Napolitano, E., Farina, V., *TL*, **42**, 3231.

Me,,,$\overset{CO_2{}^iBu}{\underset{CO_2H}{\,N H}}$ + OHC—〈benzene〉—Ph $\xrightarrow[c\text{-}C_6H_{12},\ reflux]{Ts\text{-}OH}$ Me,,,$\overset{CO_2{}^iBu}{N}$—oxazolidinone—Ph

91%

IV.K.3-27 Bertau, M. et al., *TA*, **12**, 2103.

$R^1\overset{O}{\underset{R}{-}}\overset{}{-}NHNH_2$ $\xrightarrow{NaNO_2,\ H^+,\ H_2O}$ oxazolidinone

61-89%
e.e. = 99%

IV.K.3-28 Kobayashi, S. et al., *SL*, 1140.

$\underset{Cl}{}CHO$ + $H\overset{O}{\underset{NH_2}{N}}R$ + $(CH_2=CHCH_2)_4\text{-}Sn$ $\xrightarrow[THF/H_2O,\ 30°C]{Sc(OTf)_3}$ product

63-75%

IV.K.3-29 Vilarrasa, J. et al., *TL*, **42**, 4995.

$\underset{R}{\overset{N_3}{}}\,\underset{OH}{\overset{R^1}{}}$ $\xrightarrow[THF,\ -78°C\rightarrow rt]{CO_2,\ NaH,\ PMe_3}$ oxazolidinone

91-96%

IV.K.3-30 Bach, T. et al., *ECJ*, **7**, 2581.

$\underset{R}{\overset{O}{}}O\overset{O}{-}N_3$ $\xrightarrow[EtOH,\ 0°C\rightarrow rt]{TMS\text{-}Cl,\ FeCl_3}$ product

33-72%
d.r. ≤15.7:1

IV.K.3-31 Ragaini, F. et al., *OM*, **20**, 3390.

diene + Ph-NO$_2$ $\xrightarrow{CO,\ Ru_3(CO)_{12}}$ product

≤58%

ArN—NAr

IV.K.3-32 Decicco, C.P. et al., *OL*, **3**, 1029.

Cu(OAc)$_2$, TEA

52%

IV.K.3-33 Palomo, C. et al., *JOC*, **66**, 4180.

1. H$_2$, Pd/C, EtOAc

2. NaOCl, TEMPO

95%

(similarly with other examples)

IV.K.3-34 Fuentes, J. et al., *TA*, **12**, 1267.

1. TBAF

2. Im$_2$C=S, rt

79%

IV.K.3-35 Kai, H., Nakai, T., *TL*, **42**, 6895.

DEAD, PPh$_3$, 4-NO$_2$-PhOH

THF, rt

25-92%

IV.K.3-36 Hu, L. et al., *TL*, **42**, 1449.

PhI(O$_2$CCF$_3$)$_2$

MeCN/H$_2$O

99%

IV.K.3-37 Espino, C.G. et al., *AG(E)*, **40**, 598.

44-84%

IV.K.3-38 Roumestant, M.-L., *JOC*, **66**, 6541.

62%

IV.K.3-39 Kundu, N.G. et al., *JOC*, **66**, 20.

1. Ar-I, Pd(PPh₃)₂Cl₂, CuI, TEA

2. KOH, EtOH/H₂O, 80°C, 8-10h

35-93%

IV.K.3-40 Kim, Y. et al., *OL*, **3**, 4149.

TFA, -40°C→ rt

30-90%
d.r. = 1.1-4.3:1

IV.K.3-41 Daich, A. et al., *H*, **54**, 275.

TsOH
Ph-Me, reflux

70%

IV.K.4. Heterocycles with 1 N and 1 S

IV.K.4-1 Chen, L.-C. et al., *H*, **55**, 1231.

0-80%

IV.K.4-2 Tozer, M.J. et al., *OL*, **3**, 369.

66-92% 3-6:1

IV.K.4-3 McNab, H. et al., *JCS(P1)*, 424.

750-850°C

23-54%

IV.K.4-4 Daich, A. et al., *JOC*, **66**, 4695.

TFA, rt

92-98% 1.2-15:1

IV.K.4-5 Alper, H. et al., *JOC*, **66**, 3502.

$Pd_2(dba)_3 \cdot CHCl_3$, dppp

THF

43-98%

IV.K.4-6 Zhao, R. et al., *TL*, **42**, 2101.

1. NBS, dioxane/H_2O
2. $RNHCSNH_2$, 80°C

60-98%

IV.K.4-7 Mahler, S.G. et al., *TL*, **42**, 8143.

70-93%

IV.K.4-8 Chen, Z.-C., Zhang, P.-F., *S*, 358.

$$R\!\!-\!\!\equiv\!\!-I^+Ph \; {}^-OTs + H_2N\text{-}CS_2NH_4 \xrightarrow{\text{DMF/H}_2\text{O, rt}}$$

45-73%

IV.K.4-9 Lillalgordo, J.M. et al., *S*, 2021.

38-78%

IV.K.4-10 deMeijere, A. et al., *EJOC*, 3025.

37-92%

IV.K.4-11 Takahashi, M. et al., *H*, **55**, 1759.

1. BuLi, TMEDA, THF
2. Me$_2$NCOCl
3. H$^+$
4. AcOH, reflux

20-48%

IV.K.4-12 Katritzky, A.R. et al., *JOC*, **66**, 6792.

DCC or H$^+$

63-97%

IV.K.4-12 Zhang, Y. et al., *OPP*, **33**, 181.

35-70%

IV.K.5. Heterocycles with 1 O and 1 S

No Entries

IV.K.6. Heterocycles with 3 or more N's

IV.K.6-1 Grigg, R. et al., *T*, **57**, 7729.

36-54%

IV.K.6-2 Journet, M., Cai, D. et al., *TL*, **42**, 9117.

$$R\!\!=\!\!=\!\!-CHO \xrightarrow[\text{DMSO, rt}]{NaN_3}$$

99%

IV.K.6-3 Katritzky, A.R. et al., *S*, 897.

$$\xrightarrow{R^3NHNH_2,\ CHCl_3}$$

36-95%

IV.K.6-4 Music, I., Vercek, B., *SC*, **31**, 1511.

$$\xrightarrow{PhI(OAc)_2,\ CH_2Cl_2}$$

51%

IV.K.6-5 Shawali, A.S. et al., *JPC*, **66**, 4055.

75-85%

IV.K.6-6 Migawa, M.T., Townsend, L.B., *JOC*, **66**, 4776.

45%

IV.K.6-7 Alvarez-Builla, J. et al., *JOC*, **66**, 8528.

51-76%

IV.K.6-8 Kidwai, M. et al., *SC*, **31**, 1639.

70-93%

(SiO$_2$, K-10 clay and neutral Al$_2$O$_3$ are supports used)

IV.K.6-9 Katritzky, A.R. et al., *JOC*, **66**, 6797.

51-96%

IV.K.6-10 Haddadin, M.J., Kurth, M.J. et al., *JOC*, **66**, 1310.

14-36%

IV.K.6-11 Demko, Z.P., Sharpless, K.B., *OL*, **3**, 4091 and *JOC*, **66**, 7945.

24-96%

IV.K.7. Heterocycles with 2 N's and 1 O

IV.K.7-1 Mulvihill, M.J. et al., *S*, 1965.

60-95%

IV.K.7-2 Gangloff, A.R. et al., *TL*, **42**, 1441.

5-98%

IV.K.7-3 Kang, S.-K. et al., *H*, **54**, 985.

52-77%

IV.K.7-4 Hassine, B.B. et al., *TL*, **42**, 9131.

62-82%

IV.K.8. Heterocycles with 2 N's and 1 S

IV.K.8-1 Katritzky, A.R. et al., *JOC*, **66**, 4045.

80-85%

IV.K.8-2 Katritzky, A.R., Huang, T.B., Steel, P.J., *JOC*, **66**, 5601.

15-78%

IV.K.8-3 Torroba, T. et al., *JOC*, **66**, 5766.

35-95%

IV.K.8-4 Kim, K. et al., *H*, **55**, 75; **see also:** Rees, C.W., Yue, T.-J., *JCS(P1)*, 662.

27-57%

IV.L. Other Heterocycles

IV.L-1 DuBois, J. et al., *JACS*, **123**, 6935.

60-91%

IV.L-2 Hanson, P.R. et al., *S*, 612.

99%

IV.L-3 Kohn, W.D. et al., *OL*, **3**, 971.

1. DBU, DMF
2. SnCl$_2$·2H$_2$O, DMF
3. TFA, Ph-OMe
   TIPS, H$_2$O

40-53%

IV.M. Reviews

IV.M-1 Karlsson, S., Hoberg, H.-E., *OPP*, **33**, 103.
  Review: **"Asymmetric 1,3-Dipolar Cycloadditions for the Construction of Entiomerically Pure Heterocycles. A Review."**

IV.M-2 Sessler, J.L., Davis, J.M., *ACR*, **34**, 989.
  Review: **"Sapphyrins: Verastile Anion Binding Agents."**

IV.M-3 Mamardashvili, N.Z., Gobubchikov, O.A., *RCR*, **70**, 577.
Review: **"Spectral Properties of Porphyrins and their Precursors and Derivatives."**

IV.M-4 Burrell, A.K., Officer, D.L. et al., *CRV*, **101**, 2751.
Review: **"Synthetic Routes to Multiporphyrin Arrays."**

IV.M-5 Gilchrist, T.L., *AA*, **34**, 51.
Review: **"Activated 2H-Azirines as Dienophiles and Electrophiles."**

IV.M-6 Alcaide, B., Almendros, P., *CSR*, **30**, 226.
Review: **"4-Oxoazetidine-2-carboxaldehydes as Useful Building Blocks in the Stereocontrolled Synthesis."**

IV.M-7 Alcaide, B., Almendros, P., *OPP*, **33**, 315.
Review: **"Recent Progress in the Synthesis and Reactivity of Azetidine-2,3-diones."**

IV.M-8 Sunagawa, M., Sasaki, A., *H*, **54**, 497.
Review: **"The Structural Aspects of Carbapenam Antibiotics.**

IV.M-9 Shawali, A.S., Elsheikh, S.K., *JHC*, **38**, 541.
Review: **"Annelated [1,2,4,5]Tetrazines."**

IV.M-10 Erian, A.W., *JHC*, **38**, 793.
Review: **"Recent Trends in the Chemistry of Fluorinated Five and Six-Membered Heterocycles**

IV.M-11 Zaleska, B., Lis, S., *S*, 811.
Review: **"Pyrrolidinetrione Derivatives: Synthesis and Applications in Heterocyclic Chemistry."**

IV.M-12 Ferreira, V.F. et al., *OPP*, **33**, 411.
Review: **"Recent Advances in the Synthesis of Pyrroles."**

IV.M-13 Komarova, L.G., Rusanov, A.L., *RCR*, **70**, 81.
Review: **"Rigid-Rod Poly(benzobisazoles) and Molecular Composites Based on Them.**

IV.M-14 Vaughan, K., *OPP*, **33**, 59.
Review: **"Recent Progress in the Synthesis of *bis*-Triazenes."**

IV.M-15 Chakrabarty, M. et al., *H*, **55**, 2431.
Review: **"A Sojourn in the Synthesis and Reactivity of Diindolylalkanes."**

IV.M-16 Makosza, M., Wojciechowski, K., *H*, **54**, 445.
Review: **"Nucleophilic Aromatic Substitution of Hydrogen as a Tool for the Synthesis of Indole and Quinoline Derivatives."**

IV.M-17 Moustafa, O.S., Yamada, Y., *JHC*, **38**, 809.
Review: **"Recent Trends in Isomeric Thienoquinoxalines [1980-2000]."**

IV.M-18 Litvinov, V.P. et al., *RCR*, **70**, 299.
Review: **"Pyridopyridines."**

IV.M-19 Levai, A., *JHC*, **38**, 1011.
Review: **"Synthesis and Chemical Transformations of 1,4- 4,1-, and 1,5-Benzoxazepines."**

IV.M-20 Chandrasekhar, V., Nagendran, S., *CSR*, **30**, 193.
Review: **"Phosphazenes as Scaffolds for the Construction of Multi-Site Coordination Ligands."**

IV.M-21 Bansal, R.K., Heincke, J., *CRV*, **101**, 3549.
Review: **"Annellated Heterophospholes and Phospholides and Analogues with Related Non-Phosphorus Systems."**

IV.M-22 Quintanilla-Licea, R., Teuber, H.-J., *H*, **55**, 1365.
Review: **"Review on Reactions of Acetylacetaldehyde with Aromatic and Biogenic Amines and Indoles— Synthesis of Heterocycles *via* Hydroxymethylene Ketones."**

IV.M-23 Bowman, W.R. et al., *JCS(P1)*, 2885.
Review: **"Synthesis of Heterocycles by Radical Cyclisation."**

IV.M-24 Elliott, M.C., Williams, E., *JCS(P1)*, 2303.
Review: **"Saturated Oxygen Heterocycles."**

IV.M-25 Gilchrist, T.L., *JCS(P1)*, 2491.
Review: **"Synthesis of Aromatic Heterocycles."**

IV.M-26 D'Auria, M., *H*, **54**, 475.
Review: "Photochemical Dimerization in Solution of Heterocyclic Substituted Alkenes Bearing an Electron Withdrawing Group."

IV.M-27 Fillion, H. et al., *H*, **54**, 1095.
Review: "Cycloadditions of α,β-Unsaturated N,N-Dimethylhydrazones. A Diels-Alder Strategy for the Building of Aza-Hetero Rings."

IV.M-28 Brandsma, L., *EJOC*, 4569.
Review: "Unsaturated Carbanions, Heterocumulenes and Thiocarbonyl Compounds—New Routes to Heterocycles."

IV.M-29 Hajos, G. et al., *EJOC*, 3405.
Review: "Recent Advances in Ring Transformations of Five-Membered Heterocycles and their Fused Derivatives."

IV.M-30 Mongin, F., Queguiner, G., *T*, **57**, 5059.
Review: "Advances in the Directed Metallation of Azines and Diazines (Pyridines, Pyrimidines, Pyrazines, Pyridazines, Quinolines, Benzodiazines and Carbolines). Part 1: Metallation of Pyridines, Quinolines and Carbolines."

IV.M-31 Queguiner, G. et al., *T*, **57**, 4489.
Review: "Advances in the Directed Metallation of Azines and Diazines (Pyridines, Pyrimidines, Pyrazines, Pyridazines, Quinolines, Benzodiazines and Carbolines). Part 2: Metallation of Pyrimidines, Pyrazines, Pyridazines and Benzodiazines."

IV.M-32 Aratani, N., Osuka, A., *BCJ*, **74**, 1361.
Review: "A New Strategy for Construction of Covalently Linked Giant Porphyrin Arrays with One, Two, and Three Dimensionally Arranged Architectures."

IV.M-33 Konovalova, N.V. et al., *RCR*, **70**, 939.
Review: "Synthetic Molecular Systems Based on Porphyrins as Models for the Study of Energy Transfer in Photosynthesis."

IV.M-34 Fu, G.C., *PAC*, **73**, 1113.
  Review: "Applications of 'Planar-Chiral' Heterocycles in Asymmetric Catalysis."

IV.M-35 Asakawa, Y. et al., *H*, **54**, 529.
  Review: "Biotransformations of Terpenoids from the Crude Drugs and Anikal Origin by Microorganisms."

V
PROTECTING GROUPS

V.A. Aldehyde and Ketone Protecting Groups

V.A-1 Mohajerani, B., Heravi, M.M., Ajami, D., *M*, **132**, 871; Heravi, M.M. et al., *M*, **132**, 651; Hajipour, A.R. et al., *SC*, **31**, 1625.

$$R^1\text{-}CR\text{-}OTHP \xrightarrow[\text{CH}_2\text{Cl}_2, \Delta]{\text{CrClO}_3\text{NH}_4, \text{Zeofen}} R^1\text{-}CR=O$$

78-98%

[similarly with Montmorillonite K-10 supported
NH$_4$ClCrO$_7$ or BnPPh$_3^+$HSO$_5^-$, BiCl$_3$]

V.A-2 Kaneda, K. et al., *TL*, **42**, 8329; Reddy, B.M. et al., *SC*, **31**, 1819; Jin, T.S. et al., *JCR(S)*, 289; Porta, O. et al., *T*, **57**, 217.

$$R^1\text{-}CR=O \xrightarrow[\text{Ph-Me, reflux}]{(\text{CH}_2\text{OH})_2, \text{Ti}^{4+}\text{-Montmorillonite}} R^1\text{-}CR\langle O \rangle$$

95-99%

[similar reactions reported with different alcohols and
Pt-Mo/ZrO$_2$; ZrO$_2$/SO$_4^{2-}$ or TiCl$_4$, TEA]

V.A-3 Masaki, Y. et al., *SL*, 1311.
Novel Polymer Effect in Cleavage Reactions of Acetals and Silyl Ethers in Aqueous Media Catalyzed by a Polymer-Supported Dicyanoketone Acetal.

V.A-4 Hayashi, M. et al., *SL*, 934.
TTMPP: A Novel Catalyst for Selective Deacetalation.

V.A-5 Heravi, M.M. et al., *M*, **132**, 985; Habibi, M.H. et al., *TL*, **42**, 6771; Watanabe, Y. et al., *TL*, **42**, 4641.

$$R^1\text{-}CR\langle O \rangle \xrightarrow{\text{K}_2\text{FeO}_4 \text{ supported on Montmorillonite K-10}} R^1\text{-}CR=O$$

82-95%

[similar deprotections reported with K$_5$CoW$_{12}$O$_{40}$·3H$_2$O or pyr(HF)$_x$]

V.A-6 Ballini, R., Maggi, R. et al., *S*, 1826; Yadav, J.S. et al., *SL*, 238; **see also:** Sato, T., *SL*, 1581.

$$\underset{R}{\overset{R^1}{>}}\!\!=\!\!O \quad \xrightarrow[\text{CH}_2\text{Cl}_2,\ \text{rt}]{\text{HOCH}_2\text{CH}_2\text{SH, Amberlyst 15}} \quad \underset{R}{\overset{R^1}{\diagup}}\!\!<\!\!\underset{S}{\overset{O}{\diagdown}}$$

75-95%

[similar reactions with LiBF$_4$ as catalyst or from acetals with R-SH/LiBr]

V.A-7 Mohan, R.S. et al., *TL*, **42**, 8133

$$\text{R-CHO} \quad \xrightarrow[\text{MeCN}]{\text{Ac}_2\text{O, Bi(OTf)}_3\!\cdot\!x\text{H}_2\text{O}} \quad \text{R-CH(OAc)}_2$$

64-98%

(no reaction with ketones)

V.A-8 Chavan, S.P. et al., *SL*, 1251.

$$\underset{R}{\overset{R^1}{\diagup}}\!\!<\!\!\underset{S}{\overset{O}{\diagdown}} \quad \xrightarrow[\text{Amberlyst 15}]{\text{OHC-CO}_2\text{H, }\mu\text{W}} \quad \underset{R}{\overset{R^1}{>}}\!\!=\!\!O$$

78-95%

(solvent free conditions)

V.A-9 Firouzabadi, H., Iranpoor, N., Hazarkhani, H., *JOC*, **66**, 7527.

$$\underset{R}{\overset{R^1}{>}}\!\!\underset{X}{\overset{X}{<}} + \underset{\text{HS}}{\overset{\text{HS}}{\diagup}}\!\!)_n \quad \xrightarrow[\text{CH}_2\text{Cl}_2,\ \text{rt}]{\text{I}_2} \quad \underset{R}{\overset{R^1}{>}}\!\!\underset{S}{\overset{S}{<}}\!\!)_n$$

70-96%

V.A-10 Khan, A.T. et al., *SL*, 785; Fleming, F.F. et al., *JOC*, **66**, 6502.

$$\underset{R}{\overset{R^1}{>}}\!\!\underset{SR^2}{\overset{SR^2}{<}} \quad \xrightarrow[\text{CH}_2\text{Cl}_2,\ 0\text{-}5°\text{C}]{\text{CTAB or TBATB}} \quad \underset{R}{\overset{R^1}{>}}\!\!=\!\!O$$

65-96%

[similar deprotection reported with PhI(O$_2$CCF$_3$)$_2$/TFA]

V.A-11 Quan, L.G., Cha, J.K., *SL*, 1925.

$$\text{R-CHO} \quad \xleftrightarrow[\substack{\text{aq HF} \\ \text{MeCN, rt}}]{\substack{\text{TBDMS-Cl, imidazole} \\ \text{DMF, rt}}} \quad \underset{R}{\diagup}\!\!\overset{\text{OTBDMS}}{\underset{}{}}\text{N}\diagdown$$

88-96%   85-96%

V.A-12 Jin, T.-S. et al., *SC*, **31**, 1669; Deka, N., Sarma, J.C., *CL*, 794; Muthusamy, S. et al., *TL*, **42**, 359; Firouzabadi, H. et al., *BCJ*, **74**, 2401; Firouzabadi, H., Iranpoor, N., Hazarkhani, H., *SL*, 1641; Prajapati, D. et al., *JCR(S)*, 313; Curini, M. et al., *SL*, 1182.

$$\begin{array}{c} R^1 \\ \diagdown \\ R \end{array} = O \quad \xrightarrow[\text{CH}_2\text{Cl}_2, \text{rt}]{\text{HSCH}_2\text{CH}_2\text{SH, Monmorillonite}} \quad \begin{array}{c} R^1 \\ R \end{array} \overset{S}{\underset{S}{\diagup\diagdown}}$$

75-95%

[similar dithioacetal formation reported with I_2/Al_2O_3; $InCl_3$; LiOTf;Cl_3isocyanuric acid; $CdI_2/\mu W$ or $ZrPSO_3H$ catalysts]

V.A-13 Hajipour, A.R. et al., *SC*, **31**, 1187; Heravi, M.M. et al., *JCR(S)*, 496; Hirano, M. et al., *JCR(S)*, 277; Ganguly, N.C. et al., *SC*, **31**, 1607; Harris, C.E. et al., *TL*, **42**, 4775; Mohan, R.S. et al., *S*, 1010; Tajbakhsh, M., Heravi, M.M. et al., *M*, **132**, 1229; Zhang, Q. et al., *OPP*, **33**, 87.

$$\begin{array}{c} R^1 \\ \diagdown \\ R \end{array} = NOH \quad \xrightarrow[\begin{subarray}{c} \left[\begin{array}{c} N \\ N \\ Bn \end{array}\right]_2 \end{subarray}]{} Cr_2O_7{}^{2-}, AlCl_3 \quad \begin{array}{c} R^1 \\ \diagdown \\ R \end{array} = O \quad 79\text{-}96\%$$

(also hydrolysis of semicarbazones)

[similar deprotection of oximes with $EtN^+MnO_4^-$; $(NH_4)_2S_2O_8$, $AgNO_3$, moist Momtmorillonite; pyridinium fluorchromate; $KMnO_4/Al_2O_3$; $Bi(NO_3)_3\cdot5H_2O$; N-methylpiperidinium chlorochromate/Al_2O_3; or $PEG\text{-}NO_2$]

V.A-14 Shirini, F., Azadbar, M.R., *SC*, **31**, 3775; Yadav, J.S. et al., *SL*, 1134; Chen, Z.-C. et al., *SC*, **31**, 3847; Heravi, M.M. et al., *M*, **132**, 881.

$$\begin{array}{c} Ar \\ \diagdown \\ R \end{array} = N\text{-}Y \quad \xrightarrow[\text{MeCN, 5-90min}]{[(NO_3)_3Ce]_3\cdot H_2IO_6} \quad \begin{array}{c} Ar \\ \diagdown \\ R \end{array} = O$$

70-95%

[similar deprotection of hydrazones and semicarbazones with $CeCl_3\cdot7H_2O/SiO_2$; μW; polymer-supported-$I(O_2CCF_3)_2$ or Clayfen, μW]

V.B. Amino Acid Protecting Groups

V.B-1 Taddei, M. et al., *TL*, **42**, 5191.

$$\begin{array}{c} R \\ \diagdown \\ CbzHN \end{array} - CO_2Me \quad \xrightarrow[{}^i\text{PrOH}]{\text{HCO}_2\text{NH}_4, \text{Pd/C}, \mu W} \quad \begin{array}{c} R \\ \diagdown \\ H_2N \end{array} - CO_2Me$$

90-95%

(also deprotection of benzyl amines)

V.B-2 Strazzolini, P. et al., *T*, **57**, 9033.

$$\underset{\text{BocHN}}{\overset{R}{\diagdown}}\!\!-CO_2R^1 \quad \xrightarrow[\text{CH}_2\text{Cl}_2, 0°\text{C}]{\text{HNO}_3} \quad \underset{\text{HNO}_3\cdot\text{H}_2\text{N}}{\overset{R}{\diagdown}}\!\!-CO_2R^1$$

63-92%

V.B-3 Davies, S.G. et al., *JCS(P1)*, 3106.

V.B-4 Jeyaraj, D.A., Waldmann, H., *TL*, **42**, 835.

PhAcOZ-AA⌐-Ser-Ala-OAll

H-AA⌐-Ser-Ala-OAll

penicillin G acylase
KI, buffer, MeOH

60-69%

V.B-5 Marcantoni, E., Bartoli, G. et al., *JOC*, **66**, 4430.

$$\underset{\text{BocHN}}{\overset{R}{\diagdown}}\!\!-CO_2{}^t\text{Bu} \quad \xrightarrow[\text{MeCN, reflux}]{\text{CeCl}_3\cdot7\text{H}_2\text{O, NaI}} \quad \underset{\text{BocHN}}{\overset{R}{\diagdown}}\!\!-CO_2\text{H}$$

75-99%

V.B-6 Najera, C. et al., *TL*, **42**, 7579.
A New Polymer-Supported Reagent for the Fmoc-Protection of Amino Acids

---

V.B-7 Waldmann, H. et al., *ECJ*, 1184.
Acid-Labile Protecting Groups for the Synthesis of Lipidated Peptides

V.C. Amine Protecting Groups

V.C-1 Vedejs, E. et al., *JOC*, **66**, 7542.

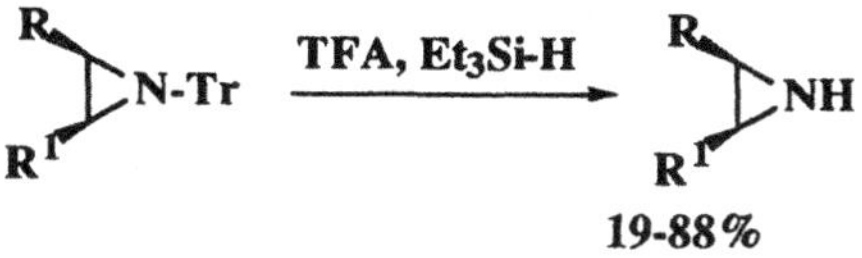

R, R'—(aziridine)—N-Tr →[TFA, Et$_3$Si-H] R, R'—(aziridine)—NH

19-88%

V.C-2 Bornaghi, L. et al., *TL*, **42**, 3129; Singh, L., Seifert, J., *TL*, **42**, 3133.

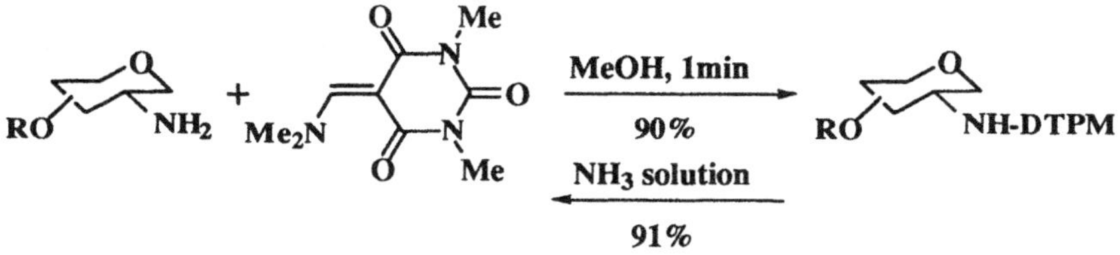

RO—(sugar)—NH$_2$ + Me$_2$N—(barbituric acid, N-Me, N-Me) →[MeOH, 1min / 90%] / ←[NH$_3$ solution / 91%] RO—(sugar)—NH-DTPM

V.C-3 Lipshutz, B.H. et al., *OL*, **3**, 4145.

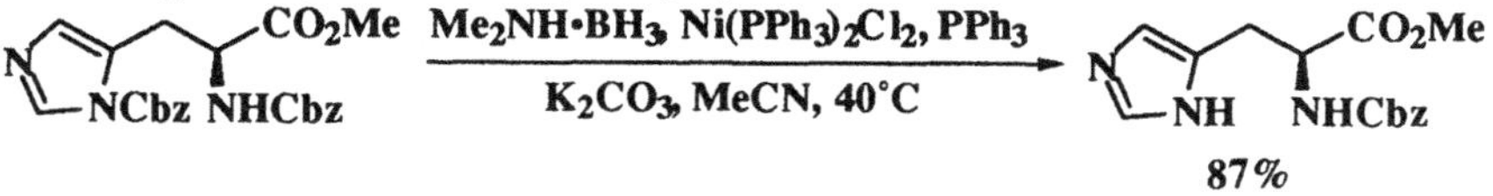

(imidazole NCbz)—CO$_2$Me NHCbz →[Me$_2$NH·BH$_3$, Ni(PPh$_3$)$_2$Cl$_2$, PPh$_3$ / K$_2$CO$_3$, MeCN, 40°C] (imidazole NH)—CO$_2$Me NHCbz

87%

(N-Alloc>Het-NCbz>RR^{1}NCbz)

V.C-4 Morvan, F. et al., *BMCL*, **11**, 2813.

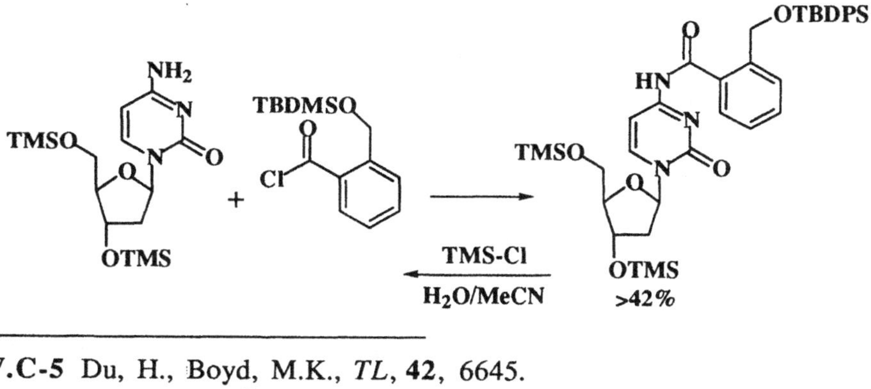

TMSO—(sugar, OTMS)—(cytosine NH$_2$) + TBDMSO—(benzoyl chloride) →[TMS-Cl / H$_2$O/MeCN] TMSO—(sugar, OTMS)—(cytosine with HN-benzamide, OTBDPS)

>42%

V.C-5 Du, H., Boyd, M.K., *TL*, **42**, 6645.

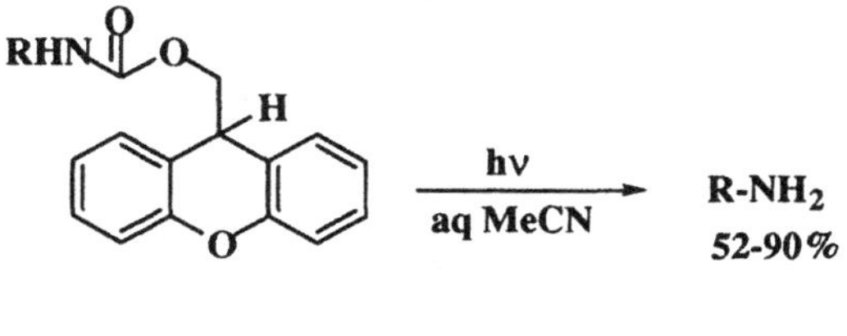

RHN—C(=O)—O—CH$_2$—(xanthene)H →[hν / aq MeCN] R-NH$_2$

52-90%

V.C-6 Curran, D.P. et al., *JOC*, **66**, 4261.
Fluorous Boc (fBoc) Carbamates: New Amine Protecting Groups for Use in Fluorous Synthesis

---

V.C-7 Chee, G.-L., *SL*, 1593.

$$RZ\text{-}CO_2{}^i Pr \xrightarrow[\text{MeNO}_2,\ 0\text{-}50°C]{\text{Al}_2\text{Cl}_3} R\text{-}ZH$$

78-92%
z = NH, NR, O

V.C-8 Lazny, R. et al., *T*, **57**, 5825.

$$RR^1N\text{-}H \xrightarrow[\text{H}_2\text{O}]{\text{PhN}_2\text{BF}_4} RR^1N\text{-}N{=}N\text{-}Ph$$

75-91%

$$\xleftarrow[\text{EtOH, rt, 5-12h}]{\text{TFA}}$$

V.C-9 Ravina, E. et al., *TL*, **42**, 8633.

90-98%

V.C-10 Casimir, J.R., *S*, 75, 1985.

53-95%

V.C-11 Konno, K. et al., *SL*, 1167; Pak, C.S., Lim, D.S., *SC*, **31**, 2209.

92-97%

(2-Ns groups removed with Ph-SH,Cs$_2$CO$_3$, DMF, rt)
[similar deprotection of a 2-pyridinylsulfonamide with Mg/MeOH]

V.D. Carboxyl Protecting Groups

V.D-1 Serrano-Wu, M.H. et al., *TL*, **42**, 8593.

$$R-CO_2H \quad 82\text{-}92\%$$

NaH / DMF, rt

V.D-2 Jackson, R.W., *TL*, **42**, 5163.

$$R\text{-}CO_2{}^tBu \xrightarrow[\text{Ph-Me, reflux}]{SiO_2} R\text{-}CO_2H$$

25-94%

V.D-3 Fuchs, P.L. et al., *OL*, **3**, 2137.

TFA / CH_2Cl_2, rt

80-99%
from syn oxime

+

80-99%
from anti oxime

V.E. Hydroxyl Protecting Groups

V.E-1 Sharma, G.V.M. et al., *TL*, **42**, 759; Oriyama, T. et al., *SC*, **31**, 2761.

$$R\text{-}OH + Ph_2CHOH \xrightarrow[\text{CH}_2\text{Cl}_2]{Yb(OTf)_3 \text{ or } FeCl_3} RO\text{—}CHPh_2$$

10-92% (Yb catalyst)
14-88% (Fe catalyst)

[diphenyl methyl ether formation alaso reported from
R-OTHP or RO-SiR^1$_3$ with HCO_2CHPh_2, TMS-OTf]

V.E-2 Sharma, G.V.M., Rakesh, *TL*, **42**, 5571.

$$R\text{-}OH + Cl_3C\text{—} \cdots \xrightarrow[]{\text{TfOH, CH}_2\text{Cl}_2} RO\text{—}CH_2\text{—}\bigcirc\text{—}Ph$$

58-75%

(deprotection with DDQ or DDQ/Mn(OAc)$_3$)

V.E-3 Rollin, P. et al., *S*, 286.
Phenylsulfonylethylidene (PSE) Acetals: A Novel Protecting Group in Carbohydrate Chemistry

V.E-3 Xiao, X., Bai, D., *SL*, 535.

$$\xrightarrow[\text{MeCN, rt}]{\text{CeCl}_3\text{·7H}_2\text{O, (CO}_2\text{H)}_2}$$

64-98%

V.E-5 Ranu, B.C., Hajra, A., *JCS(P1)*, 355, 2262; Singh, V.K. et al., *TL*, **42**, 5309; Tanemura, K. et al., *CL*, 1012.

$$\xrightarrow[\text{EtOAc, reflux}]{\text{In, I}_2}$$

R￬OAc

75-82%

[similar transformations reported with Ac_2O/Cu(OTf)$_2$ {from RO-THP or RO-TBDMS}; or Ac_2O/CAN {from RO-MEM}]

V.E-6 Habibi, M.H. et al., *TL*, **42**, 2851; Patel, B.K. et al., *TL*, **42**, 7679; Pachamuthu, K., Vankar, Y.D., *JOC*, **66**, 7511; Deka, N., Sarma, J.C., *JOC*, **66**, 1947; Das, B. et al., *SL*, 1777; Kaisalo, L.H., Hase, T.A., *TL*, **42**, 7699; Stephan, E. et al., *JCR(S)*, 518; Oriyama, T. et al., *SC*, **31**, 2305.

R-OH +

$$\xrightarrow[\substack{\text{Me}_2\text{CO, rt} \\ \text{5-99\%}}]{\text{K}_5\text{CoW}_{12}\text{O}_{40}\text{·3H}_2\text{O}}$$

$$\xleftarrow[\substack{\text{MeOH, rt} \\ \text{94-99\%}}]{\text{K}_5\text{CoW}_{12}\text{O}_{40}\text{·3H}_2\text{O}}$$

[preparation of THP ethers reported also with Bu_4NBr_3, CH_2Cl_2; CAN; I_2 μW; $SOCl_2/SiO_2$. Deprotection with Bu_4NBr_3, MeOH; Pd/C, H_2; TMS-I; Sc(OTf)$_3$]

V.E-7 Mukaiyama, T. et al., *CL*, 424.

R-OH +

$$\xrightarrow[\text{CH}_2\text{Cl}_2, \text{rt}]{\text{TMS-OTf}}$$

RO-PMB

74-99%

V.E-8 Yadav, J.S. et al., *JCR(S)*, 528; Khalafi-Nezhad, A., Alamdari, R.F., *T*, **57**, 6805; Reynolds, R.C. et al., *TL*, **42**, 7755.

$$\xrightarrow[\text{MeCN/H}_2\text{F, reflux}]{\text{InBr}_3}$$

85-95%

[trityl deprotection also reported with catalytic Ce(OTf)$_4$ or column chromatography]

V.E-9 Sawada, D., Ito, Y., *TL*, **66**, 2501; Sabitha, G. et al., *OL*, **3**, 1149; Lee, A.S.-Y. et al., *T*, **57**, 2121; Deville, J.P., Pehar, V., *JOC*, **66**, 4097.

$$R\text{-OH} + \quad MeS\diagup\!\!\diagdown OR^1 \quad \xrightarrow[\text{CH}_2\text{Cl}_2]{\text{CuBr}_2,\ \text{Bu}_4\text{NBr},\ 4\text{Å MS}} \quad RO\diagup\!\!\diagdown OR^1$$

58-99%

(also protection for acids)

[deprotection of MEM or MOM groups reported with
CeCl$_2$·7H$_2$O; CBr$_2$/^{i}PrOH; or Montmorillonite K-10]

V.E-10 Dalpozzo, R. et al., *SL*, 1897; Vatele, J.-M., *SL*, 1989; Chen, F.-E. et al., *S*, 1772; Chandrasekhar, S. et al., *T*, **57**, 3435; Kitov, P.I., Bundle, D.R., *OL*, **3**, 2835.

$$RO\diagdown\!\!\diagup\!\!\diagdown R^1 \quad \xrightarrow[\text{MeNO}_2,\ \text{reflux}]{\text{CeCl}_3\text{·7H}_2\text{O},\ \text{NaI},\ \text{HS(CH}_2)_3\text{SH}} \quad R\text{-OH}$$

15-86%

[deprotection of allyl groups reported also with I$_2$; (Bu$_4$N)$_2$S$_2$O$_8$;
PHHS, ZnCl$_2$, Pd(PPh$_3$)$_4$; NMO, OsO$_4$]

V.E-11 Falck, J.R. et al., *AG(E)*, **40**, 1281; Spencer, J.B. et al., *TL*, **42**, 4033.

$$RO\diagdown\!\!\diagup\!\!\diagdown\text{(OMe)}_n \quad \xrightarrow[\text{EtOAc/H}_2\text{O}]{\text{CrCl}_2,\ \text{LiI}} \quad R\text{-OH}$$

80-98%

(2,6-(OMe)$_2$ > 3,4-(OMe)$_2$ > 4-OMe > Bn
Bn > 2,6-(OMe)$_2$ by [H]
3,4-(OMe)$_2$ > 2,6-(OMe)$_2$ by DDQ)

[deprotection of a PMB ether in competition with a NAP ether reported with CAN]

V.E-12 Seio, K., Sekine, M., *TL*, **42**, 8657.
Synthesis of Pentathymidylate Using a 4-Monomethoxy-tritylthio (MMTrS) Group as a 5'-Hydroxyl Protecting Group: Toward Oligonucleotide Synthesis Without Acid Treatment.

V.E-13 Bandgar, B.P., Kasture, S.P., *M*, **132**, 1101; Pizzo, F. et al., *JOC*, **66**, 6734.

$$R\text{-OH} + (TMS)_2NH \quad \xrightarrow{\text{LiClO}, \mu W} \quad RO\text{-TMS}$$

58-83%

$$\xleftarrow{\text{clay}, \mu W, \text{H}_2\text{O}}$$

58-88%

[TMS ethers prepared also using TMS-N$_3$, Bu$_4$NBr]

V.E-14 Suzuki, T., Oriyama, T., *S*, 555.

$$R\text{-OTBS} \xrightarrow[\text{EtCN, -78°C}]{\text{THP-OAc, TBS-OTf}} R\text{-OTHP}$$
88-96%

(**failed for 3° and phenolic ethers**)

V.E-15 Blass, B.E. et al., *TL*, **42**. 1611.

V.E-16 Barros, M.T. et al., *SL*, 1146.

$$R\text{-OTBDPS} \xrightarrow[\text{MeOH, reflux}]{\text{Br}_2} R\text{-OH}$$
69-99%

(**at rt, TBDMS>TBDPS**)

V.E-17 Tanabe, Y. et al., *CC*, 2478; LeBideau, F. et al., *CC*, 1408.

[**silyl ethers also available from alcohols using PhR$_3$SiH, KOH, 18-C-6**]

V.E-18 Sureshan, K.M., Shashidhar, M.S., *TL*, **42**, 3037.

V.E-19 Sekine, M. et al., *TL*, **42**, 1069.

V.E-20 Silverman, R.B. et al., *OL*, **3**, 2477.

V.E-21 Avery, M.A. et al., *TL*, **42**, 7153; Love, K.R., Seeberger, P.H., *S*, 317.

$$RO \overset{O}{\underset{}{\|}} OCH_2CCl_3 \xrightarrow[\text{MeOH}]{\text{In, aq NH}_4\text{Cl}} \begin{array}{c} R\text{-OH} \\ 82\text{-}98\% \end{array}$$

(deprotection also for trichloroacetate;
Bz, TBDMS, TBDPS, CO₂Et, CONR₂ are stable)

V.E-22 Loudwig, S., Goeldner, M., *TL*, **42**, 7957.

V.E-23 Pirrung, M.C. et al., *JACS*, **123**, 3638.
Photochemically Removable Silyl Protecting Groups

V.E-24 Pirrung, M.C. et al., *OL*, **3**, 1105; **see also:** Furuta, T. et al., *OL*, **3**, 1809.
NPPOL Protecting Group for High-Fidelity Automated 5'→ 3' Photochemical DNA Synthesis

V.F. Other Protecting Groups

V.F-3 Pearson, A.J., Hwang, J.-J., *T*, **57**, 1489.
The Triphenylmethyl Group as Thiol Protection During Ruthenium-Promoted Synthesis of Tetraalkyl-*p*-phenylenediamine Systems Having Alkanethiol Side Chains

V.F-5 Guzaev, A.P., Manoharan, M., *JACS*, **132**, 783.
2-Benzamidoethyl Group—A Novel Type of Phosphate Protecting Group for Oligonucleotide Synthesis

V.F-6 Beaucage, S.L. et al., *TL*, **42**, 5635.
The 4-Oxopentyl Group as a Labile Phosphate/Thiophosphate Protecting Group for Synthetic Oligodeoxyribonucleotides

V.F-7 Choi, H.Y., Chi, D.Y., *JACS*, **132**, 9202.
A Facile Debromination Reaction: Can Bromide Now Be Used as a Protective Group in Aromatic Chemistry?

V.F-1 Lee, S.-G. et al., *SL*, 1956.

$$\underset{RS}{\overset{\displaystyle O}{\|}}\underset{}{}R^1 \xrightarrow[\text{CH}_2\text{Cl}_2, 0°\text{C}\rightarrow \text{rt}]{\text{TiCl}_4, \text{Zn}} \begin{array}{c} \text{R-SH} \\ \text{82-87\%} \end{array}$$

V.F-2 Barlos, K. et al., *TL*, **42**, 6965.

$$\text{Mmt-SH} + \text{X}\underset{n}{\diagdown}\text{CO}_2\text{R} \xrightarrow[\text{DMF/CH}_2\text{Cl}_2, \text{rt}]{^i\text{Pr}_2\text{NEt}} \text{Mmt-S}\underset{n}{\diagdown}\text{CO}_2\text{R} \\ \text{75-85\%}$$

(removal with TFA, CH_2Cl_2, Et_3SiH)

V.F-4 Rich, D.H. et al., *OL*, **3**, 1205.

$$\underset{\text{HCl·H}_2\text{N}}{\overset{\text{SH}}{\diagup}}\text{CO}_2\text{H} \xrightarrow[72\%]{\text{Fmoc-Cl, dioxane/H}_2\text{O, 0°C}} \underset{\text{FmocHN}}{\overset{\text{SFmoc}}{\diagup}}\text{CO}_2\text{H}$$

$$\underset{\text{FmocHN}}{\overset{\text{S)}_2}{\diagup}}\text{CO}_2\text{H} \xleftarrow[75\%]{\text{I}_2, \text{TEA, CH}_2\text{Cl}_2\text{/MeOH}}$$

VI
USEFUL SYNTHETIC PREPARATIONS

VI.A. Functional Group Preparations

VI.A.1. Acetals and Ketals

VI.A.1-1 Packard, G.K., Rychnovsky, S.D., *OL*, **3**, 3393; Beau, J.-M. et al., *TL*, **42**, 7567.
β-Selective Glycosylations

VI.A.1-2 Withers, S.G. et al., *AG(E)*, **40**, 417.
β-Mannosynthase: Synthesis of β-mannosides with a Mutant β-Mannosidase

---

VI.A.1-3 Crich, D. et al., *S*, 323.
2,4,6-Tri-*tert*-butylpyrimidine (TTBP): A Cost Effective, Readily Available Alternative to the Hindered Base 2,6-Di-*tert*-butylpyridine and Its 4-Substituted Derivatives in Glycosylatio and other Reactions

---

VI.A.1-4 Hirama, M. et al., *AG(E)*, **40**, 946.
AgPF$_6$ as a Remarkable Activator of 2-Deoxy Thio Glycosides

---

VI.A.1-5 Fairbanks, A.J. et al., *OL*, **3**, 2371.

VI.A.1-6 Singh, G. et al., *TA*, **12**, 1373.

213

VI.A.1-7 Wipf, P., Reeves, J.T., *JOC*, **66**, 7910.

R=OH + [structure] $\xrightarrow[\text{CH}_2\text{Cl}_2, -20°\text{C} \rightarrow \text{rt}]{\text{Cp}_2\text{ZrCl}_2, \text{AgClO}_4, 4\text{Å MS}}$ [product]

58-82%
α:β = 4.1-20.5:1

VI.A.1-8 Yamanoi, T., Yamazaki, I., *TL*, **42**, 4009.

[structure] + B(O-PhR)₃ $\xrightarrow[\text{CH}_2\text{Cl}_2]{\text{Yb(OTf)}_3, \text{CaCl}_2}$ [product]

45-83%
α:β = 1.3-2.3:1

VI.A.1-9 Yin, H., Lowary, T.L., *TL*, **42**, 5829.

[structure] $\xrightarrow[\text{CH}_2\text{Cl}_2, -78 \rightarrow 0°\text{C}]{\text{R-OH, NIS, AgOTf}}$ [product]

62-91%
α:β = 0.5-4.5:1

VI.A.1-11 Uchiyama, M. et al., *TL*, **42**, 1559; **see also:** Uchiyama, M. et al., *TL*, **42**, 1931.

[structure] + Ph—CH=CH—OMe $\xrightarrow[\text{Ph-H, reflux}]{\begin{array}{l}1.\ \text{MeOH, -78°C, 80\%}\\2.\ \text{Bu}_3\text{SnH, AIBN}\end{array}}$ [product]

90%
e.e. = 74%

VI.A.1-12 Cheng, X., Hii, K.K., *T*, **57**, 5445.

[structure] + HNRR¹ $\xrightarrow{\text{K}_2\text{Pd(SCN)}_4 \text{ or Pd(PPh}_3)(\text{SCN})_2}$ [product]

52-98%

VI.A.1-13 Tangestaninejad, S. et al., *JCR(S)*, 365.

[structure] $\xrightarrow[\text{Me}_2\text{CO}]{\text{Sn(TPP)(ClO}_4)_2}$ [product]

60-98%

VI.A.1-10 Chung, S.-K., Park, K.-H., *TL*, **42**, 4005.

59-64%
α:β = 2.9-3:1

VI.A.2. Acids and Anhydrides

see also: II.B.1, III.A, V.E, VI.A.8

VI.A.2-1 Chang, Y.-F., Tai, D.-F., *TA*, **12**, 177.
Enhancement of the Enantioselectivity of Lipase OF Catalyzed Hydrolysis

---

VI.A.2-2 Jolly, R.S. et al., *TA*, **12**, 1431; Molinari, F. et al., *TA*, **12**, 501.

15-92%
e.e. = 0-99%

VI.A.2-3 Kvittingen, L. et al., *TL*, **42**, 8543.
Regioselective Hydrolysis of Diesters of (Z)- and (E)-2-Methylbutenedioic Acids by PLE

VI.A.2-4 Mori, M. et al., *OL*, **3**, 3345.

33-82%

VI.A.2-5 Lee, J.C. et al., *SL*, 1563.

64-80%

VI.A.3. Alcohols and Related Species

see also: I.A.1, I.G.2, II.A.1

VI.A.3-1 Guanti, G., Riva, R., *TA*, **12**, 1185; **see also:** Garcia-Mera, X. et al. *TA*, **12**, 365; Kamal, A., Khanna, G.B.R., *TA*, **12**, 405; Bolte, J. et al., *TA*, **12**, 869; Fadnavis, N.W. et al., *TA*, **12**, 1695; Yasohara, Y. et al., *TL*, **42**, 3331; Tanaka, M., Suemune, H. et al., *TA*, **12**, 897.

41-68%   10-49%
e.e. = 4-99%   e.e. = 54-99%

[various other lipase-catalyzed alcohol resolutions reported]

VI.A.3-2 Perez, H.I. et al., *TA*, **12**, 1709; Allan, G.R., Carnell, A.J., *JOC*, **66**, 6495; Ferreira, E.M., Stoltz, B.M., *JACS*, **123**, 7725; Sigman, M.S. et al., *JACS*, **123**, 7475.

e.e. = 3-99%

[similarly with *sphingomonas Paucimobilis* or (-)-sparteine, O_2 with Pd catalysis]

VI.A.3-3 Faber, K. et al., *TA*, **12**, 1519.

46-90% conversion
e.e. = 59-97%   e.e. = 9-99%

VI.A.3-4 Knochel, P. et al., *AG(E)*, **40**, 1235; Hayashi, T. et al., *JOC*, **66**, 1441.

1. [Rh(cod)$_2$]BF$_4$ catecholborane
   lig, DME, rt

2. H_2O_2, KOH

50-85%
e.e. = 58-93%

lig =
P(c-hex)$_2$
P(c-hex)$_2$

[similarly with Cl$_3$Si-H/Pd followed by H_2O_2, KF]

VI.A.3-5 Muller, M. et al., *AG(E)*, **40**, 555.
Synthesis of Functionalized Cyclohexadiene-*trans*-diols with Recombinant Cells of *E. coli*

VI.A.3-6 Donohoe, T.J. et al., *TL*, **42**, 8951.
Homoallylic Alcohols and Trichloroacetamides as Hydrogen Bond Donors for Directed Dihydroxylation

---

VI.A.3-7 Otera, J. et al., *ECJ*, **7**, 3321.

$$R^1\!\!-\!\!CH(R)\!-\!OAc \xrightarrow[\text{MeOH}]{[^tBu_2Sn(Cl)OH]_2} R^1\!\!-\!\!CH(R)\!-\!OH$$

≤99%

VI.A.3-8 Ready, J.M., Jacobsen, E.N., *JACS*, **123**, 2687; see also: Trost, B.M., Tang, W., *OL*, **3**, 3409.

$$\text{R-OH} + \underset{R^1}{\triangle O} \xrightarrow[\text{MeCN, 4°C}]{\text{oligosalen, 3Å MS}} RO\!\!-\!\!CH_2\!\!-\!\!CH(OH)\!-\!R^1$$

60-99%
e.e. = 94-99%

VI.A.3-9 Yoshida, J. et al., *JOC*, **66**, 3970.

$$\text{Py-TMS} \xrightarrow[\text{2. E}^+]{\text{1. }^tBuLi, Et_2O, -78C} \text{Py-Si-E} \xrightarrow[\text{THF, MeOH}]{H_2O_2, KF, KHCO_3} \text{HO}\!\!-\!\!E$$

55-99%   90-98%

VI.A.3-10 Kanth, J.V.B., Brown, H.C., *JOC*, **66**, 5359.

$$R^1\!\!-\!\!CR\!=\!CH\!-\!R^2 + \underset{}{O\!\!-\!\!BH_2Cl} \xrightarrow[\text{2. H}_2O_2, NaOH]{\text{1. dioxane/CH}_2Cl_2, rt} R^1\!\!-\!\!CR(R)\!-\!CH(R^2)\!-\!OH$$

59-99%

VI.A.3-11 Tanaka, T. et al., *TL*, **42**, 8007.

$$\text{R-NH}_2 \xrightarrow[\text{(HOCH}_2CH_2)_2O, 210°C]{\text{KOH}} \text{R-OH}$$

42-77%

VI.A.4. Aldehydes and Ketones

VI.A.4-1 Lee, K., Im, J.-M., *TL*, **42**, 1539.

$$\text{R} \overset{\overset{O \quad O}{\parallel \quad \parallel}}{\underset{PPh_3}{\diagdown}} R^1 \xrightarrow[\text{CH}_2\text{Cl}_2, \text{ rt}]{\text{MMPP}} \text{R} \overset{\overset{O \quad O}{\parallel \quad \parallel}}{\underset{O}{\diagdown}} R^1$$

76-92%

VI.A.4-2 Mukaiyama, T. et al., *CL*, 580.

$$\text{R-CH}_2\text{OTf} \xrightarrow[\text{2. DBU, Ph-Me, rt}]{\text{1. MeS(O)NH}^t\text{Bu, CH}_2\text{Cl}_2, \text{ rt}} \text{R-CHO}$$

43-91%

VI.A.4-3 Pedro, J.R. et al., *T*, **57**, 1075.

$$\text{R} \overset{\text{CO}_2\text{H}}{\underset{R^1}{\overset{|}{\text{C}}-\text{OH}}} \xrightarrow{\text{O}_2, \text{ pivaldehyde, Co(III) complex}} \text{R} \overset{O}{\overset{\parallel}{\diagdown}} R^1$$

40-97%

VI.A.4-4 Satoh, T. et al., *T*, **57**, 493.

$$\text{R} \diagup \overset{\overset{O}{\parallel}}{\underset{R^1}{\overset{\text{Cl}}{\text{C}}-\overset{}{\text{S}}-\text{Ar}}} \xrightarrow[\text{2. aq HClO}_4]{\text{1. TFAA, NaI}} \text{R} \overset{O}{\overset{\parallel}{\diagdown}} R^1$$

68-98%

VI.A.4-5 Tanaka, K., Fu, G.C., *JOC*, **66**, 8177; Gree, R. et al., *EJOC*, 3141.

$$\text{R} \overset{R^1}{\diagup}\diagdown \text{OH} \xrightarrow[\text{THF, 100°C}]{\text{[Rh(cod)L]BF}_4} \text{R} \overset{R^1}{\diagup}\diagdown \text{CHO}$$

60-98%
e.e. = 50-92%

$$L = \text{(ferrocene ligand with } Ar_2P\text{)}$$

VI.A.4-6 Bassetti, M., Gimeno, J. et al., *TL*, **42**, 8467; Wakatsuki, Y. et al., *OL*, **3**, 735.

$$\text{R} \equiv \xrightarrow[\text{SDS or CTAB}]{\text{H}_2\text{O, [Ru(η}^5\text{-C}_9\text{H}_7)(\text{PPh}_3)_2\text{Cl}} \text{R} \diagdown \text{CHO}$$

25-97%

VI.A.4-7 Filimonov, V.D. et al., *S*, 1001.

$$Ph\text{—}\!\!\equiv\!\!\text{—}Ar \xrightarrow[\text{dioxane, rt-100°C}]{SO_3} \underset{\substack{Ph \quad Ar \\ 5\text{-}98\%}}{\overset{O \quad O}{\bigvee\bigvee}}$$

VI.A.4-8 Troupel, M. et al., *T*, **57**, 525.
Nickel Catalyzed Electrosynthesis of Ketones from Organic Halides and Iron Pentacarbonyl. Unsymmetrical Ketones

VI.A.5. Amides and Related Species

VI.A.5-1 Ren, R.X. et al., *TL*, **42**, 8441, 8445; Khodaei, M.M., Salehi, P. et al., *SC*, **31**, 2047; Chrandrasekhar, S., Gopalaiah, K., *TL*, **42**, 8123; Arisawa, M., Yamaguchi, M., *OL*, **3**, 311; Sharghi, H., Sarvari, M.H., *JCR(S)*, 446.

$$\text{(cyclohexylidene)}=N\text{-OH} \xrightarrow[\text{ionic liquid, 75°C}]{P_2O_5} \underset{95\text{-}99\%}{\text{(caprolactam)}=O, -NH}$$

[Beckmann rearrangements reported also with $FeCl_3$, neat;
in the solid state; with [RhCl(cod)]$_2$, MSA, P(tol)$_3$]

VI.A.5-2 Najera, C. et al., *TL*, **42**, 5013; Banwell, M., Smith, J., *SC*, **31**, 2011; Giacomelli, G. et al., *JOC* **66**, 2534.

$$R\text{-}CO_2H + R^1NHOR^2\cdot HCl \xrightarrow[\substack{Me \quad N_{+} \cdot R^3 \\ N_{+} \quad S \quad R^3 \\ O^- \quad Me}]{\text{TEA or DIEA, MeCN}} \underset{\substack{R^1 \\ 60\text{-}95\%}}{R\overset{O}{\underset{}{\text{C}}}N\text{-}OR^2}$$

(no racemization; Boc, Cbz, Fmoc are stable)
[Weinreb amide synthesis reported similarly with a pyridine N-oxide
disulfide, PBu$_3$ or 2-chloro-9,16-dimethoxy-1,3,5-triazine coupling reagent]

VI.A.5-3 Jung, K.W. et al., *JOC*, **66**, 1035; Salvatore, R.N., Jung, K.W. et al., *TL*, **42**, 6023.

$$R\text{-}NH_2 + R^1\text{-}X \xrightarrow[\text{DMF, 23°C}]{CO_2, Cs_2CO_3, TBAI} \underset{78\text{-}96\%}{R\text{-}NHCO_2R^1}$$

VI.A.5-4 Yamamoto, H. et al., *SL*, 1371.
3,5-Bis(perfluorodecyl)phenylboronic Acid as an Easily Recyclable Direct Amide Condensation Catalyst

VI.A.5-5 Tanaka, T. et al., *T*, **57**, 9309.

49-76%
e.e. = 4-85%

VI.A.5-6 Kunishima, M. et al., *T*, **57**, 1551; Kaminski, Z.J. et al., *JOC*, **66**, 6276.

$$R\text{-}CO_2H \ + \ R^1R^2NH \xrightarrow[]{R^3\text{-}OH \text{ or } H_2O, \text{ rt}} \quad \underset{84\text{-}99\%}{R\text{-}C(=O)NR^1R^2}$$

VI.A.5-7 Feroci, M., Inesi, A. et al., *JOC*, **66**, 6185.

57-97%

VI.A.5-8 Kunieda, T. et al., *TL*, **42**, 6565.

35-71%

VI.A.5-9 Svedas, V.K. et al., *TA*, **12**, 1645.

e.e. = 96-98%

VI.A.5-10 Yagupolskii, Y.L. et al., *TL*, **42**, 8181.

$$R\text{-}N{=}C{=}X \ + \ TMS\text{-}CF_3 \xrightarrow{Me_4NF} CF_3\text{-}C(=X)NHR$$

80-95%

VI.A.5-11 Schuemacher, A.C., Hoffmann, R.W., *S*, 243.

$$\underset{R^1}{\overset{R}{\diagdown}}\hspace{-6pt}CO_2H \;+\; O=C=N\hspace{-4pt}\underset{R^2}{\overset{R^3}{\diagup}} \quad\xrightarrow[\text{CH}_2\text{Cl}_2,\text{ rt}]{\text{DMAP}}\quad \underset{R^1}{\overset{R}{\diagdown}}\hspace{-6pt}\overset{O}{\underset{HN}{\diagdown}}\hspace{-4pt}\underset{R^2}{\overset{R^3}{\diagup}}$$

61-98%

VI.A.5-12 Kim, K. et al., *TL*, **42**, 8197.

$$\underset{S\diagdown_S\diagup N}{\overset{O}{\diagup}\overset{Cl}{}} \;+\; HNRR^1 \quad\xrightarrow[\text{CH}_2\text{Cl}_2,\text{ rt}]{}\quad R^1RN\overset{O}{\underset{NRR^1}{\diagup}}$$

37-99%

VI.A.5-13 Chandrasekhar, S. et al., *SL*, 1561.

$$\begin{matrix}\text{Ar-NHNH}_2\\\text{or}\\\text{Ar-N=N-Ar}\end{matrix}\quad\xrightarrow[\text{EtOH, rt}]{\text{PMHS, Pd/C, (Boc)}_2\text{O}}\quad \text{Ar-NHBoc}$$

76-90%

VI.A.5-14 Zwierzak, A. et al., *TL*, **42**, 5093.

$$\text{R-CHO}\quad\xrightarrow[\text{2. NaBH}_4\text{, THF, rt}]{\text{1. BocNH}_2\text{, TsNa, HCO}_2\text{H, MeOH/H}_2\text{O}}\quad R\diagup\diagdown NHBoc$$

44-70%

VI.A.5-15 Salehi, P., Khodaei, M.M. et al., *SC*, **31**, 1947.

$$\text{R-OH} + R^1\text{-CN}\quad\xrightarrow{\text{Mg(HSO}_4)_2}\quad R^1\overset{O}{\underset{NHR}{\diagup}}$$

67-97%

VI.A.5-16 Ley, S.V. et al., *JCS(P1)*, 358.

$$R\overset{O}{\underset{NR^1R^2}{\diagup}} \;+\; \underset{N\diagdown_{NH}}{\overset{S}{\diagup}\overset{OEt}{}}\quad\xrightarrow{\text{Ph-Me, 90°C, 30h}}\quad R\overset{S}{\underset{NR^1R^2}{\diagup}}$$

60-99% conversion

VI.A.5-17 Wang, Y., Ding, K., *JOC*, **66**, 3238.

$$\underset{R}{\overset{OAc}{\diagup\diagdown\hspace{-2pt}R^1}}\quad\xrightarrow[\underset{\text{PAr}_2/\text{XR}_2}{}]{[\text{Pd}(\eta^3\text{-C}_3\text{H}_5)\text{Cl}]_2,\text{ NaN(CHO)}_2,\text{ MeCN, 60°C}}\quad \underset{R}{\overset{N(CHO)_2}{\diagup\diagdown\hspace{-2pt}R^1}}$$

1-99%
e.e. = 31-96%

VI.A.5-118 Jones, W.D. et al., *OM*, **20**, 1028.

R—⟨benzene⟩—SH + HNR^{1}R^2 $\xrightarrow[\text{THF, rt}]{\text{CO, Ni(PPh}_3)_n\text{Br}_2}$ R—⟨benzene⟩—S—C(=O)—NR^{1}R^2

18-86%

VI.A.5-19 Buchwald, S.L. et al., *OL*, **3**, 3803 and *JACS*, **123**, 7727.

R—⟨benzene⟩—I + HN(Boc)(NH$_2$) $\xrightarrow[\text{DMF, 80°C}]{\text{CuI, phenanthroline, Cs}_2\text{CO}_3}$ R—⟨benzene⟩—N(Boc)(NH$_2$)

43-90%

VI.A.5-20 Cacchi, S. et al., *OL*, **3**, 2539; Lovely, C.J. et al., *TL*, **42**, 7155; Beletskaya, I.P. et al., *TL*, **42**, 4381.

Ar-Br + (oxazolidinone with R^1, R, N—H) $\xrightarrow[\text{Ph-Me, 120°C}]{\text{Pd(OAc)}_2 \text{ or Pd}_2\text{(dba)}_3, \text{'BuONa}}$ (oxazolidinone with R^1, R, N—Ar)

2-88%

[similar aryl coupling reactions reported with a chiral pyrrolidinone,
Pd$_2$(dba)$_3$, Xantphos, Cs$_2$CO$_3$ or, under similar conditions, with N-phenyl urea]

VI.A.6. Amines

VI.A.6-1 Mori, M. et al., *OL*, **3**, 1913; Widhalm, M., van Leeuwen, P.W.N.M., *JOC*, **66**, 759.

(cyclohexene with R, OCO$_2$Me) + HN(Ts)(allyl) $\xrightarrow[\text{Pd}_2\text{(dba)}_3, \text{CHCl}_3, \text{rt}]{}$ (cyclohexene with R, N(Ts)(allyl))

(binaphthyl ligand with OPPh$_2$, OPPh$_2$)

40-70%
e.e. = 5-78%

[similar amination reported with an allyl acetate, BnNH$_2$,
[Pd(η^3-C$_3$H$_5$)Cl]$_2$ and a chiral bisferrocenyl phosphine]

VI.A.6-2 Klepacz, A., Zwierzak, A., *SC*, **31**, 1681.

R-OH + HN(CO$_2$'Bu)(P(OEt)$_2$=O) $\xrightarrow[\text{2. TsOH·H}_2\text{O, EtOH, reflux}]{\text{1. DIAD, TPP, THF, rt, 2h}}$ R-NH$_3^+$ ⁻OTs

30-77%

VI.A.6-3 Hosomi, A. et al., *SL*, 1617; Allegretti, M. et al., *TL*, **42**, 4257.

$$\underset{R}{\overset{R^1}{>}}=O \; + \; Me_2HSi\text{-}NR^2R^3 \quad \xrightarrow[\text{CH}_2\text{Cl}_2, 0°C \to rt]{\text{LA}} \quad \underset{R}{\overset{R^1}{>}}-NR^2R^3$$

53-94%

[reductive amination reported also with ammonium formate, Pd/C]

VI.A.6-4 Yang, S.-C. et al., *OM*, **20**, 5326; Yang, S.-C., Tsai, Y.-C., *OM*, **20**, 763; **see also:** Mahrwald, R., Quint, S., *TL*, **42**, 1655.

<97% ≤99:1

[similar alkylations reported with alkynyl alcohols, HNRR^1, TiCl$_4$]

VI.A.6-5 Zaragoza, F., Stephensen, H., *JOC*, **66**, 2518.

$$R^1R^2NH \; + \; R^3CH_2\text{-}OH \; + \; N\equiv CCH_2P^+R_3 \quad \xrightarrow[\text{EtCN, 90°C, 2h}]{^{i}\text{Pr}_2\text{NEt}} \quad R^1R^2N\overset{}{\frown}R^3$$

68-83%

VI.A.6-6 Alper, H. et al., *JACS*, **123**, 7719.

10-87%   3-99%

VI.A.6-7 Jorgensen, K.A. et al., *CC*, 1240; Bartoli, G., Marcantoni, E. et al., *JOC*, **66**, 9052; Gomtsyan, A. et al., *JOC*, **66**, 3613; Cardillo, G. et al., *TA*, **12**, 2395.

6-93%
e.e. = 0-96%

[amino conjugate additions reported also with CeCl$_3$·7H$_2$O, NaI, SiO$_2$;
from Weinreb amides, vinyl magnesium bromide then amines or from
TMS-NHOTMSand vinyl 1,1-diesters with chiral bisoxaline copper catalyst]

VI.A.6-8 Selua, M. et al., *JOC*, **66**, 677.

$$\text{Ph-NH}_2 + \text{RO-CO}_2\text{Me} \xrightarrow[\text{triglyme, 115-130°C}]{\text{Faujasite zeolite}} \text{Ph-NHMe}$$
90-97%

VI.A.6-9 Whiting, A. et al., *TL*, **42i** 8387.

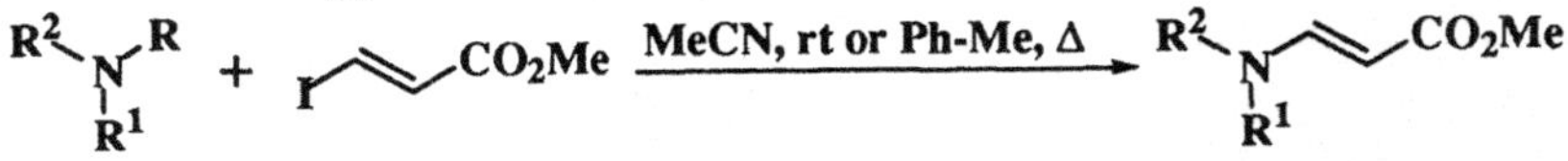

90-96%

VI.A.6-10 Singaram, B. et al., *JOC*, **66**, 1999 and *OL*, **3**, 3915.

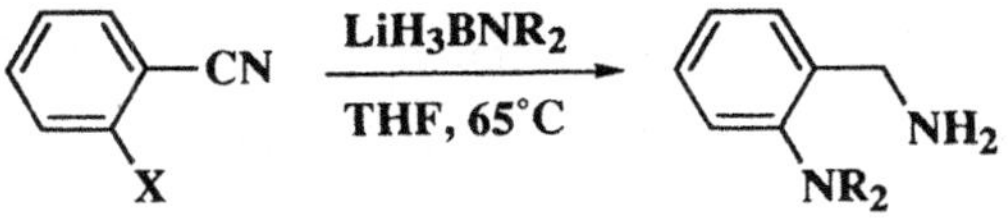

70-94%

[amines also prepared from a mesylate and LiH₃BNR₂]

VI.A.6-11 Fort, Y. et al., *T*, **57**, 7657 and *TL*, **42**, 5689; Dommisse, R. et al., *T*, **57**, 7027; **see also:** Beller, M. et al., *JOC*, **66**, 1403.

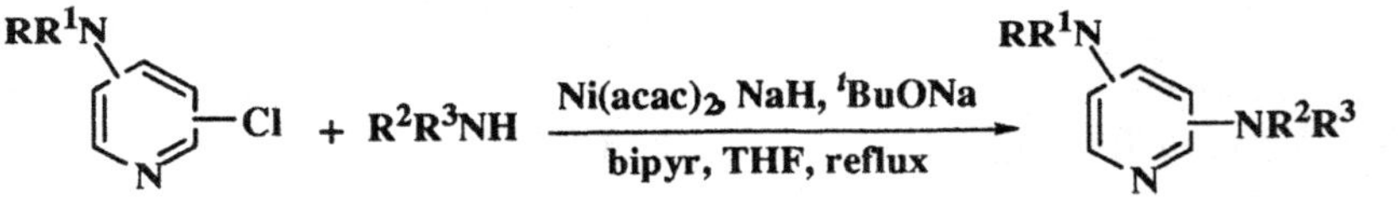

52-86%

[similarly with Pd catalyst and with ᵗBuOK]

VI.A.6-12 Antilla, J.C., Buchwald, S.L., *OL*, **3**, 2077.

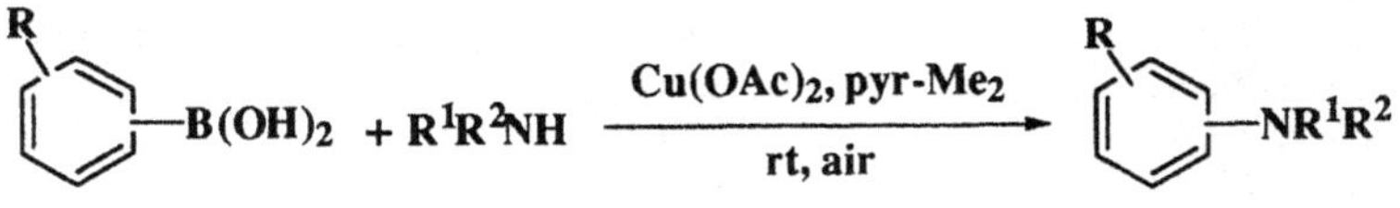

50-91%

VI.A.6-13 Hartwig, J.F. et al., *OL*, **3**, 2729; Huang, X., Buchwald, S.L., *OL*, **3**, 3417.

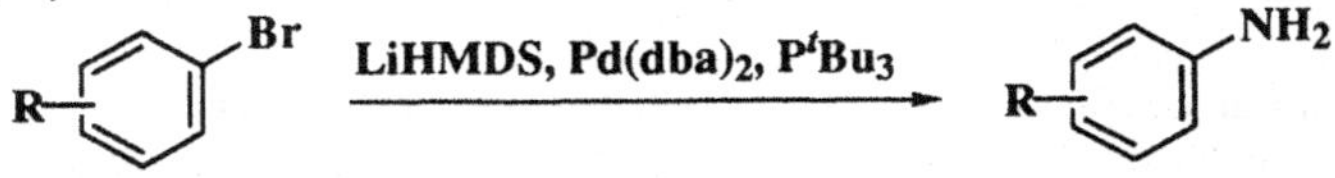

75-99%

[a different ligand used also for a similar transformation]

VI.A.6-14 Doris, E. et al., *TL*, **66**, 8301.

$$\text{Ar-NH}_2 + \text{Et}_2\text{Zn} \xrightarrow[\text{CH}_2\text{Cl}_2]{\text{Cu(acac)}} \text{Ar-NHEt}$$

35-95%

VI.A.6-15 Parrish, C.A., Buchwald, S.L., *JOC*, **66**, 3820; Ali, M.H., Buchwald, S.L., *JOC*, **66**, 2560; Nolan, S.P., *JOC*, **66**, 7729; Wong, K.-T. et al., *OL*, **3**. 2285; **see also:** Rasmussen, S.C. et al., *JOC*, **66**, 9067.

$$\xrightarrow[\text{Ph-Me, 80°C}]{\text{Pd(OAc)}_2, \text{lig, }^t\text{BuONa}}$$

79-97%

[various ligands, catalysts and substrates used for similar transformations]

VI.A.6-16 Venkataraman, D. et al., *OL*, **3**, 4315 and *TL*, **42**, 4791; **see also:** Ma, D., Xia, C., *OL*, **3**, 2583.

$$+ \text{Ph}_2\text{NH} \xrightarrow[]{^t\text{BuOK, Ph-Me, 110°C}}$$

49-88%

[N-aryl-β-amino acids prepared similarly using aryl halides, β-amino esters, CuI/K$_2$CO$_3$]

VI.A.6-17 Ishikawa, T., Saito, S. et al., *JACS*, **123**, 7734.

$$\xrightarrow[\text{THF, 0C}]{\text{BuLi}}$$

20-86%

VI.A.6-18 Kane, R.R. et al., *SL*, 643.

$$\text{Ar-NH}_2 + \xrightarrow[\text{2. NaBH}_4 \text{ rt}]{\text{1. DMF, 40°C}}$$

62-73%

VI.A.7. Amino Acid Derivatives

VI.A.7-1 Belokon, Y.N., Kagan, H.B. et al., *AG(E)*, **40**, 1948.

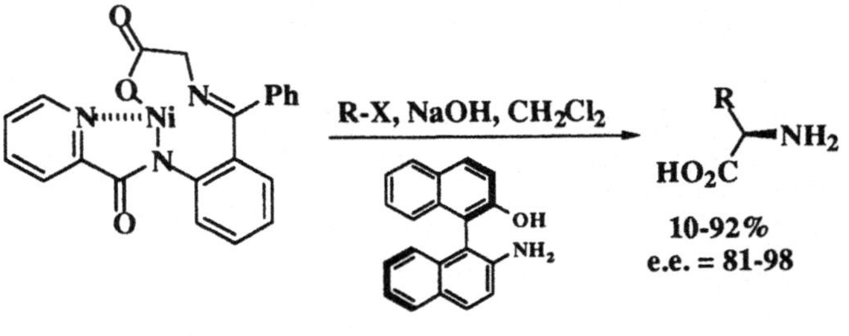

R-X, NaOH, CH$_2$Cl$_2$

10-92%
e.e. = 81-98

VI.A.7-2 Kise, N. et al., *TL*, **42**, 7637.

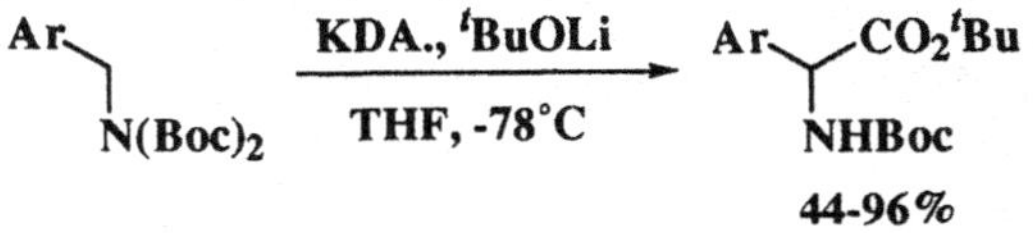

KDA., ^{t}BuOLi

THF, -78°C

44-96%

VI.A.7-3 Farras, J., Romea, P., Urpi, F. et al., *T*, **57**, 7665.

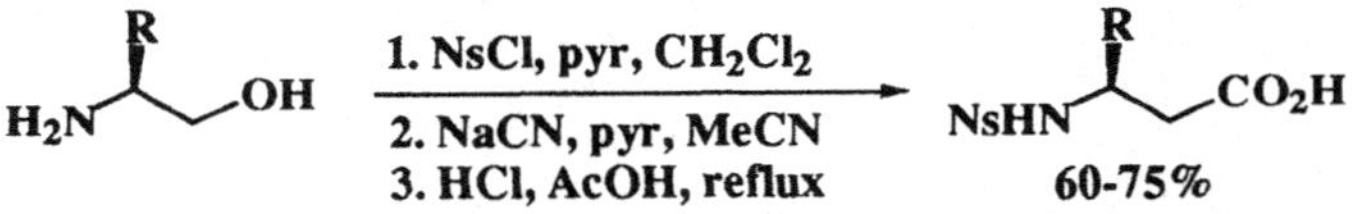

1. NsCl, pyr, CH$_2$Cl$_2$
2. NaCN, pyr, MeCN
3. HCl, AcOH, reflux

60-75%

VI.A.7-4 Jiang, B. et al., *TL*, **42**, 2545.

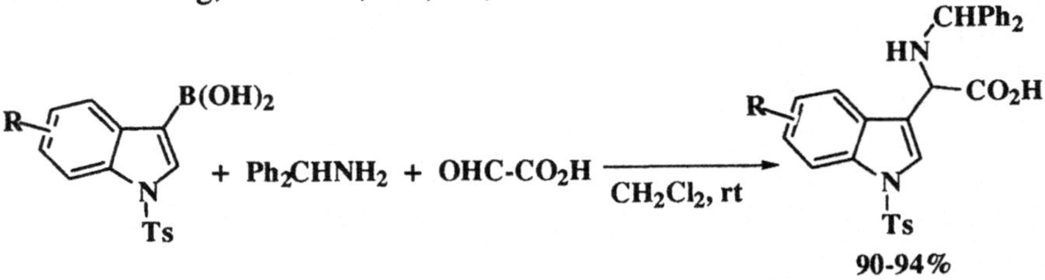

+ Ph$_2$CHNH$_2$ + OHC-CO$_2$H

CH$_2$Cl$_2$, rt

90-94%

VI.A.7-5 Harwood, L.M. et al., *JCS(P1)*, 1581.

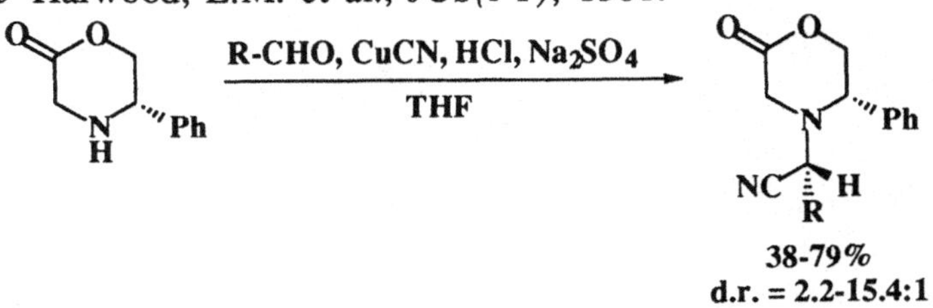

R-CHO, CuCN, HCl, Na$_2$SO$_4$

THF

38-79%
d.r. = 2.2-15.4:1

VI.A.7-6 Feng, X., Jiang, Y. et al., *SL*, 1551.
Enantioselective Strecker Reaction Promoted by Chiral N-Oxides

VI.A.7-7 Westwell, A.D. et al., *JCR(S)*, 546; Adamczyk, M., Reddy, R.E., *TA*, **12**, 1047.
Synthesis of L-Azatyrosine and Analogs

VI.A.7-8 Gellman, S.H. et al., *JOC*, **66**, 5629.
An Efficient Route to Either Enantiomer of *trans*-2-Aminocyclo-pentanecarboxylic Acid

VI.A.7-9 Chen, F.-Y., Uang, B.-J., *JOC*, **66**, 3650.
D-DOPA Synthesis

VI.A.7-10 Long, A., Baldwin, S.W., *TL*, **42**, 5343.
Homophenyl Alanine Derivatives *via* Nitrone Cycloaddition to Styrenes

VI.A.7-11 Lajoie, G.A. et al., *TL*, **42**, 3807.
Synthesis of N-Protected N-Methylserine and Threonine

VI.A.7-12 Hruby, V.J. et al., *TL*, **42**, 4601.
Synthesis of (2S,3S)-β-Methyltryptophan

VI.A.7-13 Miller J.M. et al., *JOC*, **66**, 6046; Anderson, J.C., Flaherty, A., *JCS(P1)*, 276; Koskinen, A.M.P. et al., *JOC*, **66**, 2061.
Stereocontrolled Synthesis of Proline

VI.A.7-14 Baldwin, J.E. et al., *JCS(P1)*, 668.
The Efficient Enantioselective Synthesis of Quinoxaline, Pyrazine and 1,2,4-Triazine Substituted α-Amino Acids from Vicinal Tricarbonyls

VI.A.7-15 Hernandez, N., Martin, V.S., *JOC*, **66**, 4934; Shireman, B.T., Miller, M.J., *JOC*, **66**, 4809.
Synthesis of α-Diamino Acids

VI.A.7-16 Doris, E. et al., *JOC*, **66**, 6487.
Synthesis of γ-Amino Acids by Rearrangement of α-Cyano-cyclopropanone Imidates: Application to the Regioselective Labeling of Amino Acids

VI.A.7-17 Maruoka, K. et al., *SL*, 1185; Tanaka, M., Suemume, H. et al., *JOC*, **66**, 2667.
Synthesis of α,α-Disubstituted α-Amino Acids

VI.A.7-18 Jamart-Gregoire, B. et al., *JOC*, **66**, 2869; Le Grel, P. et al., *JOC*, **66**, 4923.
Synthesis of Substituted Hydrazino Acetic Acid Derivatives

VI.A.7-19 Fustero, S. et al., *T*, **57**, 703; Williams, R.M. et al., *JACS*, **123**, 3473.
β-Amino Acid Synthesis

VI.A.7-20 Gonzalez-Muniz, R. et al., *JOC*, **66**, 3538; Madalengoitia, J.S. et al., *JOC*, **66**, 6483; Harouutounian, S.A., *JOC*, **66**, 7915; Halcomb, R.L. et al., *JOC*, **66**, 2219; Edmonds, M.K., Abell, A.D., *JOC*, **66**, 3747.
Synthesis of Conformationally Constrained Amino Acids and Peptide Surrogates

VI.A.7-21 Rich, D.H. et al., *OL*, **3**, 2309, 2313, 2317.
Synthesis of Aspartic Peptidase Inhibitors

VI.A.7-22 Nishiyama, Y. et al., *TL*, **42**, 8789.
O-(7-Azabenzotriazol-1-yl)-1,1,3,3-tetramethyluronium Hexafluorophosphate-1-hydroxy-7-azabenzotriazole-Copper(II) Chloride: A Promising Epimerization-Free Segment Coupling System for Peptide Synthesis

VI.A.7-23 Haswell, S.J. et al., *CC*, 990.
The Synthesis of Peptides Using Micro Reactors

VI.A.7-24 Scherkenbeck, J. et al., *JOC*, **66**, 3760.
Azadepsipeptides: Synthesis and Evaluation as a Novel Class of Peptidomimetics

VI.A.8. Azides

VI.A.8-1 Tamami, B., Mahdavi, H., *TL*, **42**, 8721.

$$R\text{—epoxide} \xrightarrow[\text{H}_2\text{O}]{\text{NaN}_3, \text{PTC}} R\text{—CH(OH)—CH}_2\text{—N}_3$$

87-95%

VI.A.8-2 Pizzo, F. et al., *JOC*, **66**, 3554; Fringuelli, F. et al., *JOC*, **66**, 4719.

≤99% conversion 99:1

VI.A.8-3 Deardorff, D.R. et al., *JOC*, **66**, 7191.

85-90%

VI.A.8-4 Viuf, C., Bols, M., *AG(E)*, **40**, 623.

74-98%

VI.A.9. Esters

(see also: I.G.2, IV.D, V.D)

VI.A.9-1 Yamamoto, H. et al., *SL*, 1117; **see also:** Caddick, S. et al., *T*, **57**, 6305.
A Green Method for the Selective Esterification of Primary Alcohols in the Presence of Secondary or Aromatic Alcohols

---

VI.A.9-2 Mohan, R.S. et al., *S*, 2091; Otera, J. et al., *JOC*, **66**, 8926; Sato, T. et al., *SL*, 1584; Karimi, B., Seradj, H., *SL*, 519; Mohammadpoor-Baltork, I. et al., *JCR(S)*, 280.

85-92%

[similar esterifications using Ac_2O, $LiClO_4$ or NBS or with AcOH, BiX_3]

VI.A.9-3 Siddiqui, B.S. et al., *TL* , **42**, 9059.

$$R\text{-}CO_2H + MsCl \xrightarrow{\text{pyr, 0°C}} R\text{-}CO_2Me$$

65-71%

VI.A.9-4 Venkateswarlu, Y. et al., *SC*, **31**, 2599; Lakouraj, M. et al., *JCR(S)*, 378.

$$\text{R} \overset{O}{\triangle} \xrightarrow{\text{Ac}_2\text{O, HY Zeolite, rt}} \underset{\text{R}}{\overset{\text{OAc}}{\bigvee}}\text{OAc}$$

85-95%

[ether cleavage reported also with FeCl$_3$, Monmorillonite K-10]

VI.A.9-5 Dyke, C.A., Bryson, T.A., *TL*, **42**, 3959; Lee, A.S.-Y. et al., *TL*, **42**, 301; Karmakar, D., Das, P.J., *SC*, **31**, 535.

$$\text{R-CO}_2\text{H} \xrightarrow[\text{2. R}^1\text{-OH}]{\text{1. BCl}_3} \text{R-CO}_2\text{R}^1$$

24-99%

[methyl or *tert*-butyl esters prepared similarly using hv, CBr$_4$ or S, Zn dust, respectively]

VI.A.9-6 Maruoka, K. et al., *TL*, **42**, 9245; Boyer, B. et al., *TL*, **42**, 855.

$$\text{R-CO}_2\text{H} \xrightarrow[\text{THF, rt}]{\text{R}^1\text{-X, TBAHSO}_4\text{, KF·2H}_2\text{O}} \text{R-CO}_2\text{R}^1$$

51-99%

[esters prepared similarly using BiPh$_3$ then R-Br]

VI.A.9-7 Rademann, J. et al., *AG(E)*, **40**, 381.

$$\text{R-CO}_2\text{H} \xrightarrow{\hspace{4cm}} \text{R-CO}_2\text{R}^1$$

≤80%

VI.A.9-8 Charette, A.B. et al., *JOC*, **66**, 2178.

$$\underset{\text{R}}{\overset{\text{R}^1}{\bigvee}}\text{OH} \xrightarrow{\text{4-NO}_2\text{Ph-CO}_2\text{H, R}^2\text{O}_2\text{CN=NCO}_2\text{R}^2\text{, Ph-H, rt}}$$

38-75% conversion

VI.A.9-9 Otera, J. et al., *SL*, 637.

$$\xrightarrow[\text{Ph-Me, reflux}]{\text{MeOH, LA}} \text{R-CO}_2\text{Me}$$

70-99%

VI.A.9-10 Bandgar, B.P. et al., *SL*, 1338, 1715 and *SC*, **31**, 2063; Chavan, S.P. et al., *SC*, **31**, 289.

$$\text{Me-CO-CH}_2\text{-CO}_2\text{R} + \text{R}^1\text{-OH} \xrightarrow[\text{Ph-Me, reflux}]{\text{LiClO}_4} \text{Me-CO-CH}_2\text{-CO}_2\text{R}^1$$

65-94%

[transesterifications reported also using NBS, FeSO$_4$, CuSO$_4$ or Amberlyst-15 catalysts]

VI.A.9-11 Matsuda, T. et al., *TL*, **42**, 8319.
Control on Enantioselectivity with Pressure for Lipase-Catalyzed Esterification in Supercritical Carbon Dioxide

---

VI.A.9-12 Gais, H.-J. et al., *JOC*, **66**, 3384; **see also:** Joly, S., Nair, M.S., *TA*, **12**, 2283; Nakano, S. et al., *TA*, **12**, 59; Kamal, A. et al., *JOC*, **66**, 997; Vidari, G. et al., *TA*, **12**, 1779; Ravina, E. et al., *TA*, **12**, 1723; Miyazawa, T. et al., *TA*, **12**, 1595; Chung, S.-K., Kwon, Y.-U., *OL*, **3**, 3013; Scilimati, A. et al., *TA*, **12**, 853.

$$\text{X-CH}_2\text{-CH(OH)-Ar} \xrightarrow[\text{•-NH}_2, \text{Ph-Me/H}_2\text{O}]{\substack{\text{EtCO}_2\text{CH=CH}_2 \\ \text{pig liver esterase, MeO-PEG}}} \text{X-CH}_2\text{-CH(OCOEt)-Ar} + \text{X-CH}_2\text{-CH(OH)-Ar}$$

4-51% conversion
e.e. = 61-99%   e.e. = 2-99%

[various other substrates used with lipase enzymes foe alcohol or ester resolution]

VI.A.9-13 Cordova, A., Janda, K.D., *JOC*, **66**, 1906; Gotor, V. et al., *JOC*, **66**, 4227; **see also:** Pamies, O., Backvall, J.-E., *JOC*, **66**, 4022; Kim, M.-J., Park, J. et al., *JOC*, **66**, 4736.

$$\text{R}^1\text{R-CH-OH} + \text{R}^3\text{O-CO-CH}_2\text{-CO-R}^2 \xrightarrow[\text{C. antarctica lipase B}]{\text{immobilized}} \text{R}^1\text{R-*CH-O-CO-CH}_2\text{-CO-R}^2$$

92-98%
e.e. = 90-98%

VI.A.9-14 Janda, K.D. et al., *JOC*, **66**, 868; Janda, K.D. et al., *JOC*, **66**, 5645.

$$\text{R}^1\text{R-CH-OH} \xrightarrow{\text{Bz-Cl, TEA, CH}_2\text{Cl}_2, -78°\text{C}} \text{R}^1\text{R-CH-OBz} + \text{R}^1\text{R-CH-OH}$$

15-54%   24-58%
e.e. = 0-97%   e.e. = 0-97%

VI.A.9-15 Roesky, P.W. et al., *ECJ*, **7**, 3078.

$$R\text{-}CHO \xrightarrow[\text{C}_6\text{D}_6,\text{ rt}]{Y[N(TMS)_2]_3} \underset{\textbf{35-86\%}}{R\text{-}C(=O)\text{-}O\text{-}CH_2\text{-}R}$$

VI.A.9-16 Lee, J.C. et al., *CC*, 956.

$$\underset{R}{\overset{O}{\|}}\!\!-\!\!CH_2R^1 \xrightarrow[\text{2. R}^3\text{CONMe}_2]{\text{1. Tl(OTf)}_3} \underset{\textbf{90-99\%}}{\text{product}}$$

VI.A.9-17 Katritzky, A.R. et al., *JOC*, **66**, 5606.

$$\underset{R}{\overset{OTf}{>}}\!\!-\!\!Bt \xrightarrow[\text{2. R}^1\text{-OH, HCl}]{\text{1. NaOMe, MeCN, 65°C}} \underset{\textbf{24-98\%}}{R\frown CO_2R^1}$$

VI.A.10. Ethers

VI.A.10-1 Wang, J. et al., *TL*, **42**, 8511; Moody, C.J. et al., *TA*, **12**, 1657; Chandrasekhar, S. et al., *SL*, 1779.

$$Ph\underset{N_2}{\overset{}{>}}CO_2R^* \xrightarrow[]{R\text{-OH, Rh}_2\text{(OAc)}_4} Ph\underset{OR}{\overset{}{>}}CO_2R^*$$

62-93%
d.e. = 24-90%

[similarly with chiral α-diazophosphonic acid derivatives
or tosylhydrazones, ^{t}BuOK]

VI.A.10-2 Basavaiah, D. et al., *T*, **57**, 8167.

$$\underset{Ar}{\overset{CO_2Me}{>}}\!\!-\!\!Br \;+\; \equiv\!\!-\!\!OH \xrightarrow[\text{CH}_2\text{Cl}_2,\text{ rt}]{\text{quinidine}} \text{product}$$

32-47%
e.e. = 25-40%

VI.A.10-3 Yamamoto, Y. et al., *JOC*, **66**, 270.

$$\triangleright\!\!=\!\!\underset{R}{\overset{R^1}{<}} \;+\; R^2\text{-OH} \xrightarrow[\text{Ph-H or Ph-Me, 100°C, 3d}]{\text{Pd(PPh}_3)_4,\text{ P(Ph-Me)}_3} R^2O\frown\!\!\underset{R}{\overset{R^1}{<}}$$

20-80%

VI.A.10-4 Lee, Y.R., Kim, D.H., *TL*, **42**, 6561.

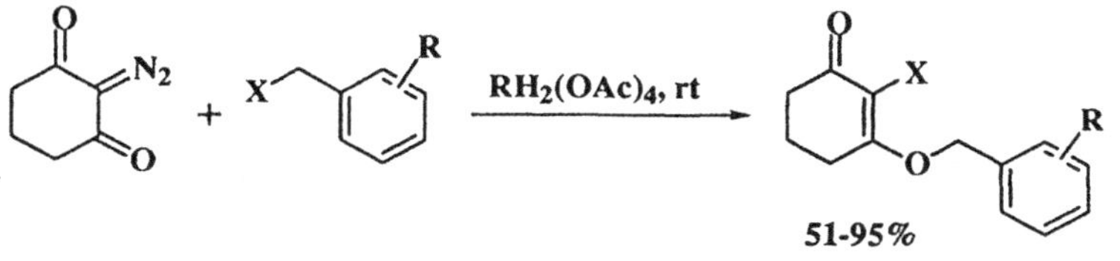

N2
X
R
RH2(OAc)4, rt
X
R
O
51-95%

VI.A.10-5 Prakash, G.K.S., Petasis, N.A. et al., *JOC*, **66**, 633; **see also:** Kelly, J.W. et al., *OL* , **3**, 139.

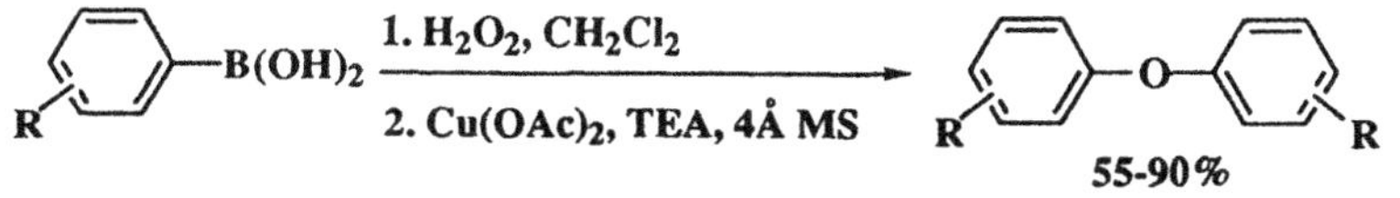

B(OH)2
R
1. H2O2, CH2Cl2
2. Cu(OAc)2, TEA, 4Å MS
O
R
R
55-90%

[a similar route used with N-hydroxy phthalimide to prepare O-arylhydroxylamines]

VI.A.10-6 Buchwald, S.L. et al., *JACS*, **123**, 10770; Parrish, C.A., Buchwald, S.L., *JOC*, **66**, 2498; Gujadhur, R., Venkataraman, D., *SC*, **31**, 2865.

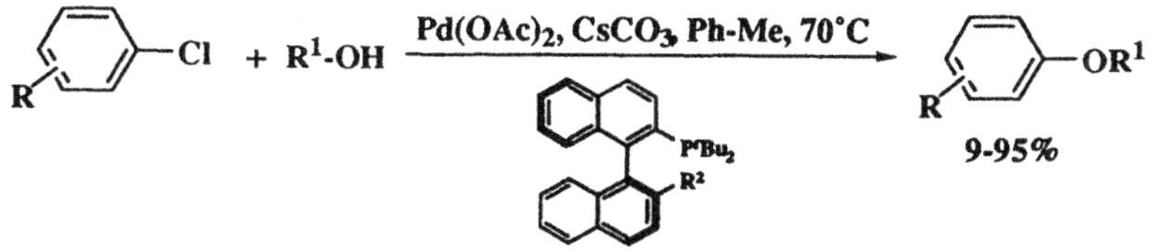

Cl
R
+ R1-OH
Pd(OAc)2, CsCO3, Ph-Me, 70°C
OR1
R
9-95%

[*tert*-butyl or bis aryl ethers prepared similarly]

VI.A.10-7 Blouin, M., Frenette, R., *JOC*, **66**, 9043.

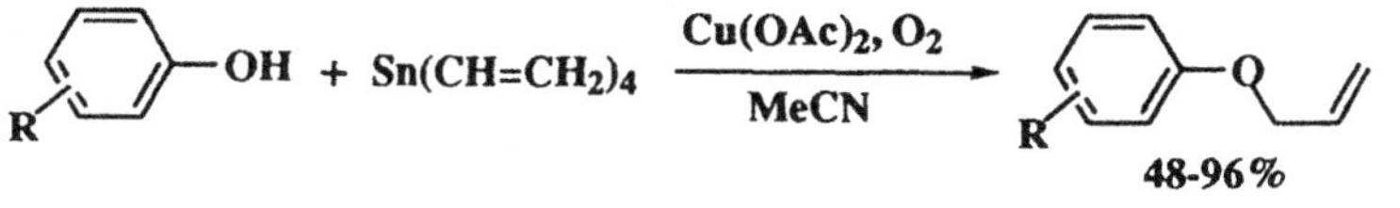

OH
R
+ Sn(CH=CH2)4
Cu(OAc)2, O2
MeCN
O
R
48-96%

VI.A.11. Halides

VI.A.11-1 Shreeve, J.M. et al., *JOC*, **66**, 6263; **see also:** Singh, R.P., Shreeve, J.M., *OL*, **3**, 2713.

R
R1
(MeOCH2CH2)NSF3
CH2Cl2, reflux, 15h
F
R
R1
F
40-80% 1:1

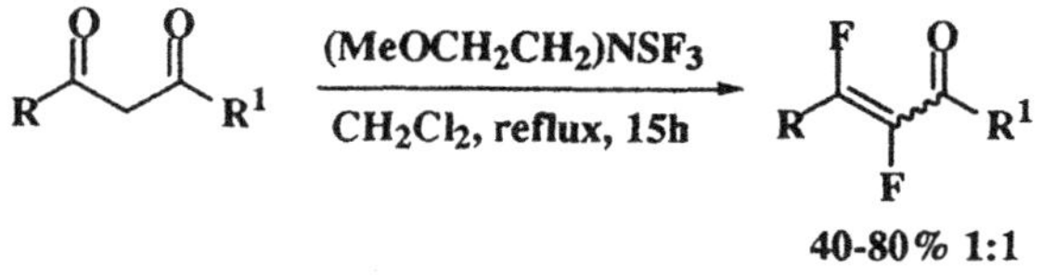

VI.A.11-2 Yoneda, N., Fukuhara, T., *CL*, 222.

$$\text{R-CH}_2\text{OH} \xrightarrow{\quad \text{IF}_5, \text{ TEA·3HF} \quad} \text{R-CH}_2\text{F}$$

32-83%

(examples of fluorination of other functional groups given)

VI.A.11-3 Singh, R.P., Shreeve, J.M., *CC*, 1196.

$$\text{R-NHF} \xleftarrow{\quad 1.2 \text{ Selectfluor}^{\text{TM}} \quad} \text{R-NH}_2 \xrightarrow[\text{DMF or DMA}]{\quad 0.45 \text{ Selectfluor}^{\text{TM}} \quad} \text{R-NF}_2$$

63-85%   72-80%

VI.A.11-4 Rozen, S. et al., *JOC*, **66**, 7464.

$$\xrightarrow[\text{CHCl}_3, \, -45^\circ\text{C}]{\quad \text{AcOF} \quad}$$

60-90%

VI.A.11-5 Konas, D.W., Coward, J.K., *JOC*, **66**, 8831.

$$+ \text{ F-N(SO}_2\text{Ph)}_2 \xrightarrow[\text{THF, -78C}]{\quad \text{LDA} \quad}$$

20-70%

VI.A.11-6 Shibata, N. et al., *JACS*, **123**, 7001; Cahard, D. et al., *AG(E)*, **40**, 4214.

$$\xrightarrow[\text{MeCN/CH}_2\text{Cl}_2, \, -80^\circ\text{C}]{\quad \text{cinchona derivative} \quad}$$

27-99%
e.e. = 1-87%

VI.A.11-7 Fuchigami, T. et al., *JOC*, **66**, 7020, 5633, 7030; Dawood, K.M., Fuchigami, T., *JOC*, **66**, 7691.

$$\xrightarrow[\text{DME or MeCN}]{\quad \text{F}^-, \text{ electrolysis} \quad}$$

24-78%
d.e. = 59-99%

[similar electrolytic fluoronations reported with other substrates]

VI.A.11-8 Kabalka, G.W. et al., *TL*, **42**, 6239.

$$X\text{-}C_6H_4\text{-}CHO \xrightarrow[\text{hex, rt}]{\text{R-BCl}_2} X\text{-}C_6H_4\text{-}CHCl\text{-}R$$

50-92%

VI.A.11-9 Lectka, T. et al., *JACS*, **123**, 1531.

$$R\text{-}COCl + \text{(pentachlorocyclohexadienone)} \xrightarrow[\text{THF, -78°C} \rightarrow \text{rt}]{\text{alkaloid base}} R\text{-}CHCl\text{-}CO_2C_5Cl_5$$

51-80%
e.e. = 80-90%

VI.A.11-10 Iranpoor, N., Firouzabadi, H., Aghapour, G., *SL*, 1176.

$$R\text{-}SH + \text{(N-X succinimide)} \xrightarrow[\text{CH}_2\text{Cl}_2, \text{rt}]{\text{PPh}_3} R\text{-}X$$

30-95%

VI.A.11-11 Roy, S.C. et al., *TL*, **42**, 9253; You, H.-W., Lee, K.-J., *SL*, 105.

$$Ar\text{-}CH=CH\text{-}CO_2H \xrightarrow[]{\text{LiX, CAN}} Ar\text{-}CH=CH\text{-}X$$

42-85%

[similarly with oxone, NaX, Na$_2$CO$_3$]

VI.A.11-12 Yadav, J.S. et al., *SL*, 1417.

$$\text{(bicyclic N-Ts aziridine)}_n \xrightarrow[\text{MeCN, rt}]{\text{InX}_3} \text{(cyclopentane NHTs, X)}_n$$

78-88%

VI.A.11-13 Pizzo, F. et al., *JOC*, **66**, 4463.

$$\text{HO,R}^1,\text{R,CO}_2\text{H,R}^2\text{I} \xleftarrow[\text{H}_2\text{O, pH 4.0}]{\text{NaI}} \text{R}^1,\text{O,R,CO}_2\text{H,R}^2 \xrightarrow[\text{H}_2\text{O, pH 1.5}]{\text{NaI, InCl}_3} \text{I,R}^1,\text{R,CO}_2\text{H,R}^2\text{OH}$$

5-93%
d.s. = 4-99:1

88-95%
d.s. = 49-99:1

VI.A.11-14 Khan, A.T. et al., *CL*, 290.

$$\underset{R}{\overset{O}{\parallel}} \diagdown \diagup R^2 \xrightarrow[\text{CH}_2\text{Cl}_2\text{F, 0°C}\rightarrow\text{rt}]{\text{CetTMATB or TBATB, K}_2\text{CO}_3} \underset{R}{\overset{O}{\parallel}}\diagdown\underset{\text{Br}}{\diagup}R^2$$

65-88%

VI.A.11-15 Jin, Z. et al., *TL*, **42**, 3771.

$$R\!\!=\!\!\!=\!\!\!-SR^1 \xrightarrow[\text{CH}_2\text{Cl}_2/\text{MeOH}]{\text{TMS-X}} \underset{X}{\overset{R}{\diagup}}\diagdown SR^1$$

96-99%
E:Z = 11-33:1

VI.A.11-16 Bandgar, B.P. et al., *TL*, **42**, 951.

$$R\text{-OH} \xrightarrow[\text{dioxane, 25°C}]{\text{KI, BF}_3\cdot\text{Et}_2\text{O}} R\text{-I}$$

0-95%

VI.A.11-17 Kosynkin, D.V., Tour, J.M., *OL*, **3**, 991; Garden, S.J., Lima, E.L.S. et al., *TL*, **42**, 2089.

$$\underset{R}{\diagup\!\!\diagdown}\text{—NH}_2 \xrightarrow[\text{CH}_2\text{Cl}_2/\text{MeOH}]{\text{BnN}^+\text{ET}_3\text{I Cl}^-, \text{NaHCO}_3} \underset{R}{\overset{I}{\diagup\!\!\diagdown}}\text{—NH}_2$$

58-99%

[aryl iodination also with KICL$_2$, H$_2$O]

VI.A.12. Nitriles and Imines

VI.A.12-1 Odom, A.L. et al., *OM*, **20**, 3967, 5011; Beller, M. et al., *JOC*, **66**, 6339.

$$R\!\!=\!\!\!=\!\!\!-R^1 \xrightarrow[\text{Ph-Me, 75°C}]{R^2\text{NH}_2, \text{Ti(NMe}_3)_4} \underset{R}{\overset{NR^2}{\diagdown}}\diagup R^1 + \underset{R}{\overset{NR^2}{\diagdown}}\diagup R^1$$

≤92% ≤100:1

VI.A.12-2 Yamamoto, T. et al., *TL*, **42**, 8653.

$$\text{ClN}\!\!=\!\!\!\underset{R}{\overset{R}{\diagdown\!\!\diagup}}\!\!=\!\!\text{NCl} + \text{Ar-MgBr} \xrightarrow[\text{THF, }\Delta]{\text{NiCl}_2(\text{dppp})} \text{ArN}\!\!=\!\!\!\underset{R}{\overset{R}{\diagdown\!\!\diagup}}\!\!=\!\!\text{NAr}$$

52-88%

VI.A.12-3 Mukaiyama, T. et al., *CL*, 390.

$$R\text{—}CH_2\text{—}\underset{H}{N}\text{—}R^1 \; + \; \underset{Cl}{\overset{Ph}{S}}{=}N^tBu \xrightarrow[\text{CH}_2\text{Cl}_2, -78°C]{\text{DBU}} R\text{—}CH{=}N\text{—}R^1$$

31-99%

VI.A.12-4 Patrick, T.B. et al., *TL*, **42**, 3553.

$$R\text{—}\underset{}{\bigcirc}\text{—}N{=}N\text{—}N\bigcirc \xrightarrow[\text{MeCN}]{\text{Zn(CN)}_2, \text{Zn(ClO}_4)_2} R\text{—}\bigcirc\text{—CN}$$

32-98%

(also aryl iodides by use of ZnI₂)

VI.A.12-5 Beller, M. et al., *TL*, **42**, 6707.

$$R\text{—}\bigcirc\text{—Cl} \xrightarrow[\text{TMEDA, Ph-Me, 160°C}]{\text{KCN, Pd(OAc)}_2, \text{dppe}} R\text{—}\bigcirc\text{—CN}$$

17-96%

VI.A.12-6 Yang, S.H., Chang, S., *OL*, **3**, 4209.

$$R\text{—}CH{=}N\text{—OH} \xrightarrow[\text{MeCN, 80°C}]{[\text{RuCl}_2(p\text{-cymene})]_2, 4\text{Å MS}} \text{R-CN}$$

80-90%

VI.A.12-7 Bose, D.S., Narsaiah, A.V., *S*, 373.

$$R\text{—}\underset{NH_2}{\overset{X}{C}} \xrightarrow[\text{CH}_2\text{Cl}_2]{\text{PyBOP, DIEA}} \text{R-CN}$$

80-95%

VI.A.12-8 Wang, B. et al., *S*, 544.

$$R\text{-CH}_2\text{NH}_2 \xrightarrow[\text{pyr/THF/H}_2\text{O, rt}]{\text{OsO}_4, \text{Me}_3\text{NO}} \text{R-CN}$$

30-60%

VI.A.12-9 Kitano, Y. et al., *S*, 437.

$$R\text{—}\underset{\text{OH}}{\overset{\text{Me}}{\bigcirc}} \xrightarrow[\text{2. aq NaHCO}_3]{\text{1. TMS-CN, AgClO}_2} R\text{—}\underset{N{\equiv}C}{\overset{\text{Me}}{\bigcirc}}$$

80-98%

VI.A.13. Other N-Containing Functional Groups

VI.A.13-1 Bak, R.R., Smallridge, A.J., *TL*, **42**, 6767; Laali, K.K., Volker, J.G., *JOC*, **66**, 35; Suzuki, H. et al., *JOC*, **66**, 4356; Moodie, R.B. et al., *JCS(P2)*, 197

$$R-\text{C}_6\text{H}_5 \xrightarrow[\text{0°C, 1min}]{\text{N}_2\text{O}_5, \text{Fe(acac)}_3} R-\text{C}_6\text{H}_4-\text{NI}_2$$

91-99%

[aromatic nitrations also reported in ionic liquids; with
NO$_2$ O$_3$, FeCl$_3$ or with N$_2$O$_5$, Faujasite]

VI.A.13-2 Ishii, Y. et al., *CC*, 1352,

$$\text{R-H} \xrightarrow{\text{HNO}_3, \text{Phth-OH}} \text{R-NO}_2$$

32-89% conversion
32-95% selecivity

VI.A.13-3 Zolfigol, M.A. et al., *JOC*, **66**, 3619.

$$\text{RR}^1\text{NH} \xrightarrow[\text{crown-NO}^+ \text{H(NO}_3)_2^-]{\text{CH}_2\text{Cl}_2, \text{rt}} \text{RR}^1\text{N-NO}$$

99%

VI.A.13-4 Metz, W.A. et al., *JOC*, **66**, 2509.

$$\underset{R}{\overset{O}{\|}}\text{C}-\text{CH}_2-\underset{R^1}{\overset{O}{\|}}\text{C} \xrightarrow[\text{CH}_2\text{Cl}_2, \text{rt}]{\text{(P)}-\text{SO}_2\text{N}_3, \text{TEA}} \underset{R}{\overset{O}{\|}}\text{C}-\underset{\text{N}_2}{\text{C}}-\underset{R^1}{\overset{O}{\|}}\text{C}$$

63-98%

VI.A.13-5 Katritzky, A.R. et al., *JOC*, **66**, 1043.

$$\underset{R^1}{\overset{R}{>}}\text{C}=\text{O} + \text{Ar-N}=\text{C}=\text{O} \xrightarrow{\text{200°C, 24h}} \underset{R^1}{\overset{R}{>}}\text{C}=\text{N-Ar}$$

71-99%

VI.A.13-6 Musiol, H.-J., Moroder, L., *OL*, **3**, 3859.
**N,N'-Di-*tert*-butoxycarbonyl-1H-benzotriazole-1-carboxamide
Derivatives Are Highly Reactive Guanidinylating Reagents**

VI.A.13-7 Sharghi, H., Sarvari, M.H., *SL*, 99; Bandgar, B.P. et al., *M*, **132**, 403.

$$\text{R-CHO} \xrightarrow[\text{90°C}]{\text{NH}_2\text{OH·HCl, CuSO}_4 \text{ or K}_2\text{CO}_3} \text{R-CH=NOH}$$

80-95%

[solventless, microwave protocols used also to prepare oximes and analogs]

VI.A.13-8 Franck, R.W., Weinreb, S.M. et al., *SL*, 232.

$$\underset{R^1}{\overset{R}{>}}\!\!=\!\!\text{NOH} \xrightarrow[\text{Et}_2\text{O, -35°C} \rightarrow \text{rt}]{R^2\text{SO}_2\text{Cl, TEA}} \underset{R^1}{\overset{R}{>}}\!\!=\!\!\text{N-SO}_2R^2$$

61-79%

VI.A.13-9 Norton, J.R. et al., *OM*, **20**, 254.

$$\text{R}\diagdown\!\!\!\diagup\text{NHR}^1 \xrightarrow{\substack{\text{1. BuLi, THF} \\ \text{2. Cp}_2\text{ZrMe(OTf), THF} \\ \text{3. R}^2\text{N=C=N}^2 \\ \text{4. HCl or H}_2\text{O}}} \underset{R}{R^2\text{HN}}\diagup\!\!\overset{NR^2}{\diagdown}\text{NHR}^1$$

35-91%

VI.A.13-10 Katritzky, A.R. et al., *JOC*, **66**, 2865, 2854.

$$R^1\diagdown\!\!\underset{N}{\overset{Bt}{>}}\!\!\underset{R}{\diagup}\text{NR}^3R^4 \xrightarrow[\text{THF, reflux}]{R^2\text{-SH, NaOMe}} R^1\diagdown\!\!\underset{N}{\overset{SR^2}{>}}\!\!\underset{R}{\diagup}\text{NR}^3R^4$$

44-92%

[similarly for preparation of thioimidates]

VI.A.13-11 Vallee, Y. et al., *SL*, 1281.

$$\text{R-CHO} + R^1\text{-NO}_2 \xrightarrow[\text{EtOH, 0°C} \rightarrow \text{rt}]{\text{Zn, AcOH}} R\diagdown\!\!\overset{}{\underset{O^-}{N^+}}\!\!\diagup R^1$$

45-96%

VI.A.13-12 Kohmura, Y., Katsuki, T., *TL*, **42**, 3339.

$$\text{+ Ph-I=NTs} \xrightarrow[\text{4Å MS, CH}_2\text{Cl}_2, \text{5°C}]{\text{Salen Mn(II) complex}}$$

71%
e.e. = 89%

(best of several examples)

VI.B. Additions to Alkenes and Alkynes

VI.B-1 Sun, Y. et al., *TL*, **42**, 8603.
Harvesting Short-Lived Hypoiodous Acid for Efficient Diastereoselective Iodohydroxylation in Crixivan® Synthesis

---

VI.B-2 Dolensky, B., Kirk, K.L., *JOC*, **66**, 4687; Hara, S. et al., *SL*, 1938.

71-80%

[iodofluorination reported with electrochemically generated iodonium cation]

VI.B-3 Nair, V. et al., *T*, **57**, 7417.

51-95%

VI.B-4 Tanaka, M. et al., *JACS*, **123**, 2899; **see also:** Tanaka, T. et al., *OL*, **3**, 3627.

$$R\!\!\equiv\!\!\ + PhS\text{-}CO_2Me \xrightarrow[\text{Ph-Me/octane, 110°C}]{Pd(PCy_3)_2}$$

32-96%

VI.B-5 Arisawa, M., Yamaguchi, M., *OL*, **3**, 763; Gareau, Y. et al., *T*, **57**, 5739.

$$R\!\!\equiv\!\!\ + (BuS)_2 \xrightarrow[\text{Me}_2\text{CO, reflux, 10h}]{RhH(PPh_3)_4, P(Ph\text{-}OMe)_3, CF_3SO_3H}$$

62-99%

VI.B-6 Liepins, V., Backvall, J.-E., *CC*, 265; Gonzalez-Nogal, A.M. et al., *JOC*, **66**, 1961.

1. DMPS-CuCNLi
2. E⁺

44-73%

VI.B-7 Tiecco, M. et al., *SL*, 1767 and *TA*, **12**, 1493.

45-98%

VI.B-8 Fokin, V.V., Sharpless, K.B., *AG(E)*, **40**, 3455; Joullie, M.M. et al., *JOC*, **66**, 7223; **see also:** Li, G. et al., *T*, **57**, 8407.

88-96% 1.6-3:1

VI.B-9 Zwierzak, A., Klepacz, A., *TL*, **42**, 4539.

55-83%

VI.B-10 Shibasaki, M. et al., *JACS*, **123**, 1256.

53-99%

VI.C. Boron Compounds

VI.C-1 Yang, F.-Y., Cheng, C.-H., *JACS*, **123**, 761.

52-93%
E:Z = 1:13.3-19

VI.C-2 Shimizu, M., Hiyama, T. et al., *AG(E)*, **40**, 4283, 790.

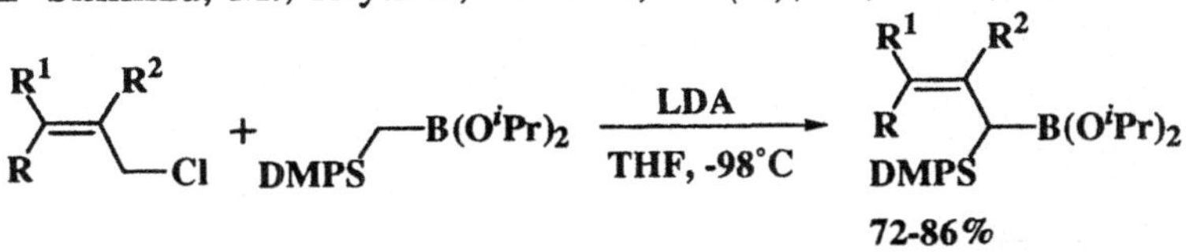

72-86%

VI.C-3 Le Floch, P. et al., *JOM*, **640**, 197; Ishiyama, T., Ishida, K., Miyaura, N., *T*, **57**, 9813; **see also:** Song, Y.-L., Morin, C., *SL*, 266.

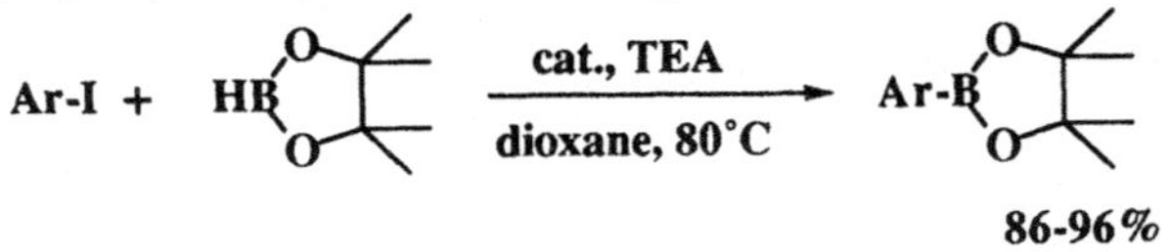

86-96%

[boronic acids prepared similarly]

VI.C-4 Vedso, P. et al., *OL*, **3**, 1435.

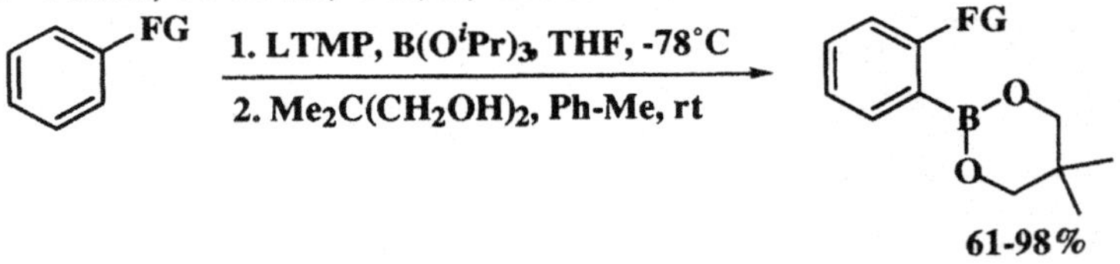

1. LTMP, B(O^{i}Pr)$_3$, THF, -78°C
2. Me$_2$C(CH$_2$OH)$_2$, Ph-Me, rt

61-98%

VI.C-5 Falck, J.R. et al., *JOC*, **66**, 7148.

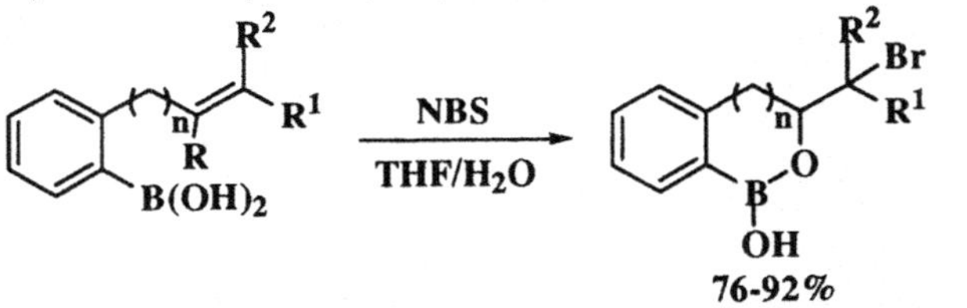

NBS
THF/H$_2$O

76-92%

VI.D. Nucleotides and Related Compounds

VI.D-1 Janeba, Z. et al., *CCC*, **66**, 517.
   Synthesis of 8-Amino- and N-Substituted 8-Aminoadenine Derivatives of Acyclic Nucleoside and Nucleotide Analogs

---

VI.D-2 Olgen, S., Chu, C.K., *Z*, **56b**, 804.
   Synthesis and Antiviral Activity of 2'-Deoxy-2'-fluoro-L-arabinofuranosyl-1,2,3-triazole Derivatives

VI.D-3 Gotor, V. et al., *JOC*, **66**, 4079.

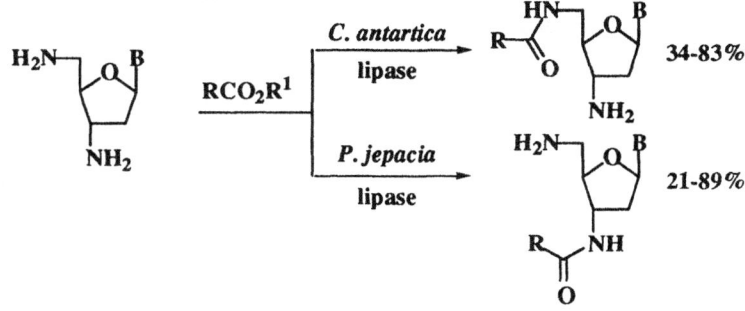

VI.D-4 Chattopadhyaya, J. et al., *JOC*, **66**, 6560.
Total Synthesis of 2',3',4',5',5"-^{2}H$_5$-Ribonuclosides: The Key Building Blocks for NMR Structure Elucidation of Large RNA

VI.D-5 Nielsen, P. et al., *JOC*, **66**, 4878; Wengel, J. et al., *JOC*, **66**, 5106; Kværno, L., Wengel, J., *JOC*, **66**, 5498; Makara, G.M. et al., *JOC*, **66**, 5783; Koshkin, A.A. et al., *JOC*, **66**, 8504.
Bicyclic Nucleosides

VI.D-6 Cushman, M. et al., *JOC*, **66**, 8320.
Design, Synthesis and Evaluation of 9-D-Ribityl-1,3,7-trihydro-2,6,8-purinetrione, a Potent Inhibitor of Riboflavin Synthase

VI.D-7 Gurjar, M.K., Maheshwar, K., *JOC*, **66**, 7552; Herdewijn, P. et al., *JOC*, **66**, 8478; Chu, C.K. et al., *JOC*, **66**, 4852; Hrebabecky, H., Holy, A., *CCC*, **66**, 785; Mehta, G. et al., *TL*, **66**, 7663.
Carboxylic Nucleosides

VI.D-8 Umezawa, K. et al., *JCS(P1)*, 298.
Synthesis of Sugar-Modified Derivatives of the Unususal Nucleoside Oxanosine and its Carbocyclic Analogs as Potential Inhibitors of HIV

VI.D-9 Izawa, K. et al., *TL*, **42**, 7605.
Radical Deoxygenation and Dehalogenation of Nucleoside Derivatives with Hypophosphorous Acid and Dialkyl Phosphites

VI.D-10 Liu, F., Austin, D.J., *JOC*, **66**, 8643 and *OL*, **3**, 2273.
A General Synthesis of 5'-Azido-5'-deoxy-2',3'-O-isopropylidene Nucleosides

---

VI.D-11 Veliz, E.A., Beal, P.A., *JOC*, **66**, 8592.
6-Bromopurine Nucleosides as Reagents for Nucleoside Analogue Synthesis

---

VI.D-12 Pirrung, M.C. et al., *JOC*, **66**, 2067.
A Universal, Photocleavable DNA Base: Nitropiperonyl 2'-Deoxyriboside

---

VI.D-13 Izawa, K. et al., *JOC*, **66**, 7469.
Synthesis of 9-(2,3-Dideoxy-2-fluoro-*p*-D-*threo*-penofuranosyl)adenine (fddA) *via* a Purine 3'Deoxynucleoside

---

VI.D-14 Marcotte, S., Quirion, J.-C. et al., *S*, 929.
Synthesis of 3'-Deoxy-3'-difluoromethyluridine and 2'-Deoxy-2'-difluormethyluridine

---

VI.D-15 Tyler, P.C. et al., *JOC*, **66**, 5723.

41-62%

VI.D-16 Hocek, M. et al., *CCC*, **66**, 483.
Synthesis and Structure-Activity Relationship Study in Cytostatic Activity of 6-Aryl, 6-Heteraryl-, and 6-Benzylpurine Ribonuclesides

---

VI.D-17 Manoharan, M. et al., *TL*, **42**, 8777; Kumar, V.A. et al., *T*, **57**, 1311; Pedroso, E. et al., *T*, **57**, 179; Seela, F., Debelak, H., *JOC*, **66**, 3303; An, H. et al., *JOC*, **66**, 2789.
Oligonucleotide Synthesis

VI.D-18 Aso, M., Suemune, H. et al., *JOC*, **66**, 3513.
Molecular Design of a New Class of Spin-Labeled Ribonucleosides with N-*t*-Butylaminoxyl Radicals

---

VI.D-19 Jung, M.E., Toyota, A., *JOC*, **66**, 2624.
Preparation of 4'-Substituted Thymidines by Substitution of the Thymidine S^1-Esters

---

VI.D-20 Shuto, S., Matsuda, A. et al., *ECJ*, **7**, 2332.
1'α-Branched-Chain Sugar Pyrimidine Ribonucleosides from Uridine

---

VI.D-21 Rao, P., Benner, S.A., *JOC*, **66**, 5012.
Fluorescent Charge-Neutral Analogs of Xanthosine: Synthesis of a 2'-Deoxyribonucleoside Bearing a 5-Aza-7-deazaxanthine Base

---

VI.D-22 Marlow, A.L., Kiessling, L.L., *OL*, **3**, 2517.
Improved Chemical Synthesis of UDP-Galactofuranose

VI.D-23 Kraszewski, A. et al., *TL*, **42**, 8055.
A New Method for the Synthesis of Nucleoside 2',3'-O,O-Cyclic Phosphorodithioates *via* Aryl Cyclic Phosphites as Intermediates

---

VI.D-24 Lu, Y., Just, G., *T*, **57**, 1677.
Stereoselective Synthesis of Dithymidine Phosphorothioates Using D-Xylose Derived Chiral Auxiliaries

VI.E Phosphorus, Selenium and Tellurium Compounds.

VI.E-1 Guilleman, J.-C., Janati, T., Denis, J.-M., *JOC*, **66**, 7864.

$$\underset{Cl}{\overset{R}{\diagdown}}\underset{Cl}{\overset{PH_2}{\diagup}} \quad \xrightarrow[\text{Et}_2\text{O, -60°C}]{\text{DBU}} \quad \text{R-C}\equiv\text{P}$$
20-81%

VI.E-2 Gauvry, N., Mortier, J., *S*, 553.

$$\underset{}{\overset{O}{\overset{\|}{R-P(OR^1)_2}}} \quad \xrightarrow[\text{2. MeOH}]{\text{1. BBr}_3\text{, Ph-Me/hex, -30} \rightarrow 70°\text{C}} \quad \underset{}{\overset{O}{\overset{\|}{R-P(OH)_2}}}$$
77-95%

VI.E-3 Deprele, S., Montchamp, J.-L., *JOC*, **66**, 6745; **see also:**
Montchamp, J.-L., Dumond, Y.R., *JACS*, **123**, 510; Mirzaei, F., Han, L.-
B., Tanaka, M., *TL*, **66**, 297; Han, L.-B., Zhao, C.-Q., Tanaka, M., *JOC*,
66, 5929.

$$MO-PH_2 \quad + \quad \diagup\!\!\!\diagdown R \quad \xrightarrow{BEt_3, MeOH} \quad \underset{MO}{\overset{H}{\diagdown}}\!\!\overset{O}{\underset{}{\overset{\|}{P}}}\!\!\diagup\!\!\diagdown R$$

10-92%

[similar additions used for the preparation of phosphonates or
vinyl phosphine oxides using PdMe₂(dppb) or RuBr(PPh₃)₃ catalysts]

VI.E-4 Stelzer, O. et al., *TA*, **12**, 1159.

$$\underset{PPh_2}{\overset{P(NMe_2)_2}{\diagup}} \quad \xrightarrow[\text{heptane, reflux}]{R^*\text{-OH}} \quad \underset{PPh_2}{\overset{P(OR^*)_2}{\diagup}}$$

64-87%

VI.E-5 Timperley, C.M. et al., *JCS(P1)*, 26.

$$P(OMe)_3 \xrightarrow[\text{6-64\%}]{Na, R\text{-OH}} \underset{RO}{\overset{OMe}{\overset{\|}{P}}}OMe \xrightarrow[\text{66-95\%}]{Me\text{-I, }\Delta} \underset{RO}{\overset{O}{\overset{\|}{P}}}\overset{Me}{OMe} \xrightarrow[\text{2. MeOH}]{1.\ TMS\text{-Br}} \underset{RO}{\overset{Me}{\diagdown}}PO_2H$$

60-97%

VI.E-6 Lee, S. et al., *CC*, 1698.

$$Ar\text{-H} + Ph\text{-NH}_2 \xrightarrow[\underset{Me^-N\diagdown\diagup N-Bu}{}\ PF_4^-]{MeP(O)(OEt)_2,\ Sc(OTf)_3} \quad \underset{Ar}{\overset{PhHN}{\diagdown}}\!\!\underset{O}{\overset{}{\overset{\|}{P}}}(OEt)_2$$

90-99%

VI.E-7 Yadav, J.S. et al., *S*, 2277 and *SL*, 1131; Kaboudin, B., Nazari,
R., *TL*, **42**, 8211.

$$R\text{-CH=N-R}^1 + HO\text{-P(OEt)}_2 \xrightarrow[\text{MeCN, rt}]{ZrCl_4} \quad \underset{R}{\overset{R^1HN}{\diagdown}}\!\!\underset{O}{\overset{}{\overset{\|}{P}}}(OEt)_2$$

78-93%

[similarly with different catalysts]

VI.E-8 Chan, K.S. et al., *TL*, **42**, 4883.

$$\underset{R}{\diagup\!\!\!\diagdown}\!\!-X \xrightarrow[\text{DMF, 160-165°C}]{PPh_3, Pd/C} \underset{R}{\diagup\!\!\!\diagdown}\!\!-PPh_2$$

16-53%

VI.E-9 Davis, F.A. et al., *OL*, **3**, 1757.

71-97%
d.e. = 64-95%

VI.E-10 Tedeschi, L., Enders, D., *OL*, **3**, 3515.
Asymmetric Synthesis of β-Phosphono Malonates *via* Fe₂O₃-Mediated Phospha-Michael Addition to Knoevenagel Acceptors

---

VI.E-11 Kaboudin, B., Balakrishna, M.S., *SC*, **31**, 2773; Zhou, T., Chen, Z.C., *SC*, **31**, 3289.

$$R\text{-Cl} + P(OR^1)_3 \xrightarrow[\text{neat}]{Al_2O_3, \mu W} R\text{--}P(OR^1)_2$$

70-90%

[aryl phosphonates prepared from Ar₂IBF₄, HP(O)(OR)₂, Pd(PPh₃)₄]

VI.E-12 Keay, B.A. et al., *JOC*, **66**, 7478.
A Simple Resolution Procedure Using the Staudinger Reaction for the Preparation of P-Stereogenic Phosphine Oxides

---

VI.E-13 Juge, S. et al., *TA*, **12**, 1441.

85-94%
d.e. = 45-99%

VI.E-14 Masson, S. et al., *S*, 1623.

40-91%

VI.E-15 Ishihara, H., Koketsu, M. et al., *JACS*, **123**, 8408.
Reaction of Lithium Aluminum Hydride with Elemental Selenium: Its Application as a Selenating Reagent into Organic Molecules

VI.E-16 Schmitt, A., Reissig, H.-U., *S*, 867.

$$R \overset{O}{\underset{}{\diagdown}} O \xrightarrow[\text{2. Ph-SeH, BF}_3\text{·Et}_2\text{O}]{\text{1. }^i\text{Bu}_2\text{AlH}} R \overset{O}{\underset{}{\diagdown}} SePh$$

83-91%

VI.F. Silicon Compounds.

VI.F-1 Bienz, S. et al., *OM*, **20**, 1849.
New Silicon Groups as Potential Chiral Auxilliaries. Synthesis and Highly Selective Chiral 1,6-Induction in 1,2-Additions to Acylsilanes

---

VI.F-2 Takaki, K. et al., *TL*, **42**, 9211; Kayran, C., Rouzi, P., *Z*, **56b**, 1138; Perales, J.B., van Vranken, D.L., *JOC*, **66**, 7270.

$$\overset{R^1}{\underset{R}{\diagup}}\overset{R^2}{=} + \text{Ph-SiH}_3 \xrightarrow[\text{THF, rt}]{\text{Ph(Bn)N-YbNPh}_2\text{(hmpa)}} \overset{R^1 \quad R^2}{\underset{R \quad \text{SiH}_2\text{Ph}}{\diagup}}$$

35-95%

[hydrosilylation also reported with triethylsilane,
tricarbonyl(*o*-xylene) metal complexes or with PhMe₂SiH, H₂PtCl₆]

VI.F-3 Lee, T.W., Corey, E.J., *OL*, **3**, 3337.
(2-Methoxyphenyl)dimethylsilyl Lithium and Cuprate Reagents Offer Unique Advantages in Multistep Synthesis

---

VI.F-4 Purido, F.J. et al., *CC*, 1606.
A Tandem Allylsilane-Vinylsilane Difunctionalization by Silylcupration of Allene Followed by Reaction with α,β-Unsaturated Nitriles

---

VI.F-5 Yoo, B.R., Jung, I.N. et al., *JOM*, **631**, 36.
Friedel-Crafts Alkylation of Benzene with (Polychloromethyl)silanes

---

VI.F-6 Manoso, A.S., Deshong, P., *JOC*, **66**, 7449.

$$\text{Ar-X} + (\text{EtO})_3\text{SiH} \xrightarrow[\text{NMP}]{\text{Pd(dba)}_2, \text{P}^t\text{Bu}_2\text{(Ph-Ph)}, {}^i\text{Pr}_2\text{NEt}} \text{Ar-Si(OEt)}_3$$

0-92%

VI.G. Sulfur Compounds

VI.G-1 Atwood, D., Laitinen, R., guest eds., *JOM*, **623**, 1-201.
Special Issue: Group 16 Chemisrtry

VI.G-2 Node, M. et al., *OL*, **3**, 3121 and *TL*, **42**, 9207.

1. CH_2Cl_2, rt,
2. $BF_3 \cdot Et_2O$, CH_2Cl_2, rt
3. $^nC_{12}H_{25}$-SH

74-91%
e.e. = 98%

VI.G-3 Cai, Y., Roberts, B.P., *TL*, **42**, 8235.

R-CHO + Ph_3Si-SH $\xrightarrow{\text{imidazole, } CH_2Cl_2}$

70-80%

VI.G-4 Mukaiyama, T. et al., *CL*, 638.

CuO, $TiCl_4$, Ph-Me, 0°C, 1h
or CuO, TiF_4, CH_2Cl_2, rt, 12h

58-99%

VI.G-5 Zaragoza, F. et al., *T*, **57**, 5451.

R-CH_2OH + HS-R^1 $\xrightarrow[\text{EtCN, 90°C, 2-15h}]{Me_3P^+CH_2CN \ I^-, \ DIPEA}$

38-84%

VI.G-6 Kobayashi, S. et al., *SL*, 983; **see also:** Kamimura, A. et al., *TL*, **42**, 8497.

$Hf(OTf)_4$ CH_2Cl_2, 4Å MS, 0°C

53-92%
e.e. = 43-94%

VI.G-7 Kita, Y. et al., *JOC*, **66**, 2434.

$$\text{TMS-OTf}, \quad \text{MeCN}, -30°C \to rt, 10min$$

18-99%

VI.G-8 Zhong, P., Guo, M.-P., *JCR(S)*, 370.

$$R\!\!\equiv\!\!-SeR^1 \xrightarrow[\text{2. } R^2\text{-SCl}]{\text{1. } Cp_2Zr(H)Cl, THF, rt}$$

69-82%

VI.G-9 Sharghi, H. et al., *JOC*, **66**, 7287; Renard, P.-Y., Mioskowski, C. et al., *TL*, **42**, 8479.

$$\xrightarrow{NH_4SCN, \text{ macrocyclic diamide}}$$

0-93%

[thiocyanates prepared also from R-Br, TMS-NCS, Bu₄NF]

VI.G-10 Salunkhe, M.M. et al., *SC*, **31**, 3041; **see also:** Prakash, O., Moriarty, R.M. et al., *JOC*, **66**, 2019.

$$H_2N-\!\!\!\bigcirc\!\!\!- + NH_4SCN \xrightarrow{NaBO_3 \cdot H_2O, AcOH} H_2N-\!\!\!\bigcirc\!\!\!-SCN$$

0-95%

VI.G-11 Naso, F. et al., *JOC*, **66**, 5933.
Highly Stereoselective Route to Dialkyl Sulfoxides Based on Sequential Displacement of Oxygen and Carbon Leaving Groups by Grignard Reagents on Carbon Leaving Groups by Reagents on Sulfinyl Compounds

VI.G-12 Olah, G.A., Mathew, T., Prakash, G.K.S., *CC*, 1696; Dubac, J. et al., *JOC*, **66**, 421; Salunkhe, M.M. et al., *JOC*, **66**, 8616; Zhang, J., Zhang, Y., *JCR(S)*, 516; Steensma, R.W. et al., *TL*, **42**, 2281.

$$R\text{-}SO_3H + Ar\text{-}H \xrightarrow{\text{Nafion-H, reflux}} RSO_2\text{-}Ar$$

30-82%

[sulfones also prepared from arenes and sulfonyl chlorides with FeCl₃ under microwave heating or in an ionic liquid; *via* R-X, Ar-SO₂Cl, Sm/HgCl₂ or Ar-SO₂CF₃, RMgX]

VI.G-13 Karade, N.N. et al., *SL*, 1573.

$$R\text{—}\langle\!\!\bigcirc\!\!\rangle + SOCl_2 \xrightarrow[\text{CHCl}_3]{\text{montmorillonite K-10}} R\text{—}\langle\!\!\bigcirc\!\!\rangle\text{—}\overset{\displaystyle O}{\underset{}{S}}\text{—Cl}$$

61-91%

VI.G-14 Katsuki, T. et al., *TL*, **42**, 7071; Marzinzik, A.L., Sharpless, K.B., *JOC*, **66**, 594.

$$Ar^{\diagdown S}\diagup R + Ts\text{-}N_3 \xrightarrow{\text{4Å MS, CH}_2\text{Cl}_2,\ rt} Ar\diagup\overset{\displaystyle N^-Ts}{\underset{}{S^+}}\diagdown R$$

36-99%
e.e. = 66-99%

[N-tosylsulfinimines prepared also by reaction of sulfides
with chloramine T in MeCN]

VI.G-15 Percel, V. et al., *JOC*, **66**, 2104.
Synthesis of Functional Aromatic Multisulfonyl Chlorides and their Masked Precursors

---

VI.G-16 Katohgi, M., Togo, H., *T*, **57**, 7481.

$$R\text{-}SO_2NHCH_2R^1 \xrightarrow{\text{PhI(OAc)}_2,\ I_2,\ (((\bullet}} R\text{-}SO_2NH_2$$

42-81%

VI.G-17 Benicewicz, B.C. et al., *TL*, **42**, 3791.

$$\overset{\|}{\underset{Ph}{}} \xrightarrow[\text{2. Lawesson's reagent, Ph-Me, reflux}]{\text{1. Ph-COSH, montmorillonite K-10, Ph-H, reflux}} Ph\diagup\overset{\displaystyle Me}{\underset{}{}}\diagdown S\overset{\displaystyle S}{\underset{}{}}Ph$$

70%

VI.G-18 Harpp, D.N. et al., *TL*, **42**, 8607.

$$Ph_3CSCl + (RR^1N\text{-}S)_2 \longrightarrow (RR^1N\text{-}S)_2S$$

75-85%

VI.H. Tin Compounds

VI.H-1 Capperucci, A., Degl'Innocenti, A. et al., *T*, **57**, 6267.

51-83%

VI.H-2 Maleczka, R.E., Jr. et al., *S*, 1495.
The Regiochemical Influence of Oxo-Substitution in Palladium-Mediated Hydrostannations of 1-Alkynes

---

VI.H-3 Kocienski, P. et al., *S*, 331.

91%

VI.H-4 Pippel, D.J. et al., *JACS*, b123, **4919**.

79%
e.r. = 15.7:1

VII
REVIEWS

VII.A Techniques

VII.A-1 Anslyn, E.V. et al., *ACR*, **34**, 963.

Review: "Teaching Old Indicators New Tricks."

VII.A-2 Bierbaum, V.M., guest ed., *CRV*, **101**, 209-606.

Reviews: "Frontiers in Mass Spectrometry."

VII.A-3 Riguera, R. et al., *TA*, **12**, 2915.
Review: "A Practical Guide for the Assignment of the Absolute Configuration of Alcohols, Amines and Carboxylic Acids by NMR."

VII.A-4 Lloyd-Jones, G.C., *SL*, 161.
Account: "Isotopic Desymmetrisation as a Stereochemical Probe."

VII.A-5 Williams, C.M., Mander, L.N., *T*, **57**, 425.
Report: "Chromatography with Silver Nitrate."

VII.A-6 Curran, D.P., *SL*, 1488.
New Tools "Fluorous Reverse Phase Silica Gel. A New Tool
Synthesis: for Preparative Separations in Synthetic Organic and Organofluorine Chemistry."

VII.A-7 Fuchs, P.L., *T*, **57**, 6855
Report: "Increase in Intricacy—A Tool for Evaluating Organic Syntheses."

VII.A-8 Rowlands, G.J., *T*, **57**, 1865.
Report: "Ambifunctional Cooperative Catalysts."

VII.A-9 Jarowicki, K., Kocienski, P., *JCS(P1)*, 2109.
Review: "Protecting Groups."

VII.A-10 Lidstrom, P. et al., *T*, **57**, 9225.
  Report: "Microwave Assisted Organic Synthesis—A Review."

VII.A-11 Perreux, L., Loupy, A., *T*, **57**, 9199.
  Report: "A Tentative Rationalization of Microwave Effedts in Organic Synthesis According to the Reaction Medium and Mechanistic Considerations."

VII.A-12 Ogibin, Y.N., Nikishin, G.I., *RCR*, **70**, 543.
  Review: "Electrochemical Reactions of Alkenes Induced by Anodic Oxidation and their Applications in Organic Synthesis."

VII.A-13 Kadereit, D., Waldmann, H., *CRV*, **101**, 3367.
  Review: "Enzymatic Protecting Group Techniques."

VII.A-14 Hansch, C. et al., *CRV*, **101**, 2727.
  Review: "Comparative QSAR: Angiotensin II Antagonists."

VII.A-15 Hansch, C. et al., *CRV*, **101**, 619.
  Review: "Chem-Bioinformatics and QSAR: A Review of QSAR Lacking Positive Hydrophobic Terms."

VII.A-16 Roberts, S.M., *JCS(P1)*, 1475.
  Review: "Preparative Biotransformations."

VII.A-17 Davis, B.G., Jones, J.B. et al., *TA*, **12**, 249.
  Article: "Expanding the Utility of Proteases in Synthesis: Broadening the Substrate Acceptance in Non-Coded Amide Bond Formation Using Chemically Modified Mutants of Subtilisin."

VII.A-18 Sood, A., Panchagnula, R., *CRV*, **101**, 3275.
  Review: "Preoral Route: An Opportunity for Protein and Peptide Drug Delivery."

VII.A-19 Amery, G., *AA*, **34**, 61.
  Review: "Reactive Chemical Hazard Evaluation in the Scale-Up of Chemical Processes."

VII.A-20 Fulop, F., conference ed., *PAC*, **73**, 1387-1509.
Lectures: "Hungarian-German-Italian-Polish Joint Meeting on Medicinal Chemistry, Budapest, Hungary, September 2001."

VII.A-21 Clark, J.H., symposium ed., *PAC*, **73**, 77-203.
Lectures: "International Symposium on Green Chemistry."

VII.A-22 Hjeresen, D.L., conference ed., *PAC*, **73**, 1229-1330.
Lectures: "IUPAC CHEMRAWN XIV Conference on Green Chemistry: Toward Environmentally Benign Processes and Products, Boulder, Colorado, June 2001."

VII.A-23 Thomas, J.M. et al., *PAC*, **73**, 1087.
Lecture: "Benign by Design. New Catalysts for an Environmentally Conscious Age."

VII.A-24 Sheldon, R., *CC*, 2399.
Feature Article: "Catalytic Reactions in Ionic Liquids."

VII.A-25 de Miguel Y. et al., *JCS(P1)*, 3085.
Review: "Supported Catalysts and their Applications in Synthetic Organic Chemistry."

VII.A-26 Engberts, J.B.F.N., Blandamer, M.J., *CC*, 1701.
Feature Article: "Understanding Chemical Reactions in Water: From Hydrophobic Encounters to Surfactant Aggregates."

VII.A-27 Siskin, M., Katritzky, A.R., *CRV*, **101**, 825.
Review: "Reactivity of Organic Compounds in Superheated Water: General Background."

VII.A-28 Katritzky, A.R. et al., *CRV*, **101**, 837.
Review: "Reactions in High Temperature Aqueous Media."

VII.A-29 Bren, V.A., *RCR*, **70**, 1017.
Review: "Flourescent and Photochromic Chemosensors."

VII.A-30 Banerjee, A.K. et al., *RCR*, **70**, 971.
Review: "Silica Gel in Organic Synthesis."

VII.A-31 Raston, C.L. et al., *CC*, 2159.
Feature   "**Recent Advances in Solventless Organic Reactions:**
Article: **Towards Benign Synthesis with Remarkable**
          Versatitility."

VII.B Asymmetric Synthesis and Molecular Recognition

VII.B-1 Christoffers, J., Mann, A., *AG(E)*, **40**, 4591.
Minieview: "Enatioselective Construction of Quaternary
Stereocenters."

VII.B-2 Kim. Y.H., *ACR*, **34**, 955.
Review: "Dual Enantioselective Control in Asymmetric
          Synthesis."

VII.B-3 Maruoka, K., guest ed., *T*, **57**, 805-913.
Symposium "Lewis Acid Control of Asymmetric Synthesis."
in Print:

VII.B-4 O'Brien, P., *JCS(P1)*, 95.
Review: "Stoichiometric Asymmetric Processes."

VII.B-5 Tye, H., Comina, P.J., *JCS(P1)*, 1729.
Review: "Catalytic Asymmetric Processes."

VII.B-6 Shibasaki, M., Kanai, M., *CPB*, **49**, 511.
Review: "Multifunctional Asymmetric Catalysis."

VII.B-7 Dalko, P.I., Moisan, L., *AG(E)*, **40**, 3726.
Review: "Enantioselective Organocatalysis."

VII.B-8 List, B., *SL*, 1675.
Account: "Asymmetric Aminocatalysis."

VII.B-9 Kondepudi, D.K., Asakura, K., *ACR*, **34**, 946.
Review: "Chiral Autocatalysis, Spontaneous Symmetry
          Breaking, and Stochastic Behavior."

VII.B-10 Noyori, R., Ohkuma, T., *AG(E)*, **40**, 40.
Review: "Asymmetric Catalysis by Architectural and
          Functional Molecular Engineering: Practical Chemo-
          and Stereoselective Hydrogenation of Ketones."

VII.B-11 Backvall, J.-E. et al., *CSR*, **30**, 321.
Review: "Racemisation in Asymmetric Synthesis. Dynamic Kinetic Resolution and Related Processes in Enzyme and Metal Catalysis."

VII.B-12 Bolm, C. et al., *AG(E)*, **40**, 3284.
Review: "Catalyzed Asymmetric Arylation Reactions."

VII.B-13 Cardona, F., Goti, A., Brandi, A., *EJOC*, 2999.
Microreview: "Kinetic Resolutions by Means of Cycloaddition Reactions."

VII.B-14 Backvall, J.-E. et al., *CSR*, **30**, 321.
Review: "Racemisation in Asymmetric Synthesis. Dynamic Kinetic Resolution and Related Processes in Enzyme and Metal Catalysis."

VII.B-15 Mikami, K., Maryanoff, B.E. et al., *T*, **57**, 2917.
Report: "Acyclic Stereocontrol Between Remote Atom Centers *via* Intramolecular and Intermolecular Stereo-Communication."

VII.B-16 McCarthy, M., Guiry, P.J., *T*, **57**, 3809.
Report: "Axially Chiral Bidentate Ligands in Asymmetric Catalysis."

VII.B-17 Plaquevent, J.-C., Cahard, D. et al., *S*, 1742.
Special Topic: "Solution and Solid-Phase Approaches in Asymmetric Phase-Transfer Catalysis by Cinchona Alkaloid Derivatives."

VII.B-18 Kacprzak, J., Gawronski, J., *S*, 961.
Review: "Cinchona Alkaloids and their Derivatives: Versatile Catalysts and Ligands in Asymmetric Synthesis."

VII.B-19 Groger, H., Wilken, J., *AG(E)*, **40**, 529.
Review: "The Application of L-Proline as an Enzyme Mimic and Further New Asymmetric Syntheses Using Small Organic Molecules as Chiral Catalysts."

VII.B-20 Vedejs, E. et al., *SL*, 1499.
Account: "Enantioselective Acyl Transfer Using Chiral Phosphine Catalysts."

VII.B-21 Spivey, A.C., Andrews, B.I., *AG(E)*, **40**, 3131.
Highlights: "Catalysis of the Asymmetric Desymmerization of
Cyclic Anhydrides by Nucleophilic Ring-Opening
with Alcohols."

VII.B-22 Seebach, D. et al., *AG(E)*, **40**, 93.
Review: "Taddols, their Derivatives and Taddol Analogues:
Versatile Chiral Auxiliaries."

VII.B-23 Alcaide, B., Almendros, P., *CSR*, **30**, 226.
Review: "4-Oxo-azetidine-2-carbaldehydes as Useful
Building Blocks in Stereocontrolled Synthesis."

VII.B-24 Hruby, V.J., Soloshono, K., eds., *T*, **57**, 6329-6650.
Symposium "Asymmetric Synthesis of Novel Sterically
in Print 88: Constrained Amino Acids."

VII.B-25 O'Donnell, M.J., *AA*, **34**, 3.
Review: "The Preparation of Optically Active α-Amino Acids
from the Benzophenone Imines of Glycine
Derivatives."

VII.B-26 Chen, Y., Reymond, J.-L., *S*, 934.
Special   "Enantioselective Epoxidation with a Library of
Topic: Catalytic Antibodies."

VII.B-27 Daverio, P., Zanda, M., *TA*, **12**, 2225.
Review: "Enantioselective Reductions by Chirally Modified
Alumino- and Borohydrides."

VII.B-28 Csaky, A.G., Plumet, J., *CSR*, **30**, 313.
Review: "Stereoselective Coupling of Ketone and
Carboxylate Enolates."

VII.B-29 Eames, J., Weerasooriya., N., *TA* , **12**, 1.
Review: "Recent Advances into the Enantioselective
Protonation of Prostereogenic Enol Derivatives."

VII.B-30 Gong, B., *SL*, 582.
Review: "Specifying Non-Covalent Interactions: Sequence-
Specific Assembly of Hydrogen-Bonded Molecular
Duplexes."

VII.B-31 Herrmann, W.A., ed., *JOM*, **621**, 1-358.
Special "Perspectives on Organometallic Stereochemistry."
Issue:

VII.B-32 Timmerman, P. et al., *AG(E)*, **40**, 2383.
Review: "Noncovalent Synthesis Using Hydrogen Bonding."

VII.B-33 Ghadiri, M.R. et al., *AG(E)*, **40**, 988.
Review: "Self-Assembling Organic Nanotubes."

VII.B-34 Toda, F., *PAC*, **73**, 1137.
Lecture: "Crystalline Inclusion Complexes as Media of
Molecular Recognitions and Selective Reactions."

VII.B-35 Krische, M.J. et al., *T*, **57**, 1139.
Report: "Hydrogen Bonding in Noncovalent Synthesis:
Selectivity and the Directed Organization of
Molecular Strands."

VII.B-36 Motherwell, W.B. et al., *T*, **57**, 4663.
Report: "Recent Progress in the Design and Synthesis of
Artificial Enzymes."

VII.B-37 Timmerman, P., Prins, L.J., *EJOC*, 3191.
Microreview: "Noncovalent Synthesis of Melamine-Cyanuric/
Barbituric Acid Derived Nanostuctures: Regio- and
Stereoselection."

VII.B-38 Jones, G.B., *T*, **57**, 7999.
Report: "π Shielding in Organic Synthesis."

VII.B-39 Lyndsey, G.M., Philp, D., *CSR*, **30**, 287.
Review: "Applying Biological Principles to the Assembly and
Selection of Synthetic Superstructures."

VII.B-40 Matile, S., *CSR*, **30**, 158.
Review: "En Route to Supramolecular Functional Plasticity:
Artificial β-Barrels, the Barrel-Stave Motif and
Related Approaches.

VII.C Reactions

VII.C-1 Kumar, A., *CRV*, **101**, 1.
Review: "Salt Effects on Diels-Alder Reaction Kinetics."

VII.C-2 Faber, K. et al., *CSR*, **30**, 332.
Review: "Enzyme-Initiated Domino (Cascade) Reactions."

VII.C-3 Elliott, M.C. et al., *T*, **57**, 6651.
Report: "Annulation Reactions of Azoles and Azolines with Heterocumulenes."

VII.C-4 Grossman, R.B., *SL*, 13.
Account: "The Double Annulation of Tethered Diacids and Alkynones: History and Scope."

VII.C-5 Katritzky, A.R., Toader, D., *SL*, 458.
Account: "The Preparartion of Mono-1,1-trans-1,2-di- and trisubstituted Ethylenes by Benzotriazole Methodology."

VII.C-6 Miura, T. et al., *SL*, 1055.
Review: "Recent Advances of BINAP Chemistry in the Industrial Aspects."

VII.C-7 Moser, W.H., *T*, **57**, 2065.
New Tools "The Brook Rearrangement in Tandem Bond Synthesis: Formation Strategies."

VII.C-8 Lu, X. et al., *ACR*, **34**, 535.
Review: "Reactions of Electron-Deficient Alkynes and Allenes Under Phosphine Catalysis."

VII.C-9 Fujiwara, Y. et al., *ACR*, **34**, 633.
Review: "Catalytic Functionalization of Arenes and Alkanes *via* C-H Bond Activation."

VII.C-10 Pavlov, V.A., *RCR*, **70**, 1037.
Review: "Structural and Configurational Relationships 'Metal-Complex-Substrate-Product' in Asymmetric Catalytic Hydrogenation, Hydrosilylation, and Cross-Coupling Reactions."

VII.C-11 Schneider, C., *SL*, 1079.
Account: "The Silyloxy-Cope Rearrangement of *syn*-Aldol Products: Evolution of a Powerful Synthetic Strategy."

VII.C-12 Kotyatkina, A.I. et al., *RCR*, **70**, 641.
Review: "1,3-Dipolar Cycloaddition Reactions of Nitrile Oxides in the Synthesis of Natural Compounds and their Analogues."

VII.C-13 Tai, C.-H., Cook, P.F., *ACR*, **34**, 49.
Review: "Pyridoxyl 5'-Phosphate-Dependent α,β-Elimination Reactions: Mechanism of O-Acetylserine Sulfhydrylase,"

VII.C-14 Litvinovskaya, R.P., Khripach, V.A., *RCR*, **70**, 405.
Review: "Regio- and Stereochemistry of 1,3-Dipolar Cycloaddition of Nitrile Oxides to Alkenes."

VII.C-15 Harmata, M., *ACR*, **34**, 595.
Review: "Exploration of Fundamental and Synthetic Aspects of the Intramolecular 4+3 Cycloaddition Reaction."

VII.C-16 Shea, K.J. et al., *AG(E)*, **40**, 820.
Review: "The Type 2 Intramolecular Diels-Alder Reaction: Synthesis and Chemistry of Bridgehead Alkenes."

VII.C-17 Deslongchamps, P. et al., *T*, **57**, 4243.
Report: "The Transannular Diels-Alder Strategy: Applications to Total Synthesis."

VII.C-18 Oh, T. et al., *T*, **57**, 6099.
Report: "Recent Delepolments in Imino Diels-Alder Reactions."

VII.C-19 Christoffers, J., *SL*, 723.
Account: "Catalysis of the Michael Reaction and the Vinylogous Michael Reaction by Ferric Chloride Hexahydrate."

VII.C-20 de Vries, J.G., *CJC*, **79**, 1086.
Review: "The Heck Reaction in the Production of Fine Chemicals."

VII.C-21 Stavropoulos, P. et al., *ACR*, **34**, 745.
Review: **"The Gif Paradox."**

VII.C-22 Gibson, S.E. et al., *T*, **57**, 7449.
Report: **"Advances in the Heck Chemistry of Aryl Bromides and Chlorides."**

VII.C-23 Luzzio, F.A., *T*, **57**, 915.
Report: **"The Henry Reaction: Recent Examples."**

VII.C-24 Breit, B., Seiche, W., *S*, 1.
Review: **"Recent Advances on Chemo-, Regio- and Stereoselective Hydroformylation."**

VII.C-25 Ceyer, S.T., *ACR*, **34**, 737.
Review: **The Unique Chemistry of Hydrogen Beneath the Surface: Catalytic Hydrogenation of Hydrocarbons."**

VII.C-26 Bunz, U.H.F., *ACR*, **34**, 998.
Review: **"Poly(*p*-phenyleneethynylene)s by Alkene Metathesis."**

VII.C-27 Krause, N., Hoffmann-Roder, A., *S*, 171.
Review: **"Recent Advances in Catalytic Enantioselective Michael Additions."**

VII.C-28 Mateos, A.F. et al., *T*, **57**, 1049
Article: **"The Nazarov Cyclization of β-Carbonyl-β'-furyl-divinyl Ketones and Related Compounds as Induced by Perchloric Acid."**

VII.C-29 Ganeshpure, P.A. et al., *CRV*, **101**, 3499.
Review: **"Synthetic Applications of Nonmetal Catalysts for Homogeneous Oxidations."**

VII.C-30 Reedijk, J. et al., *CSR* , **30**, 376.
Review: **"Homogeneous Bio-Inspired Copper-Catalyzed Oxidation Reactions."**

VII.C-31 Quideau, S., Feldman, K.S., guest eds., *T*, **57**, 265-423.
Symposium "Oxidative Activation of Aromatic Rings: An
in print: Efficient Strategy for Arene Functionalization."

VII.C-32 Arterburn, J.B., *T*, **57**, 9765.
Report: "Selective Oxidation of Secondary Alcohols."

VII.C-33 Kiss, G., *CRV*, **101**, 3435.
Review: "Palladium-Catalyzed Reppe Carbonylation."

VII.C-34 Yet, L. *AG(E)*, **40**, 875.
Review: "Recent Devepolments in Catalytic Asymmetric
Strecker-Type Reactions."

VII.C-35 Lloyd-Williams, P., Giralt, E., *CSR*, **30**, 145.
Review: "Atropisomerism, Biphenyls and the Suzuki
Coupling: Peptide Antibiotics."

VII.C-36 Jedlinski, Z., *JHC*, **38**, 1249.
Review: "Nucleophilic Substitution and Electron Transfer in
the Ring-Opening Reactions of β-Lactones: A Short
Review."

VII.C-37 Suwinski, J., Swierczek, K., *T*, **57**, 1639.
Report: "*Cine-* and *Tele*-Substitution Reactions."

VII.C-38 Danishefsky, S.J. et al., *AG(E)*, **40**, 4545.
Review: "The β-Alkyl Suzuki-Miyaura Cross-Coupling
Reaction: Development, Mechanistic Study, and
Applications in Natural Product Synthesis."

VII.D Reactive Intermediates

VII.D-1 Richard, J.P. et al., ACR, *34, 981.*
Review: "Formation and Stability of Carbocations and
Carbanions in Water and Intrinsic Barriers to their
Reactions."

VII.D-2 Somsak, L., *CRV*, **101**, 81.
Review: "Carbanionic Reactivity of the Anomeric Center in
Carbohydrates."

VII.D-3 Langer, P., *ECJ*, **7**, 3858.
Review: "Regio- and Diastereoselective Cyclization Reactions
of Free and Masked 1,3-Dicarbonyl Dianions with
1,2-Dielectrophiles."

VII.D-4 Minami, T. et al., *S*, 349.
   Review: "α-Phsophonovinyl Carbanions in Organic
   Synthesis."

VII.D-5 Bertrand,G., ed., *JOM*, **617-618**, 3-754.
   Special   "Transition Metal Complexes of Carbenes and
   Issue: Related Species in 2000."

VII.D-6 Zhang, W., *T*, **57**, 7237.
   Report: "Intramolecular Free Radical Conjugate Additions."

VII.D-7 Robertson, J. et al., *CSR*, **30**, 94.
   Review: "Radical Translocation Reactions in Synthesis."

VII.D-8 Boche, G., Lohrenz, J.C.W., *CRV*, **101**, 697.
   Review: "The Electrophilic Nature of Carbenoids,
   Nitrenoids, and Oxenoids."

VII.D-9 Hodgson, D.M. et al., *CSR*, **30**, 50.
   Review: "Catalytic Enantioselective Rearrangements and
   Cycloadditions Involving Ylids from Diazo
   Compounds."

VII.D-10 Tolstikov, G.A. et al., *RCR*, **70**, 655.
   Review: "Sulfur Ylids in the Synthesis of Heterocycles and
   Carbocyclic Compounds.

VII.D-11 Dakternieks, D., Schiesser, C.H., *AJC*, **54**, 89.
   Current   "The Quest for Single-Enantiomer Outcomes in Free-
   Chemistry: Radical Chemistry."

VII.D-12 McCarroll, A., Walton, J.C., *JCS(P1)*, 3215.
   Review: "Organic Synthesis Through Free-Radical
   Annulations and Related Cascade Sequences."

VII.D-13 Doyle, M.P., Hu, W., *SL*, 1364.
   Account: "Macrocycle Formation from Catalytic Metal Carbene
   Transformations."

VII.D-14 Li, J.J., *T*, **57**, 1.
   Report: "Free Radical Chemistry of Three-Membered
   Heterocycles."

VII.D-15 Studer, A., Bossart, M., *T*, **57**, 9649.
Report: "Radical Aryl Migration Reactions."

VII.D-16 Ollivier, C., Renaud, P., *CRV*, **101**, 3415.
Review: "Organoboranes as a Souce of Radicals."

VII.D-17 Paquette, L.A., *SL*, 1.
Account: "The Electrophilic and Radical Behavior of α-Halosulfonyl Systems."

VII.D-18 Friestad, G.K., *T*, **57**, 5461.
Report: "Addition of Carbon-Centered Radicals to Imines and Related Compounds."

VII.D-19 Togo, H., Katohgi, M., *SL*, 565.
Review: "Synthetic Uses of Organohypervalent Iodine Compounds Through Radical Pathways."

VII.D-20 Ryu, I., *CSR*, **30**, 16.
Review: "Radical Carboxylations of Iodoalkanes and Saturated Alcohols Using Carbon Monoxide."

VII.D-21 Yoon, U.C., Mariano, P.S., *ACR*, **34**, 523.
Review: "The Synthetic Potential of Phthalimide SET Photochemistry."

VII.D-22 Wagner, P.J., ACR, *34, 1*.
Review: "Photoinduced Ortho [2+2] Cycloaddition of Double Bonds to Triplet Benzenes."

VII.D-23 Mehta, G., Kotha, S., *T*, **57**, 625.
Report: "Recent Chemistry of Benzocyclobutenes."

VII.D-24 Shereshovets, V.V., Khursan, S.L., Komissarov, V.D., Tolstikov, G.A., *RCR*, **70**, 105.
Review: "Organic Hydrotrioxides."

VII.D-25 Morkin, T.L., Leigh, W.J., *ACR*, **34**, 129.
Review: "Substituent Effects on the Reactivity of the Silicon-Carbon Double Bond."

VII.E. Organometallics and Metalloids

VII.E-1 Qian, C.T., Hou, X.L., symposium eds., *PAC*, **73**, 205-376.
Lectures: "XIXth International Conference on Organometallic
Chemistry."

VII.E-2 Adams, R.D., guest ed., *JOM*, **630**, 5-138.
Special   "Metals in Organic Synthesis—In Honor of Myron
Issue: Rosenblum."

VII.E-3 Shinokubo, H., Oshima, K., *SL*, 322.
Account: "Stereospecific or Stereoselective Elimination of
Vicinal-Alkoxyiodoalkanes by Means of
Organometallic Reagents."

VII.E-4 Schlosser, M., *EJOC*, 3975.
Microreview: "The Organometallic Approach to Molecular
Diversity—Halogens as Haptens.

VII.E-5 Normant, J.F., *ACR*, **34**, 640.
Review: "1,1-Dimetallic Reagents for the Elaboration of
Stereoselectively Di- or Trisubstituted Linear
Substrates."

VII.E-6 Corey, E.J., Lee, T.W., *CC*, 1321.
Feature   "The Formyl C-H•••O Hydrogen Bond as a Critical
Article: Factor in Enantioselective Lewis-Acid Catalyzed
Reactions of Aldehydes."

VII.E-7 Fallis, A.G., Forgione, P., *T*, **57**, 5899.
Report: "Metal Mediated Carbometallation of Alkynes and
Alkenes Containing Adjacent Heteratoms."

VII.E-8 Westerhausen, M., *AG(E)*, **40**, 2974.
Review: "100 Years After Grignard: Where Does the
Organometallic Chemistry of the Heavy Alkaline
Earth Metals Stand Today?"

VII.E-9 Marks, T.J. et al., *CRV*, **101**, 953
Review: "Catalysis Research of Relevance to Carbon
Management: Progress, Challenges, and
Opportunities."

VII.E-10 Friesen, R.W., *JCS(P1)*, 1969.
Review: "Generation and Reactivity of α-Metalated Vinyl Ethers."

VII.E-11 Fletcher, A.J., Christie, S.D.R., *JCS(P1)*, 1.
Review: "Applications of Stoichiometric Transition Metal Complexes in Organic Synthesis."

VII.E-12 Andersen, N.G., Keay, B.A., *CRV*, **101**, 997.
Review: "2-Furyl Phosphines as Ligands for Transition-Metal-Mediated Organic Synthesis."

VII.E-13 Knolker, H.-J. et al., *PAC*, **73**, 1075.
Lecture: "Recent Applications of Tricarbonyliron-Diene Complexes to Organic Synthesis."

VII.E-14 Adams, R.D., ed., *JOM*, **637-639**, 1-875.
Special Edition: "50th Anniversary of the Discovery of Diferrocene."

VII.E-15 Steel, P.G., *JCS(P1)*, 2727.
Review: "Recent Developments in Lanthanide Mediated Organic Synthesis."

VII.E-16 Elliot, G.I., Konopelski, J.P., *T*, **57**, 5683.
Report: "Arylation with Organolead and Organobismuth Reagents."

VII.E-17 Yus, M., *SL*, 1197.
Account: "From Arene-Catalyzed Lithiation to Other Synthetic Adventures."

VII.E-18 Henderson, K.W., Kerr, W.J., *ECJ*, **7**, 3430.
Review: "Magnesium Bisamides as Reagents in Synthesis."

VII.E-19 Harman, W.D. et al., *T*, **57**, 8203.
Report: "Synthetic Applications of the Dearomatization Agent Pentaammineosmium(II)."

VII.E-20 Trost, B.M. et al., *CRV*, **101**, 2067.
Review: "Non-Metathesis Ruthenium Catalyzed C-C Bond Formation."

VII.E-21 Hayashi, T., *SL*, 879.
Account: "Rhodium-Catalyzed Asymmetric 1,4-Addition of Organoboronic Acids and their Derivatives to Electron Deficient Olefins."

VII.E-22 Shibasaki, M. et al., *ECJ*, 7, 4066.
Review: "Zirconium Alkoxides in Catalysis."

VII.E-23 Stefani, H.A. et al., *T*, **57**, 1411.
Report: "Recent Advances in Selenocyclofunctionalization Reactions."

VII.E-24 Pu, L., Yu, H.-B., *CRV*, **101**, 757.
Review: "Catalytic Asymmetric Organozinc Additions to Carbonyl Compounds."

VII.E-25 Mugesh, G., du Mont, W.-W., Sies, H., *CRV*, **101**, 2125.
Review: "Chemistry of Biologically Important Synthetic Organoselenium Compounds."

VII.E-26 Trnka, T.A., Grubbs, R.H., *ACR*, **34**, 18.
Review: "The Development of $L_2X_2Ru=CHR$ Olefin Metathesis Catalysts: An Organometallic Success Story."

VII.E-27 Rossen, K., *AG(E)*, **40**, 4611.
Highlight: "Ru- and Rh-Catalyzed Asymmetric Hydrogenations: Recent Surprises from an Old Reaction."

VII.F. Halogen Compounds and Halogenation

(see also: VI.A.10.)

VII.F-1 Ma, S., Li, L., *SL*, 1206.
Account: "An Efficient New Methodology for the Synthesis of 1-Functionalized 2-Halo-2-alkenes *via* Hydrohalogenation Reaction of Electron-Deficient Allenes."

VII.F-2 Caine, D., *T*, **57**, 2643.
Report: "Reactions of Conjugated Haloenoates with Nucleophilic Reagents."

VII.F-3 Brel. V.K. et al., *RCR*, **70**, 231.
    Review: "Chemistry of Xenon Derivatives. Synthesis and Chemical Properties."

VII.F-4 Carmen. R.M. et al., *AJC*, **54**, 117.
    Review: "Towards the *cis*-Bromination of Double Bonds."

VII.F-5 Koser, G.F., *AA*, **34**, 89.
    Review: "[Hydroxy(tosyloxy)iodo]benzene and Closely Related Iodanes: The Second Stage of Development."

VII.F-6 Lemal, D.M., *ACR*, **34**, 662.
    Review: "Hexafluorobenzene Photochemistry: Wellspring of Fluorocarbon Structures."

VII.F-7 Jolliffe, K.A., *AJC*, **54**, 75.
    Review: "N,N-Diethylaminosulfur Trifluoride (DAST, Et_2NSF_3) and Related Reagents."

VII.G Natural Products

VII.G-1 Gottlieb, O.R., DaSilva, M.F.D.G.F., symposium eds., *PAC*, **73**, 549-626.
    Lectures: "22nd IUPAC International Symposium on the Chemistry of Natural Products."

VII.G-2 Majinda, R.R.T. et al., *PAC*, **73**, 1197.
    Lecture: "Recent Results from Natural Product Research at the University of Botswana."

VII.G-3 Fulop, F., *CRV*, **101**, 2181.
    Review: "The Chemistry of 2-Aminocycloalkenecarboxylic Acids."

VII.G-4 Donaldson, W.A., *T*, **57**, 8589.
    Report: "Synthesis of Cyclopropane Containing Natural Products."

VII.G-5 Bringmann, G., Menche, D., *ACR*, **34**, 615.
    Review: "Stereoselective Total Synthesis of Axially Chiral Natural Products *via* Biaryl Lactones."

VII.G-6 Bols, M. et al., *JCS(P1)*, 2136.
Review: **"Garner's Aldehyde."**

VII.G-7 Mehta, G., Nandakumar, J., *TL*, **42**, 7667.
Atricle: **"Restructuring α-Pinene: Novel Entry into Diverse Polycarbocyclic Frameworks."**

VII.G-8 Ghosh, A.K. et al., *S*, 1281.
Review: **"Tartaric Acid and Tartrates in the Synthesis of Bioactive Molecules."**

VII.G-9 Bringmann, G. et al., *S*, 155
Feature **"From Dynamic to Non-Dynamic Kinetic Resolution**
Article **of Lactone-Bridged Bioaryls: Synthesis of Mastigophorene B."**

VII.G-10 Hodgkinson, T.J., Shipman, M., *T*, **57**, 4467.
Report: **"Chemical Synthesis and Mode of Action of the Azinomycins."**

VII.G-11 Silva, L.P., Jr., *S*, 671.
Review: **"Total Synthesis of Bakkanes."**

VII.G-12 Murphy, P.J., Thomas, C.W., *CSR*, **30**, 303.
Review: **"The Synthesis and Biological Activity of the Marine Metabolite Cylindrospermopsin."**

VII.G-13 Nicolaou, K.C. et al., *CC*, 1523.
Review: **"Recent Developments in the Chemistry, Biology and Medicine of the Epothilones."**

VII.G-14 Kingston, D.G.I., *CC*, 867.
Feature **"Taxol, a Molecule for All Seasons."**
Article:

VII.G-15 Funayama, S. et al., *H*, **54**, 1139.
Review: **"Quinoline Alkaloids of *Orixa Japonica*."**

VII.G-16 Asakawa, Y. et al., *H*, **54**, 1057.
Review: **"Sesquiterpene Lactones and Acetogenin Lactones from the Hepaticae and Chemosystematics of the Liverworts *Frullania, Plagiochila* and *Porella*."**

VII.G-17 Wright, P.C. et al., *T*, **57**, 9347.
Report: "Marine Cyanobacteria—A Prolific Source of Natural Products."

VII.G-18 Palomo, C. et al., *SL*, 1813.
Account: "β-Lactams as Versatile Intermediates in α- and β-Amino Acid Synthesis."

VII.G-19 Peterlin-Masic, L., Kikelj, D., *T*, **57**, 7073.
Report: "Arginine Mimetics."

VII.G-20 Najera, C. et al., *OPP*, **33**, 203.
Review: "New Trends in Peptide Coupling Reagents."

VII.G-21 Shioiri, T., Hamada, Y., *SL*, 184.
Account: "Efficient Syntheses of Biologically Active Peptides of Aquatic Origin Involving Unusual Amino Acids."

VII.G-22 Pieroni, O., Lenci, F. et al., *ACR*, **34**, 9.
Review: "Photoresponsive Polypeptides."

VII.G-23 Halcrow, M.A., *AG(E)*, **40**, 346.
Review: "Chemically Modified Amino Acids in Copper Proteins that Bind or Activate Dioxygen."

VII.G-24 Seitz, O. et al., *T*, **57**, 2247.
Report: "Synthetic Peptide Conjugates—Tailor-Made Probes for the Biology of Protein Modification and Protein Processing."

VII.G-25 DeGrado,,W.F., guest ed., *CRV*, **101**, 3025-3232.
Reviews: "Protein Design."

VII.G-26 Ahn,.N., guest ed., *CRV*, **101**, 2207-2600.
Reviews: "Protein Phosphorylation and Signaling."

VII.G-27 Muir, T.W., *SL*, 733.
Account: "Development and Application of Expressed Protein Ligation."

VII.G-28 Kirschning, A. et al., *S* , 507.
Review: "Concepts for the Total Synthesis of Deoxy Sugars."

VII.G-29 Hang, H.C., Bertozzi, C.R., *ACR*, **34**, 727.
Review: "Chemoselective Approaches to Glycoprotein Assembly."

VII.G-30 Stutz, A.E., ed., *TCC*, **215**, 1-345.
Reviews: "Epimerisation, Isomerisation and Rearrangement Reactions of Carbohydrates."

VII.G-31 Kvaerno, L., Wengel, J., *CC*, 1419.
Feature   "Antisense Molecules and Furanose Confirmations—
Article   Is it Really that Simple?"

VII.G-32 Dubreuil, D. et al., *PAC*, **73**, 1189.
Lecture: "Synthesis of Nucleoside Analogs and New Tat Protein Inhibitors."

VII.G-33 Nicolaou, K.C., Mitchell, H.J., *AG(E)*, **40**, 1576.
Review: "Adventures in Carbohydrate Chemistry: New Synthetic Technologies, Chemical Synthesis, Molecular Design, and Chemical Biology."

VII.G-34 Hayakawa, Y., *BCJ*, **74**, 1547.
Account: "Toward an Ideal Synthesis of Oligonucleotides: Development of a Novel Phosphoramidite Method with High Capability."

VII.G-35 Gallego, J., Varani, G., *ACR*, **34**, 836.
Review: "Targeting RNA with Small-Molecule Drugs: Therapeutic Promise and Chemical Challenges."

VII.G-36 Reese, P.B., *ST*, **66**, 481
Review: "Remote Functionalization Reactions in Steroids."

VII.G-37 Kobayashi, S. et al., *BCJ*, **74**, 613.
Account: "Enzymatic Polymerization for Precision Polymer Synthesis."

VII.G-38 Stonik, V.A., *RCR*, **70**, 673.
Review: "Marine Polar Steroids."

VII.G-39 Suh, J., *SL*, 1343.
Account: "Synthesis of Polymeric Enzyme-Like Catalysts."

VII.G-40 Bugg, T.D.H., Lin, G., *CC*, 941.
Feature   "Solving the Riddle of the Intradiol and Extradiol
Article: Catechol Dioxygenases: How Do Enzymes Control
             Hydroperoxide Rearrangements?"

VII.G-41 Wolfenden, R., Snider, M.J., *ACR*, **34**, 938.
Review: "The Depth of Chemical Time and the Power of
            Enzymes as Castalysts."

VII.G-42 Pelliesser, H., Santelli, M., *OPP*, **33**, 1.
Review: "Chemical and Biochemical Hydroxylations of
            Steroids."

VII.G-43 Pellissier, H., Santelli, M., *OPP*, **33**, 455.
Review: "Functionalization of the 18-Methyl Group of
            Steroids."

VII.G-44 Prilezhaeva, E.N., *RCR*, **70**, 897.
Review: "Rearrangements of Sulfoxides and sulfones in the
            Total Synthesis of Natural Products."

VII.H. Others

VII.H-1 Jung, K.W. et al. *T*, **57**, 7785.
Report: "Synthesis of Secondary Amines."

VII.H-2 Lee, S.H., Cheong, C.S., *T*, **57**, 4801.
Report: "Selective Reactions of Reactive Amino Groups in
            Polyamino Compounds by Metal-Chelated or
            -Mediated Methods."

VII.H-3 Meunier, B. et al., *PAC*, **73**, 1173.
Lecture: "From Classical Antimalarial Drugs to New
            Compounds Based on the Mechanism of Action of
            Artemisinin."

VII.H-4 Hunter, C.A. et al., *JCS(P2)*, 651.
Review: "Aromatic Interactions.

VII.H-5 Butkus, E., *SL*, 1827.
Account: "Stereocontrolled Synthesis and Reactions of
            Bicyclo[3.3.1]nonanes."

VII.H-6 Schleyer, P.R. guest ed., *CRV*, **101**, 1115-1566.
Review: "Aromaticity."

VII.H-7 Gokel, G.W., Mukhopadhyay, A., *CSR*, **30**, 274.
Review: "Synthetic Models of Cation-Conducting Channels."

VII.H-8 Boger, D.L., ed., *BMCL*, **11**, 1477-1613.
**Symposium "Recent Advances in Bioorganic and Medicinal
in Print: Chemistry."**

VII.H-9 Bruckner, R., *CC*, 141.
**Feature "The β-elimination Route to Stereodefined γ-
Alkylidenebutenolides."**

VII.H-10 Elgemeie, G.H., Sayed, S.H., *S*, 1747.
Review: "Synthesis and Chemistry of Dithiols."

VII.H-11 Corsgro, A. et al., *S*, 1903.
**Review: "Regeneration of Carbonyl Compounds from the
Corresponding Oximes."**

VII.H-12 Ley, S.V. et al., *CRV*, **101**, 53.
**Review: "1,2-Diacetals: A New Opportunity for Organic
Synthesis."**

VII.H-13 Vasil'ev, A.A., Serebryakov, E.P., *RCR*, **70**, 735.
**Review: "Synthetic Methodologies for Carbo-Substituted
Conjugated Dienes."**

VII.H-14 Milata, V., *AA*, **34**, 20.
Review: "Alkoxymethylenemalonates in Organic Synthesis."

VII.H-15 Wirth, T., *AG(E)*, **40**, 2812.
Highlight: "IBX—New Reactions with an Old Reagent."

VII.H-16 Percec, V., guest ed., *CRV*, **101**, 3579-4192.
Reviews: "Frontiers in Polymer Chemistry."

VII.H-17 Komatsu, K., *BCJ*, **74**, 407.
**Review: "Cyclic π-Conjugated Systems Annelated with
Bicyclo[2.2.2]octene: Synthesis, Structure, and
Properties."**

VII.H-18 Tundo, P., *PAC*, **73**, 1117.
Lecture: "New Developments in Dimethyl Carbonate Chemistry."

VII.H-19 Besenhard, J.O., Sitte, W., Stelzer, F., Gamsjager, H., eds., *M*, **132**, 421-549.
Special Issue: "Electroactive Materials."

VII.H-20 Bartsch, R.A., Elshani, S. et al., *JHC*, **38**, 311.
Review: "Synthesis of Lariat Ether Carboxylic Acids Based on Dibenzo-16-crown-5."

VII.H-21 Laurent, C., *TA*, **12**, 2359.
Review: "Epoxy Ketones as Versatile Building Blocks in Organic Synthesis."

VII.H-22 Klingebiel, U., Schmatz, S., *M*, **132**, 1105.
Review: "Cyclization and Isomerization Reactions in Silylhydrazine."

VII.H-23 Ragnarsson, U., *CSR*, **30**, 205.
Review: "Synthetic Methodology for Alkyl Substituted Hydrazines."

VII.H-24 Krakowiak, K.E. et al., *JHC*, **38**, 1239.
Review: "One-Step Syntheses of Macrocyclic Compounds: A Short Review."

VII.H-25 Ronsisvalle, G., symposium ed., *PAC*, **73**, 66-75.
Lecture: "XVIth International Symposium on Medicinal Chemistry."

VII.H-26 Gaber, A., McNab, H., *S*, 2059.
Review: "Synthetic Applications of the Pyrolysis of Medium Acid Derivatives."

VII.H-27 Riess, J.G., *CRV*, **101**, 2797.
Review: "Oxygen Carriers ('Blood Substitutes')—Raison d'Etre, Chemistry, and Some Physiology."

VII.H-28 Gross, R.A. et al., *CRV*, **101**, 2097.
Review: "Polymer Synthesis by *In Vitro*, Enzyme Catalysis."

VII.H-29 Iorga, B., Savignac, P., *SL*, 447.
Account: **"Controlled Reactivity of Phosphonates by Temporary Silicon Connection."**

VII.H-30 Zefirova, O.N., Zefirov, N.S., *RCR*, **70**, 333.
Review: **"Physiologically Active Compounds Interacting with Serotonin (5-Hydroxytryptamine) Receptors."**

VII.H-31 Braunstein, P., Boag, N.M., *AG(E)*, **40**, 2427.
Minireview: "Alkyl, Silyl, and Phosphane Ligands—Classical Ligands in Nonclassical Bonding Modes."

VII.H-32 Inoue, K., Itaya, T. et al., *BCJ*, **74**, 1381.
Review: **"Synthesis and Functionality of Cyclophosphazene-Based Polymers."**

VII.H-33 Ghosh, A.K. et al., *S*, 2203.
Review: **"Syntheses of FDA Approved HIV Protease Inhibitors."**

VII.H-34 Silnikov, V.N., Vlassov, V.V., *RCR*, **70**, 491.
Review: **"Design of Site-Specific RNA-Cleaving Reagents."**

VII.H-35 Hudlicky, T. et al., *S*, 952.
Special **"Novel O- and N-Linked Inositol Oligomers: A New**
Topic: **Class of Unnatural Saccharide Mimics."**

VII.H-36 Katz, H.E. et al., *ACR*, **34**, 359.
Review: **"Synthetic Chemistry for Ultrapure, Processable, and High-Mobility Organic Transitor Semiconuctors."**

VII.H-37 Back, T.G., *T*, **57**, 5263.
Report: **"The Chemistry of Acetylenic and Allenic Sulfones."**

VII.H-38 Hsung, R.P. et al., *T*, **57**, 7575.
Review: **"Recent Advances in the Chemistry of Ynamines and Ynamides."**

VII.H-39 Makarov, S.V., *RCR*, **70**, 885.
Review: **"Recent Trends in the Chemistry of Sulfur-Containing Reducing Agents."**

VII.H-40 Perevalov, S.G. et al., *RCR*, **70**, 921.
   Review: "(Het)aroylpyruvic Acids and their Derivatives as Promising Building Blocks for Organic Synthesis."

VII.H-41 Komarov, I.V., *RCR*, **70**, 991.
   Review: "Organic Molecules with Abnormal Geometric Parameters."

VIII
SELECTED TOPICAL AREAS

VIII.A. Fullerene Chemistry

VIII.A.1 Diels-Alder-Type Cycloadditions

VIII.A.1-1 Cross, R.J. et al., *JACS*, **123**, 256.
Reversible Diels-Alder Addition to Fullerenes: A Study of Equilibria Using ^{3}H NMR Spectroscopy

---

VIII.A.1-2 Wu, Y.-L. et al., *JCS(P1)*, 617.
Steroid-Fullerene Adducts from Diels-Alder Reactions.

VIII.A.1-3 Krautler, B. et al., *ECJ*, **7**, 3223 and *T*, **57**, 3709.
Efficient Preparation of Monoadducts of [60]Fullerrene and Anthracenes by Solution Chemistry and their Thermolytic Decomposition in the Solid State

---

VIII.A.1-4 Chronakis, N., Orfanopoulos, M., *OL*, **3**, 545.
[4+2] Cycloadditions of Rigid *s-cis* Dienes to C_{60}. A Synchronous Diels-Alder Reaction

---

VIII.A.1-5 Hudhomme, P., Gorgues, A. et al., *TL*, **42**, 3717, 3447.
The Dumbell Bis Diels-Alder Adduct Between Tetramethylinden[4H]tetrathiafulvalene and Two C_{60}

VIII.A.2 Other Cycloadditions

VIII.A.2-1 Nishimura, J. et al., *OL*, **3**, 1193.
First Isolation and Characterization of Eight Regioisomers for [6]Fullerene-Benzyne Bisadducts

---

VIII.A.2-2 Mamane, V., Riant, O., *T*, **57**, 2555.
Asymmetric Synthesis of Chiral Ferrocenyl Fulleropyrrolidines as Potential Building Blocks for New Materials

VIII.A.2-3 Ishida, H., Itoh, K., Ohno, M., *T*, **57**, 1737.
1,3-Dipolar Cycloaddition Reaction of [60]Fullerene with Thiocarbonyl Ylides and Synthetic Applications of the Cycloadduct

VIII.A.2-4 Diederich, F. et al., *HCA*, **84**, 1207.
Hexakis-Adducts of [60]Fullerene with Different Addition Patterns: Templated Synthesis, Physical Properties, and Chemical Reactivity

VIII.A.3 Photochemical Reactions

VIII.A.3-1 Shevlin, P.B. et al., *JACS*, **123**, 1349.
Observation of Both Thermal First-Order and Photochemical Zero-Order Kinetics in the Rearrangement of [6,5] Open Fulleroids to [6,6] Closed Fullerenes

VIII.A.3-2 Birkett, P.R., Taylor, R. et al., *JCS(P2)*, 68.
Preparation and Characterisation of Two [70]Fullerene Diols, $C_{70}Ph_8(OH)_2$

VIII.A.3-3 Kabe, Y. et al., *JOM*, **636**, 82.
The Photochemical Reaction of 1,2-Digermacyclobutane with C_{60}: Possible Example of a Closed [6,5]-Bridged Fullerene Derivative of Germacyclopropane

VIII.A.3-4 Murata, S. et al., *TL*, **42**, 895.
Photooxygenative Partial Ring Cleavage of Bis(fulleroid): Synthesis of a Novel Fullerene Derivative with a 12-Membered Ring

VIII.A.3-5 Murata, Y., Komatsu, K., *CL*, 896.
Photochemical Reaction of the Open-Cage Fullerene Derivative with Singlet Oxygen

VIII.A.4 Other Fullerene Chemistry

VIII.A.4-1 de La Vaissiere, B. et al., *JCS(P2)*, 821.
Regioselectivity in Radical Reactions of C_{60} Derivatives

VIII.A.4-2 Wang, N.-X. et al,, *TL*, **42**, 7911.
Some Thermal Decomposition Reactions of $C_{60}H_{36}$

VIII.A.4-3 Hirsch, A., *AG(E)*, **40**, 1235.
New Cages and Unusual Guests: Fullerene Chemistry Continues to Excite

VIII.A.4-4 Siegel, J.S. et al., *T*, **57**, 3737.
Baskets, Covered Baskets, and Basket Balls: Corannulene Based Cyclophanes as Fullerene Mimics

VIII.A.4-5 Kitagawa, T. et al., *T*, **57**, 3537.
Reaction of Cyclopropenylium Ions with the *tert*-Butyl-C_{60} Anion: Carbocation-Carbanion Coordination *vs* Salt Formation

VIII.A.4-6 Rabinovitz, M. et al., *JOC*, **66**, 6004.
Lithium Reduction of the Bowl-Shaped C_{60} Fragment Diindeno[1,2,3,4-defg:1',2',3',4'-maop]chrysene: An Interplay Between Experiment and Calculation

VIII.A.4-7 Martin, N., Echegoyen, L. et al., *JOC*, **66**, 4393.
Reductive Electrochemistry of Spiromethano Fullerenes

VIII.A.4-8 Guldi, D.M., Maggini, M., Paulucci, F. et al., *ECJ*, **7**, 1597.
A Photosensitizer Dinuclear Ruthenium Complex: Intramolecular Energy Transfer to A Covalently Linked Fullerene Acceptor

VIII.A.4-9 Ohno, T. et al., *JOC*, **66**, 3397.
Intramolecular Charge-Transfer in a New Dyad Based on C_{60} and Bis(4'-*t*-butyldiphenyl-4-yl) Analine (BBA) Donor

VIII.A.4-10 Prato, M. et al., *JOC*, **66**, 4915.
Novel Versatile Fullerene Synthons

VIII.A.4-11 Prato, M., Guldi, D. et al., *M*, **132**, 63.
Novel Functional Fullerrene Materials: Fullerenes as Energy Acceptors

VIII.A.4-12 Prato, M. et al., *AJC*, **54**, 223.
Fullerene Derivatives as Potential DNA Photoprobes

VIII.A.4-13 Boltalina, O.V., Taylor, R. et al., *AG(E)*, **40**, 787.
$C_{60}F_{20}$: 'Saturnene,' an Extraordinary Squashed Fullerene

VIII.A.4-14 Taylor, R. et al., *JCS(P2)*, 550.
Isolation and Characterisation of Bis(oxahomo)fullerene Derivatives of $C_{60}H_{18}$

VIII.A.4-15 Meier, M.S., Kiegiel, J., *OL*, **3**, 1717.
Preparation and Characterization of the Fullerene Diols 1,2-$C_{60}(OH)_2$, 1,2-$C_{70}(OH)_2$, and 5,6-$C_{70}(OH)_2$

VIII.A.4-16 Weisman, R.B., Heymann, D., Bachilo, S.M., *JACS*, **123**, 9720.
Synthesis and Characterization of the 'Missing' Oxide of C_{60}: [5,6]-Open $C_{60}O$

VIII.A.4-17 Taylor, R. et al., *JCS(P2)*, 1038.
Some 4-Fluorophenyl Derivatives of [60]Fullerene Spontaneous Oxidation and Oxide-Induced Fragmentation to C_{58}

VIII.A.4-18 Taylor, R. et al., *JCS(P2)*, 782.
Formation and Characterisation of Alkoxy Derivatves of [60]Fullerene

VIII.A.4-19 Warkentin, J. et al., *JOC*, **66**, 7496.
Mechanism of Migration of the Trimethylsilyl Group During Reactions of Methoxy[(trimethylsilyl)ethoxy]carbene with N-Phenylmaleimide and C_{60}

VIII.A.4-20 Murthy, C.N., Geckeler, K.E., *CC*, 1194; Yuan, D.-Q., Fujita, K. et al., *TL*, **42**, 6727.
The Water-Soluble β-Cyclodextrin-[60]Fullerene Complex

VIII.A.4-21 Schuster, D.I., et al., *JOC*, **66**, 5449.
Synthetic Approaches to a Variety of Covalently Linked Porphyrin-Fullerene Hybrids

VIII.A.4-22 Prato, M., Pasimeni, L. et al., *ECJ*, **7**, 816.
Efficient Charge Separation in Porphyrin-Fullerene-Ligand Complexes

VIII.A.4-23 Van Koten, G. et al., *OM*, **20**, 3993.
Methanofullerene-Based Palladium Bis(amino)aryl Complexes and Application in Lewis Acid Catalysis

VIII.A.4-24 Van Koten, G. et al., *OM*, **20**, 4198.
C,N-2-[(Dimethylamino)methyl]phenylplatinum Complexes Functionalized with C$_{60}$ as Macromolecular Building Blocks

---

VIII.A.4-25 Tossi, A., Prato, M., *OL*, **3**, 1845.
A Novel [60]Fullerene Amino Acid for Use in Solid-Phase Peptide Synthesis.

VIII.A.4-26 Hauke, F., Hirsch, A., *CC*, 1316.
Regioselective Formation of Highly Functionalised Heterofullerenes: Pentamalonates of RC$_{59}$N Involving an Octahedral Addition Pattern

---

VIII.A.4-27 Haucke, F., Hirsch, A., *T*, **57**, 3697.
C$_{59}$N: A Key Intermediate in Azaheterofullerene Chemistry

---

VIII.A.4-28 Rasmussen, P.G., Strongin, R.M. et al., *TL*, **42**, 6823.
The Reaction of [60]Fullerene with 2-Diazo-4,5-Dicyanoimidazole

---

VIII.A.4-29 Prato, M. et al., *JOC*, **66**, 2802.
Isolation and Characterization of all Eight Bisadducts of Fulleropyrrolidine Derivatives

---

VIII.A.4-30 Langa, F., Martin, N. et al., *JOC*, **66**, 5033.
C$_{60}$-Based Triads with Improved Electron-Acceptor Properties: Pyrazolylpyrazolino[60]fullerenes

---

VIII.A.4-31 Komatsu, K. et al., *JOC*, **66**, 8187.
The Reaction of Fullerene C$_{60}$ with 4,6-Dimethyl-1,2,3-triazine: Formation of an Open-Cage Fullerene Derivative

---

VIII.A.4-32 Tour, J.M. et al., *JACS*, **123**, 6536.
Functionalization of Carson Nanotubes by Electrochemical Reduction of Aryl Diazonium Salts: A Bucky Paper Electrode

---

VIII.A.4-33 Komatsu, K. et al., *JOC*, **66**, 7235.
The Reaction of Fullerene C$_{60}$ with Phthalazine: The Mechanochemical Solid-State Reaction Yielding a New C$_{60}$ Dimer Versus the Liquid Phase Reaction Affording an Open-Cage Fullerene

VIII.A.4-34 Bickett, P.R., Zerbetto, F. et al., *JCS(P2)*, 140.
Saturation versus Inductive Effects: The Electrochemistry of the $C_{70}Ph_{2n}$ (n = 1-5) Series

---

VIII.A.4-35 Shinohara, H. et al., *CC*, 1366.
A Catalytic Synthesis and Structural Characterization of a New [84]Fullerene Isomer

---

VIII.A.4-36 Shinohara, H. et al., *AG(E)*, **40**, 397.
A Scandium Carbide Endoheral Metalofullerene: (Sc_2C_2) @ C_{84}

VIII.A.4-37 Fujiwara, K., Komatsu, K., *CC*, 1986.
First Synthesis of a Highly Symmetrical Deakis-Adduct of Fullerene Dimer C_{120}

VIII.A.4-38 Dragoe, N., Kitazawa, K. et al., *JACS*, **123**, 1294.
First Unsymmetrical Bisfullerene, C_{121}: Evidense for the Presence of Both Homofullerene and Methanofullerene Cages in One Molecule

VIII.A.5 Reviews

VIII.A.5-1 Taylor, R., *ECJ*, **7**, 4074.
Review: "Fluorinated Fullerenes."

VIII.A.5-2 Tarasov, B.P. et al., *RCR*, **70**, 131.
Review: "Hydrogen-Containing Carbon Nanostructures: Synthesis and Properties."

VIII.A.5-3 Kitagawa, T., Takeuchi, K., *BCJ*, **74**, 785.
Review: "Monofunctionalized C_{60} Ions: Their Generation, Stability, and Reactions."

VIII.A.5-4 Rakov, E.G., *RCR*, **70**, 827.
Review: "The Chemistry and Applications of Carbon Nanotubes."

VIII.B Taxol and Related Taxane Chemistry

VIII.B-1 Zhou, Z. et al., *SC*, **31**, 3609.
A Practical and Efficient Synthesis of Taxol C-13 Side Chain

VIII.B-21 Mandai, T. et al., *H*, **54**, 561
Synthesis and Biological Evaluation of Water Soluble Taxoids Bearing Sugar Moieties

VIII.B-3 Ihara, M. et al., *JOC*, **66**, 2394.
A Novel Approach to the Taxane BC Ring System Through Formation of α-Ketol by Oxidative Removal of the Phenylsulfonyl Group with Subsequent *in situ* Oxidation

VIII.B-4 Shing, T.K.M. et al., *TL*, **42**, 8361.
Synthesis of the CD Ring in Taxol from (S)-(+)-Carvone

VIII.B-5 DeMattei, J.A. et al., *JOC*, **66**, 3330.
An Efficient Synthesis of the Taxane-Derived Anticancer Agent ABT-271

VIII.B-6 Zamir, L.D. et al., *CJC*, **79**, 1381.
Reanalysis of the Biotransformation of 4(20),11(12)-Taxadiene Derivatives

VIII.B-7 Dubois, J. et al., *JOC* **66**, 5058.
Semisynthesis of D-Ring Modified Taxoids: Novel Thia Derivatives of Docetaxel

VIII.B-8 Soga, T., *BCML*, **11**, 497.
Synthesis of 10-Deoxy-10-C-morpholinoethyl Docetaxel Analogues

VIII.B-9 Fang, W.-S. et al., *TL*, **42**, 1331.
Synthesis of the 2α-Benzoylamido Analogue of Docetaxel

VIII.B-10 Takahashi, T. et al., , *TL*, **42**, 7855; 7859.
Stereo- and Regio- Selective Ti-Mediated Radical Cyclization of Epoxy-Alkenes: Synthesis of the A and C Ring Synthons of Paclitaxel

VIII.B-11 Walker, M.A. et al., *BMCL*, **11**, 1683.
Synthesis of a Novel C-10 Spiro-Epoxide of Paclitaxol

VIII.B-12 Bane, S.L. et al., *BMCL*, **11**, 2249.
Synthesis and Microtubule Binding of Fluorescent Paclitaxol Derivatives

VIII.B-13 Mastalerz, H. et al., *OL*, **3**, 1613.
Synthesis of 7β-Sulfur Analogues of Paclitxel Utilizing a Novel Epimerization of the 7α-Thiol Group

VIII.B-14 Georg, G.I. et al., *JOC*, **66**, 3321.
Novel D-Seco Paclitaxel Analogues: Synthesis, Biological Evaluation, and Model Testing

VIII.B-15 Georg, G.I. et al., *JOC*, **66**, 8211.
Efficient Synthessi of the 3'-Phenolic Metabolite of Paclitaxel

VIII.C Dendrimers, Calixeranes and Other Unnatural Products

VIII.C.1 Dendrimers

VIII.C.1-1 Wang, S., Advincula, R.C., *OL*, **3**, 3831; Parquette, J.R. et al., *T*, **57**, 9393 and *JOC*, **66**, 6440; Caminade, A.-M., Majoral, J.-P. et al., *JACS*, **123**, 6698; Cao, D., Meier, H., *AG(E)*, **40**, 186; Meijer, E.W. et al., *JOC*, **57**, 2136; Sarracino, D.A., Richert, C., *BMCL*, **11**, 1733; Martin, N., Guldi, D.M. et al., *OL*, **3**, 2645; McElhanon, J.R., Wheeler, D.R, *OL*, **3**, 2681; Martin, I.K., Twyman, L.J., *TL*, **42**, 1119, 1123; Wang, X. et al., *TL*, **42**, 2181; Jones, D.S. et al., *TL*, **42**, 2069; Sengupta, S., Sadhukhan, S.K., *TL*, **42**, 3659 and *OM*, **20**, 1889; Zhang, J. et al., *TL*, **42**, 3599; Nierengarten, J.-F. et al., *JACS*, **123**, 9743; Mestres, R. et al., *T*, **57**, 3397; Brana, M.F. et al., *BMCL*, **11**, 3027; Pieters, R.J. et al, *EJOC*, 4685; Pu, L. et al., *JOC*, **66**, 2358, 6136; Simanek, E.E. et al., *JACS*, **123**, 8914; Taylor, P.C. et al., *JOC*, **66**, 8687; Verheyde, B., Dehaen, W., *JOC*, **66**, 4062, Raymond, K.N. et al., *ECJ*, **7**, 272; Diez-Barra, E. et al., *JOC*, **66**, 5664; van Veggel, F.C.J.M., Reinhoudt, D.N. et al., *JOC*, **66**, 4643; Tsuji, Y. et al., *OM*, **20**, 5342; Verboom, W., Reinhoudt, D.N. et al., *JOC*, **66**, 4663, 5405; Tsuji, Y. et al., *OM*, **20**, 5342; Serrano, J.L. et al., *ECJ*, **7**, 1006; Smith, D.K. et al., *ECJ*, **7**, 979; Bronk, K., Thayumanavan, S., *OL*, **3**, 2057; Mullen, K. et al., *JACS*, **123**, 8101; Raymond, K.N. et al., *ECJ*, **7**, 272; Cordova, A., Janda, K.M., *JACS*, **123**, 8248; Cai, C. et al., *CC*, 1442; Hannon, M.J., Mayers, P.C., Taylor, P.C., *AG(E)*, **40** 1081; Chow, H.-F., *ECJ*, **7**, 686; Balasubramanian, R., Maitra, U., *JOC*, **66**, 3035; Thayumanavan, S. et al., *OL*, **3**, 1961; Kim, K. et al., *AG(E)*, **40**, 746; de la Mata, F.J., Gomez, R. et al., *OM*, **20**, 2583; Vargas-Berenuel, A. et al., *JOC*, **66**, 7786; Majoral, J.-P. et al., *AG(E)*, **40**, 224; Newcome, G.R. et al., *TL*, **42**, 7537.
Dendrimer Synthesis

VIII.C.1-2 Seebach, D. et al., *ECJ*, **7**, 2873; Van Koten, G. et al., *ECJ*, **7**, 181, 1289; Meijer, E.W. et al., *JACS*, **123**, 8453.
Dendrimer Catalysts

VIII.C.1-3 Kim, C., Park, J., *JOM*, **629**, 194; Matsuda, K. et al., *TL*, **42**, 3327; Caminade, A.-M., Balavoine, G., Majoral, J.-P. et al., *T*, **57**, 2521; Caminade, A.-M., Majoral, J.-P. et al., *AG(E)*, **40**, 2626 and *TL*, **42**, 3587; Mihara, H. et al., *ECJ*, **7**, 2449; Parquette, J.R. et al., *OL*, **3**, 3129.
Hetero-Containing and Miscellaneous Dendrimers

VIII.C.2 Calixeranes

VIII.C.2-1 Abidi, R., Thuery, P., Vincens, J. et al., *TL*, **42**, 1685; Lemaire, M. et al., *JCS(P1)*, 1426; Stibor, I. et al., *CCC*, **66**, 641; Sessler, J.L. et al., *JACS*, **123**, 2099; Dospil, G., Schatz, J., *TL*, **42**, 7837; Asfari, Z. et al., *TL*, **42**, 8285; Weng, L., Zhang, Z.-Z. et al., *JCS(P2)*, 545; Pochini, A. et al., *JOC*, **66**, 8302; Fuji, K. et al., *JOC*, **66**, 4083; Van Veggel, F.C.J.M., Reinhoudt, D.N. et al., *ECJ*, **7**, 4878; Deligoz, H., Cetisli, H., *JCR(S)*, 427; Liu, Y., Inoue, Y. et al., *JOC*, **66**, 7209; Leray, I. et al., *ECJ*, **7**, 4590.
Calixarene Receptors

VIII.C.2-2 Komatsu, N., Chishiro, T., *JCS(P1)*, 1532; Miyano, S. et al., *JACS*, **123**, 779; Kim, J.S. et al., *JOC*, **66**, 5976; Mattay, J. et al., *ECJ*, **7**, 465; Ungaro, R. et al., *S*, 2105; Meth-Cohn, O. et al., *JCS(P1)*, 3297; Lhotak, P., *T*, **57**, 4775; Maas, G. et al., *T*, **57**, 4161; Dondoni, A. et al., *TL*, **42**, 3295; Komatsu, N., *TL*, **42**, 1733; Coleman, A.W. et al., *TL*, **42**, 577; Fuji, K. et al., *CPB*, **49**, 507; Makha, M., Raston, C.L., *CC*, 2470; Shao, S. et al., *SC*, **31**, 1421.
Synthesis of Calixarenes

VIII.C.2-3 Dumazet-Bonnamour, I. et al., *TL*, **42**, 8177; Chowdhury, S., Georghiou, P.E., *JOC*, **66**, 6257; Agbaria, K., Biali, S.E., *JOC*, **66**, 5482; Biali, S.E. et al., *JOC*, **66**, 2891, 7059; Nierengarten, J.-F. et al., *JOC*, **66**, 6432; Kim, J.S. et al., *JCS(P1)*, 31; Harvey, P.D. et al., *OM*, **20**, 273; Zavada, J. et al., *JOC*, **66**, 4595; Lu, G.-Y. et al., *S*, 1023; Shuker, S.B. et al., *SL*, 210; Hatsui, T. et al., *TL*, **42**, 6855; Kuhnert, N., Le-Gresley, A., *JCS(P1)*, 3393; Hioki, H. et al., *JCS(P1)*, 3265; Shorthill, B.J., Glass, T.E., *OL*, **3**, 577.
Synthesis on Calixarene

VIII.C.2-4 Nishimura, J. et al., *T*, **57**, 1219; Gerkensmeier, T. et al., *Z*, **56b**, 1063; Biali, S.E. et al., *JOC*, **66**, 2900; Tomapatanaget, B., Tuntulani, T., *TL*, **42**, 8105; Gloede, J. et al., *TL*, **42**, 9139; Nomura, E. et al., *JOC*, **66**, 8030; Timmerman, P., Reinhoudt, D.N. et al., *JOC*, **66**, 8297; Black, D.StC. et al., *T*, **57**, 2203.
Isolation, Characterization and Conformational Characteristics of Calixarenes

VIII.C.3 Rotaxanes

VIII.C.3-1 Becher, J., Stoddart, J.F., et al., *JOC*, **66**, 3559 and *T*, **57**, 947; Takata, T. et al., *BCJ*, **74**, 139, 149; Anderson, H.L. et al., *CC*, 1046 and *AG(E)*, **40**, 1071.; Kim, K., et al., *CC*, 1042 and *AG(E)*, **40**, 399; Hunter, C.A. et al., *AG(E)*, **40**, 2678; Abraham, W. et al., *EJOC*, 3921; Zehnder, D.W., II, Smithrud, D.B., *OL*, **3**, 2485; Puddephatt, R.J. et al., *CC*, 1310; Doddi, G., Ercolani, G., Franconerri, S., Mencarelli, P., *JOC*, **66**, 4950; Li, Z.-T. et al., *JOC*, **66**, 7035; Ueno, A. et al., *ECJ*, **7**, 1390.
Synthesis of Rotaxanes

VIII.C.4 Supramolecules

VIII.C.4-1 Osuka, A. et al., *TL*, **42**, 3617; Tykwinski, R.R. et al., *OL*, **3**, 1045; Aakeroy, C.B. et al., *AG(E)*, **40**, 3240; Mullen, K., Rabe, J.P. et al., *JACS*, **123**, 11462; Tour, J.M. et al., *T*, **57**, 9055; Goshe, A.J., Bosnich, B., *SL*, 941; Gossauer, A. et al., *JOC*, **66**, 4973; Stoddart, J.F. et al., *JOC*, **66**, 6857; Metrangolo, P., Resnati, G., *ECJ*, **7**, 2511; Fitzmaurice, D. et al., *ECJ*, **7**, 1309; Drain, C.M. et al., *JOC*, **66**, 6513; McMenimen, K.A., Hamilton, D.G., *JACS*, **123**, 6453; Kondo, T. et al., *AG(E)*, **40**, 894; Ganapathy, S., Ganesh, K.N. et al., *OL*, **3**, 1921; Ermer, O., Neudorfl, J., *ECJ*, **7**, 4961; Smith, D.K. et al., *ECJ*, **7**, 4730; Middel, O., Verboom, W., Reinhoudt, D.N., et al., *JOC*, **66**, 3998; Roelens, S. et al., *JOC*, **66**, 4930; Hunter, C.A. et al., *ECJ*, **7**, 4863; Hannon, M.J. et al., *AG(E)*, **40**, 1079; Zaman, M.B., Tomura, M., Yamashita, Y., *JOC*, **66**, 5987; Brandt, K. et al., *JOC*, **66**, 5701; Raymond, K.N. et al., *AG(E)*, **40**, 157.
Synthesis of Supramolecules

VIII.C.5 Others

VIII.C.5-1 Bauer, H. et al., *EJOC*, 3255; Nishimura, J. et al., *TL*, **42**, 25; Moriguchi, T., Tsuge, A. et al., *JCS(P2)*, 2084; Gleiter, R. et al., *JOC*, **66**, 3416; Minuti, L., Taticchi, A. et al., *TA*, **12**, 1179; Park, K.K. et al., *JOC*, **66**, 6800; Mallouk, T.E. et al., *JOC*, **66**, 3027; Tessier, C.A., Youngs, W.J., *OM*, **20**, 1276; Dolbier, W.R., Jr., et al., *JOC*, **66**, 7055; Sankararaman, S. et al., *JOC*, **66**, 4299; Bryce, M.R., *JOC*, **66**, 3313; Bodwell, G.J. et al., *OL*, **3**, 2093; Uneyama, K. et al., *JOC*, **66**, 7216; Matsuda-Sentou, W., Shinmyozu, T., *TA*, **12**, 839; Tsuge, A. et al., *JOC*, **66**, 9023; Pelter, A. et al., *TL*, **66**, 8391.
Synthesis of Cyclophanes

VIII.C.5-2 Raston, C.L. et al., *EJOC*, 3227; Paek, K. et al., *JOC*, **66**, 5736; Kobuke, T. et al., *AG(E)*, **40**, 457; Wan, W.B., Haley, M.M., *JOC*, **66**, 3893; Tanaka, M., Kimura, K. et al., *JOC*, **66**, 7008; Yu, L., Lindsey, J.S., *JOC*, **66**, 7402, Lindsey, J.S., *JOC*, **66**, 7342; Sherburn, M.S. et al., *JOC*, **66**, 8227; Baxter, P.N.W., *JOC*, **66**, 4170; Scheeren, H.W., Nolte, R.J.M., *JOC*, **66**, 2643; Hoye, R.C. et al., *JOC*, **66**, 2722; Maia, A. et al., *JOC*, **66**, 3249; Jurczak, J. et al., *TA*, **12**, 1763; Ishida, H., Sokabe, M. et al., *JOC*, **66**, 2978; Weber, E. et al., *JCS(P2)*, 1212; Colquhoun, H.M., Arico, F., Williams, D.J., *CC*, 2574.
Synthesis of Other Macrocycles

VIII.C.6 Reviews

VIII.C.6-1 Bradley, M. et al., *AA*, **34**, 75
  Review: "Solid-Phase Dendrimer Chemistry: Synthesis and Applications."

VIII.C.6-2 Haag, R., *ECJ*, **7**, 327;
  Review: "Dendrimers and Hyperbranched Polymers as High-Loading Supports for Organic Synthesis."

VIII.C.6-3 Tully, D.C., Frechet, J.M.J., *CC*, 1229.
  Feature Article: "Dendrimers at Surfaces and Interfaces: Chemistry and Applications."

VIII.C.6-4 Campagna, S. et al., *CSR*, **30**, 367.
  Review: "Dendrimers Based on Ruthenium(II) and Osmium(II) Polypyridine Complexes and the Approach of Using Complexes as Ligands and Complexes as Metals."

VIII.C.6-5 Hecht, S., Frechet, J.M.J., *AG(E)*, **40**, 75.
Review: **"Dendrimer Encapsulation of Function: Applying Nature's Site Isolation Principle from Biomimetics to Materials Science."**

VIII.C.6-6 Astruc, D., Chardac, F., *CRV*, **101**, 2991.
Review: **"Dendritic Catalysts and Dendrimers in Catalysis."**

VIII.C.6-7 Crooks, R.M. et al., *ACR*, **34**, 181.
Review: **"Dendrimer-Encapsulated Metal Nanoparticles: Synthesis, Characterization, and Applications to Catalysis."**

VIII.C.6-8 Gorman, C.B., Smith, J.C., *ACR*, **34**, 60.
Review: **"Structure-Property Relationships in Dendritic Encapsulation."**

VIII.C.6-9 Nierengarten, J.-F., *AG(E)*, **40**, 2973.
Highlights: **"Ring-Opened Fullerenes: An Unprecedented Class of Ligands for Supramolecular Chemistry."**

VIII.C.6-10 Cotton, F.A., Lin, C., Murillo, C.A., *ACR*, **34**, 759.
Review: **"Supramolecular Arrays Based on Dimetal Building Units."**

VIII.C.6-11 Beek, P.D., Gale, P.A., *AG(E)*, **40**, 487.
Review: **"Anion, Recognition and Sensing: The State of the Art and Future Perspectives."**

VIII.C.6-12 Sherrington, D.C., Taskinen, K.A., *CSR*, **30**, 83.
Review: **"Self-Assembly in Synthetic Macromolecular Systems *via* Multiple Hydrogen Bonding Interactions,"**

VIII.C.6-13 Osuka, A. et al., *SL*, 1663.
Account: **"Discrete Giant Porphyrin Arrays: Challenges to Molecular Size, Length and the Extent of Electronic π-Conjugation."**

VIII.C.6-14 Stoddart, J.F., guest ed., *ACR*, **34**, 409-522.
Reviews: **"Special Issue on Molecular Machines."**

VIII.C.6-15 Panova, I.G., Topchieva, I.N., *RCR*, **70**, 23.
Review: "Rotaxanes and Polytaxanes. Their Synthesis and the Supramolecular Devices Based on Them."

VIII.C.6-16 Albrecht, M., *CRV*, **101**, 3457.
Review: "Let's Twist Again—Double-Stranded, Triple-Stranded, and Circular Helicates."

VIII.C.6-17 Kim, B.H. et al., *S*, 2191.
Feature "Multiple Cycloadditive Macrocyclization: An
Article: Efficient Method for Crown Ether-Type Cyclo-phanes, Bis-Calix[4]arenes and Silamacrocycles."

VIII.C.6-18 Sliwa, W., *H*, **55**, 181.
Review: "Calixarenes Bearing Azaaromatic Moieties."

VIII.C.6-19 Sessler, J.L. et al., *PAC*, **73**, 1041.
Review: "Calixphyrins. Hybrid Macrocycles at the Structural Crossroads Between Porphyrins and Calixpyrroles."

VIII.C.6-20 Warmuth, R., Yoon, J., *ACR*, **34**, 95.
Review: "Recent Highlights in Hemicarcerand Chemistry."

VIII.D. Total Syntheses of Selected Natural Products

(see also VIII.B)

VIII.D.-A

Davis, F.A., *OL*, **3**, 3169	(+)-**241D**
Chang, S., *TA*, **12**, 29	(-)-**A26771B**
Theodorakis, E.A., *JOC*, **66**, 8843	(-)-**Acanthoic Acid**
Adamczyk, M., *TA*, **12**, 2385	(S)-(-)-**Acromelobinic Acid**
Gung, B.W., *TL*, **42**, 4761	(+) & (-)-**Adociaacetylene B**
Bonjoch, J., *ECJ*, **7**, 3446	**Aeruginosin 298-A & 298-B**
Kobayashi, M., *ECJ*, **7**, 2663	**Agosterol A**
Ghosh, A.K., *OL*, **3**, 2677	**AI-77-B**
Chern, J.-W., *TL*, **42**, 1111	**Ailanthoidol**
Martin, S.F., *JACS*, **123**, 8003	(±)-**Akuammicine**
Hart, D.J., *JACS*, **123**, 5892	*ent*-**Alantrypinone**
Ma, D., *TL*, **42**, 6929	(+)-α-**Allokainic Acid**
Comins, D.L., *OL*, **3**, 469	(+)-**Allopumiliotoxin 267A**
Holmes, A.B., *ECJ*, **7**, 1845	(+)-**Allopumiliotoxin 323B'**
Paterson, I., *AG(E)*, **40**, 4055	(+)-**Altohytrin A**
Jacobsen, E.N., *JACS*, **123**, 10772	(+)-**Ambruticin**

Martin, S.F., *JACS*, **123**, 12432   (+)-**Ambruticin S**
Williams, D.R., *JACS*, **123** 765   (+)-**Amphidinolide K**
Bermejo, F.A., *JOC*, **66**, 8287   (+)-**Ampullicin**
Wahala, K., *JCS(P1)*, 642   **Angolensin**
Wu, Y.-L., *JOC*, **66**, 853   **Annonacin**
Stenstrom, Y., *TA*, **12**, 1407   (-)-**Aplyolide A**
Ogasawara, K., *TL*, **42**, 1049   (-)-**Aphanorphine**
Funk, R.L., *OL*, **3**, 3923   (±)-**Aphanorphine**
Kitahara, T., *H*, **55**, 1   **Apicidin**
Singh, G., *TL*, **42**, 6603   **Apicidin A**
De Brabander, J.K., *TL*, **42**, 1217   **Apicularen A**
Spinella, A., *SL*, 1971   **Aplyolide A**
Koert, U., *AG(E)*, **40**, 2063   **Apoptolodinone**
Voelter, W., *ZN*, **56B**, 325   S-(+)-**Argentilactone**
Enders, D., *SL*, 1796   S-(+)-**Argentilactone**
Ogasawara, K., *OL*, **3**, 291   (+)-**Arnicenone**
Barriault, L., *OL*, **3**, 1925   (+)-**Arteannium M**
Brown, R.C.D., *JOC*, **66**, 6719   (±)-**Asarinin**
Overman, L.E. *JACS*, **123**, 9468   **Asperazine**
Mehta, G., *TL*, **42**, 8101   (±)-**Asterisca-3(15),6-diene**
Krafft, M.E., *JOC*, **66**, 7443   (±)-**Asteriscanolide**
Taber, D.F., *JOC*, **66**, 944   (-)-**Astrogorgiadiol**
Suenaga, K., *OL*, **3**, 527   **Attenol A & B**
Coleman, R.S., *AG(E)*, **40**, 1736   **Azinomycin A**

VIII.D.-B

Hanessian, S., *JACS*, **123**, 10200   **Bafilomycin A$_1$**
Back, T.G., *JOC*, **66**, 4361   (±)-**Bakkenolide A**
Li, W.-D.Z., *OL*, **3**, 2555   (±)-**Balanitol**
Gerwick, W.H., *CC*, *1934*   **Barbamide**
Overman, L.E., *JACS*, **123**, 10782   **Batzelladine F**
LeDian, C., *HCA*, **84**, 890   (-)-**Bauhinin**
Boger, D.L., *JACS*, **123**, 561   **BE-22179**
Kinder, F.R. Jr., *JOC*, **66**, 2118   **Bengamide B & E**
Banwell, M.G., *JOC*, **66**, 6768   *ent*-**Bengamide E**
Chavan, S.P., *JOC*, **66**, 6197   D-(+)-**Biotin**
Woerpel, K.A., *OL*, **3**, 675   (+)-**Blastmycinone**
Suh, Y.-G., *TL*, **42**, 1691   (+)-**Brefeldin A**
Parsons, P.J., *SL*, 257   **Brevioxime**
Sha, C.-K., *OL*, **3**, 2177   (-)-**Brunsvigine**
Hesse, M., *HCA*, **84**, 180   (±)-**Budmunchiamine A, B & C**

VIII.D.-C

Charlton, J.L., *JOC*, **66**, 8606   (±)-**Cagayanin**
Nicolaou, K.C., *TA*, **12**, 937   D-**Callipeltose**

VIII.D.-D

Wang, J.-J., *JOC*, **66**, 2881
Lobo, A.M., *TL*, **42**, 6663

Massanet, G.M., *T*, **57**, 2171
Kim, S., *TL*, **42**, 7641
Han, H., *TL*, **42**, 4001
Kim, S., *TL*, **42**, 7641
Han, H. *TL*, **42**, 4001
White, J.D., *JACS*, **123**, 5407
Pan, X., *JCR(S)*, 328

Kraus, G.A., *TL*, **42**, 6649
Taber, D.F., *JOC*, **66**, 3423
Bailey, P.D., *TL*, **42**, 113
Hsung, R.P., *JOC*, **66**, 1049

Smith, C.D., *JOC*, **66**, 3459
Padwa, A., *JOC*, **66**, 3119
Jung, M., *TL*, **42**, 3997
Datta, A., *T*, **57**, 1169
Zhu, J., *TL*, **42**, 3431
Comins, D.L., *JOC*, **66**, 6829
Dhavale, D.D., *TL*, **42**, 747
Pattenden, G., *SL*, 1869
O'Doherty, G.A., *OL*, **3**, 401
Comins, D.L., *TL*, **42**, 6839
Marshall, J.A., *JOC*, **66**, 8037
Durand, P., *T*, **57**, 2757
Alvarez, M., *TL*, **42**, 315
Argade, N.P., *JOC*, **66**, 9038
Steglich, W., *SL*, 759
Rychnovsky, S.D., *AG(E)*, **40**, 3224
Trost, B.M., *JACS*, **123**, 9449
Zhang, W., *OL*, **3**, 2353

Harrowven, D.C., *T*, **57**, 9157
Ward, D.E., *JOC*, **66**, 7832
Isobe, M., *T*, **57**, 4543
Durand, P., *TL*, **42**, 2121
Tomioka, K., *TL*, **42**, 8493
Tietze, L.F., *AG(E)*, **40**, 901
Langlois, Y., *TL*, **42**, 8297
Boojamra, C.G., *JACS*, **123**, 870
Cossy, J., *EJOC*, 2841
Rokach, J., *TL*, **42**, 8277
Bringmann, G., *T*, **57**, 1253

DC-81
(+) & (-)-Debromoflustramine B
(+)-Decipienin A
(+)-*trans*-Decursidinol
Decursin
(+)-Decursinol
Decursinol Angelate
9,10-Dehydroepothiolone D
Δ^5-Dehydrosugiyl Methyl Ether
Deliquinone
(-)-Delobanone
Demethoxy Fumitremorgin C
(±)-7-Demethoxyzanthodioline
Dendroamide A
(±)-Dedrobine
(+)-Deoxoartemisitene
(-)-Deoxoprosophylline
(-)-Deoxoprosophylline
(+)-Deoxoprospinine
1-Deoxy Castanospermine
***bis*-Deoxylophotoxin**
D & L-Deoxymannojirimycin
1-Deoxynorjirimycin
(-)-Deoxypukalide
(±)-15-Deoxyspergualin
Deoxyvariolin B
Deoxyvaseinone
Deoxyviolacein
Dermostatin A
Deschlorocallipeltoside A
(±)-Desmethylamino FR 901 483
(±)-1-Desoxyhypnophilin
Destruxin B
(-)-5,11-Dideoxytetrodtoxin
Diheteropeptin
Dihydrexidine
5,6-dihydrocineromycin B
6',7'-Dihydrokeramine C
Dihydropacidamycin D
(-)-4a,5-Dihydrostreptazolin
2,3-Dinor *i*PF$_{2\alpha}$-III
Dioncophylline B

Paterson, I., *JACS*, **123**, 9535   (+)-Discodermolide
Overman, L.E., *JACS*, **123**, 9465   Ditryptophenaline
Ghosh, A.K., *OL*, **3**, 635   (-)-Doliculide
Snider, B.B., *OL*, **3**, 1761   (+) & (-)-Dysibetaine
Yamada, Y., *JOC*, **66**, 1429   Dysididiolide

VIII.D.-E

Wee, A.G.H., *JOC*, **66**, 8935   (-)-Eburnamonine
Wee, A.G.H., *JOC*, **66**, 8935   (+)-*epi*-Eburnamonine
Kim, D., *T*, **57**, 1247   (±)-β-Elemene
Denmark, S.E., *JOC*, **66**, 4276   (+)-1-Epiaustraline
Evans, D.A., *OL*, **3**, 3009   (-)-Epibatidine
Kitahara, T., *H*, **55**, 861   (±)-Epibatidine
Weinreb, S.M., *JACS*, **123**, 8851   7-Epicylindrospermopsin
Okamura, H., *TL*, **42**, 8649   (+)-Epiepoformin
Chan, T.H., *OL*, **3**, 739   Epigallocatechin-3-Gallate
Sulikowski, G.A., *TL*, **42**, 6621   (+)-5-Epiindolizidine 167B
Greene, A.E., *JOC*, **66**, 5438   (-)-Epilentiginosine
Brown, R.C.D., *TL*, **42**, 473; *JOC*, **66**, 6719 (±)-Epimagnolin A
Flachsmann, F., *HCA*, **84**, 416   (+)-7-Epispirojatamol
Ho, T.-L., *T*, **57**, 507   3-Epitacamonine
Ho, T.-L., *T*, **57**, 507   14-Epitacamonine
Dobler, M.R., *TL*, **42**, 215   (+)-Epopromycin B
Hatakeyama, S., *TL*, **42**, 7867   Epopromycin B
Avery, M.A., *TL*, **42**, 7341   Epothilone A
Carreira, E.M., *JACS*, **123**, 3611   Epothilone A & B
Furstner, A., *CC*, 1057; *ECJ*, **7**, 5299   Epothilone A & C
Sinha, S.C., *ECJ*, **7**, 1691   Epothilone A & F
Martin, H.J., *ECJ*, **7**, 2261   Epothilone B
Avery, M.A., *OL*, **3**, 3607   Epothilone B
Taylor, R.E., *OL*, **3**, 2221   Epothilone B & D
White, J.D., *JACS*, **123**, 5407   Epothilone B & D
Thomas, E.J. *TL*, **42**, 8373   Epothilone B & D
Nicolaou, K.C., *AG(E)*, **40**, 207   Epoxyquinomycin B
Li, W.Z., *TA*, **12**, 95   (+)-3,4-Epoxycenbrene-A
Pettus, T.R.R., *OL*, **3**, 905   (±)-Epoxysorbicillinol
Wood, J.L., *JACS*, **123**, 2097   (±)-Epoxysorbicillinol
Shishido, K., *TL*, **42**, 2517   (-)-Equisetin
Flachsmann, F., *HCA*, **84**, 416   (-)-Erythrodiene
Aoyama, Y., *SL*, 1452   (-)-α-Eudesmol
Corey, E.J., *JACS*, **123**, 1872   Eunicenone A
Funk, R.L., *JACS*, **123**, 9455   (±)-Euplotin A
Couture, A., *JOC*, **66**, 8064   Eupolauramine
Okamura, H., *T*, **57**, 1903   (-)- & (+)-Eutipoxide B

VIII.D.-F

Brown, R.C.D., *JOC*, **66**, 6719   (±)-Fargesin
Hatakeyama, S., *OL*, **3**, 953   (+)-Febrifugine
Takeuchi, Y., *T*, **57**, 1213   (+)-Febrifugine
Lu, Z.-H., *TL*, **42**, 8919   (R)-FluoxetIne
Boger, D.L., *JACS*, **123**, 4161   Fostriecin
Kitahara, T., *TL*, **42**, 8207   FR 901464
Jacobsen, E.N., *JACS*, **123**, 9974   FR 901464
Ciufolini, M.A., *OL*, **3**, 765; *JACS*, **123**, 7534   FR 901483
Funk, R.L., *OL*, **3**, 1125   FR 901483
Kita, T., *JACS*, **123**, 3214   Fredericamycin A
Danishefsky, S.J., *JACS*, **123**, 1878   Frondosin
Watanabe, H., *TL*, **42**, 917   Fudecalone
Simpkins, N.S., *SL*, 661   Fumagillol
Eustache, J., *OL*, **3**, 2737   (-)-Fumagillol
Avendano, C., *SL*, 1387   (-)-Fumiquinazoline F
Oshima, Y., *JOC*, **66**, 6982   Furanodictine A
Josepth-Nathan, P., *JOC*, **66**, 1186   Furstramide A & B
Josepth-Nathan, P., *JOC*, **66**, 1186   Furstramine A & B
Song, J.J., *JOC*, **66**, 605   Fusaric Acid
Song, J.J., *JOC*, **66**, 605   (S)-(+)-Fusarindlic Acid
Snider, B.B., *SC*, 2667   (±)-Fusaricide

VIII.D.-G

Guillou, C., *AG(E)*, **40**, 4745   (±)-Galathamine
Charlton, J.L., JOC, *66*, 8606   (±)-Galbulin
Johansson, M., *OL*, **3**, 2843   (+)-Galiellalactone
Avendano, C., *SL*, 1387   (-)-Glyantrypine
Zhao, G., *T*, **57**, 2147   Goniofufurone
Enders, D., *SL*, 1796   S-(-)-Goniothalamin
Etoh, H., *CL*, 210   Grandinal
Cha, J.K., *JACS*, **123**, 3243   Grandirubrine
Knapp, S., *OL*, **3**, 3583   Griseolic Acid B

VIII.D.-H

Rodrico, R.G.A., *JOC*, **66**, 3639   (±)-Halenaquinone
Takemoto, Y., *JOC*, **66**, 81   Halicholactone
Jiang, B., *JOC*, **66**, 4865   (-)-Hamacanthin
Kerr, M.A., *OL*, **3**, 3189   (±)-Hapalindole Q
Larson, D.S., *JOC*, **66**, 7427   (+)-Hatomarubicin B
Shishido, K., *JOC*, **66**, 309   (-)-Heliannuol E
Danishefsky, S.J., *AG(E)*, **40**, 4713   Heliquinomycinone
Vedejs, E., *JOC*, **66**, 7355   (-)-Hemiasterlin
Rainier, J.D., *JOC*, **66**, 1380   (±)-Hemibrevetoxin B
Yokokawa, F., *T*, **57**, 6311   (-)-Hennoxazole A

Furstner, A., *CC*, 671
Spur, B.W., *T*, **57**, 25
Kodama, M., *T*, **57**, 1235
Pilli, R.A., *TL*, **42**, 6999
Palmisano, G., *JOC*, **66**, 8447
Hsung, R.P., *JOC*, **66**, 1049
Martin, O.R., *TA*, **12**, 1807
Nicolaou, K.C., *AG(E)*, **40**, 761
Hesse, M., *HCA*, **84**, 1253
Voelter, W., *ZN*, **56B**, 689
Li, Y., *TL*, **42**, 275
Kinoshita, T., *JHC*, **38**, 165
Williams, R.M., *OL*, **3**, 4287

Herbarumin I
12(R) & (S)-HETE
Hippospongic Acid A
(±)-Homopumiliotoxin 223G
(-)-Horsfiline
Huajiaosimuline
(+)-Hyacinthacine A₂
Hybocarpone
Hydroxycelacinnine
ω-Hydroxyisochrysophanol
(-)-13-Hydroxyneocembrene
Hyperolactone A & C
(+)-Hypusine

VIII.D.-I, -J

Mukai, C., *JOC*, **66**, 5875
Noda, Y., *H*, **55**, 1839
Funk, R.L., *OL*, **3**, 2611
Cha, J.K., *JACS*, **123**, 3243
Shattuck, J.C., *OL*, **3**, 3021
Remuson, R., *TL*, **42**, 4617
Kim, G., *TA*, **12**, 2073
Liebeskind, L.S., *JACS*, **123**, 12477
Kim, G., *OL*, **3**, 2985; *TA*, **12**, 2073
Sulikowski, G.A., *TL*, **42**, 6621
Kamiyama, K., *CPB*, **49**, 1604
Ogasawara, K., *OL*, **3**, 679
Bermejo, F.A., *JOC*, **66**, 8287
Voelter, W., *ZN*, **56B**, 689
Lu, X., *JOC*, **66**, 6545
Lu, X., *JOC*, **66**, 6545
Ganem, B., *OL*, **3**, 201
Hatakeyama, S., *OL*, **3**, 953
Takeuchi, Y., *T*, **57**, 1213
Cha, J.K., *JACS*, **123**, 3243
Bols, M., *ECJ*, **7**, 3744
Crimmins, M.T., *JACS*, **123**, 1533
Kende, A.S., *OL*, **3**, 2505
Voelter, W., *ZN*, **56B**, 689

(-)-Ichthyothereol
(-)-Idiadione
(±)-Illudin C
Imerubrine
Imperanene
(±)-Indolizidine 167B
(-)-Indolizidine 167B
(-)-Indolizidine 209B
(-)-Indolizidine 209D
Indolizidine 223AB
Indolmycin
(-)-Iridolactone
(+)-Isoampullicin
Isochrysophanol
(±)-Isocynodine
(±)-Isocynometrine
Isofagomine
(+)-Isofebrifugine
(+)-Isofebrifugine
Isoimerubrine
Isolagomine
(-)-Isolaurallene
(±)-Isostemonamide
Isozyganein

Kawamoto, H., *T*, **57**, 981
Porco, J.A., Jr., *OL*, **3**, 1649
Clive, D.L.J., *JOC*, **66**, 4841
Harrowven, D.C., *TL*, **42**, 6973

J-113397
(-)-Jesterone
(+)-Juruenolide C
Justicidin B

VIII.D.-K

Giralt, E., *JACS*, **123**, 11398   **Kahalalide F**
Ogasawara, K., *TL*, **42**, 7587   **(-)-Kainic Acid**
Ganem, B., *OL*, **3**, 485   **(-)-α-Kainic Acid**
Ma, D., *JACS*, **123**, 9706   **Kaitocephalin**
Kakinuma, K., *JCS(P1)*, 569   **Kedarosamine**
Piers, E., *S*, 2138   **(±)-Kelsoene**
Mehta, G., *TL*, **42**, 2855   **(+)-Kelsoene**
Kobayashi, S., *JACS*, **123**, 1372; *JOC*, **66**, 5580 **Khafrefungin**
Jiang, B., *TA*, **12**, 2835   **Kurzilactone**

VIII.D.-L

Ruchirawat, T., *TL*, **42**, 1205   **Lamellarin G Trimethyl Ether**
Guitian, E., *SL*, 1164   **Lamellarin I & K**
Monti, H., *TL*, **42**, 6125   **Lancifolol**
Pearson, W.H., *JOC*, **66**, 6724   **(±)-Lapidlectene B**
Ma, D., *OL*, **3**, 3927   **(-)-Lasubine II**
Wood, J.L., *JOC*, **66**, 7025   **(+)-Latifolic Acid**
Wood, J.L., *JOC*, **66**, 7025   **(+)-Latifoline**
Mulzer, J., *AG(E)*, **40**, 3842   **Laulimalide**
Enev, V.S., *JACS*, **123**, 10764   **Laulimalide**
Ghosh, A.K., *TL*, **42**, 3399; *JOC*, **66**, 8973 **(-)-Laulimalide**
Paterson, I., *OL*, **3**, 3149   **(-)-Laulimalide**
Kobayashi, M., *TL*, **42**, 1941   **Lembehyne A**
Funk, R.L., *OL*, **3**, 3923   **(±)-Lennoxamine**
Greene, A.E., *JOC*, **66**, 5438   **(+)-Lentiginosine**
Funk, R.L., *OL*, **3**, 3511   **Lepadiformine**
Kibayashi, C., *JOC*, **66**, 3338   **(-)-Lepadin A, B & C**
Yamamoto, H., *SL*, 1113   **(R)-Limonene**
Marcos, I.S., *T*, **57**, 713   **(+)-Limonidilactone**
Moeller, K.D., *OL*, **3**, 2685   **Linaool Oxide**
Ohta, S., *OL*, **3**, 1359   **(±)-Linderol A**
Vannieuwenhze, M.S., *JACS*, **123**, 6983 **Lipin I**
Spur, B.W., *TL*, **42**, 6057   **15-*epi*-Lipoxin A$_4$**
Pathak, T., *T*, **57**, 1093   **D-Lividosamine**
Tatsuta, K. *CL*, 172   **LL-Z1640-2**
Spur, B.W., *T*, **57**, 25   **LTB$_4$**
Tatsuta, K., *TL*, **42**, 7625   **Luminacin C$_1$ & C$_2$**
Ciufolini, M.A., *TL*, **42**, 1907   **Luzopeptin C**
Yokokawa, F., *TL*, **42**, 4171   **Lyngbyabellin A**

VIII.D.-M

Kobayashi, Y., *JOC*, **66**, 2011   **Macrosphelide A & B**
Kobayashi, Y., *TL*, **42**, 2817   **Macrosphelide C & F**
Tamai, Y., *JCS(P1)*, 645   **(-)-Malyngolide**

Suzuki, K., *OL*, **3**, 1741   (-)-**Malyngolide**
Ghosh, A.K., *TL*, **42**, 6231   (-)-**Malyngolide**
Voelter, W., *ZN*, **56B**, 689   **Morindaparvin**
Ma, D., *OL*, **3**, 2189   **Martinellic Acid**
Snider, B.B., *OL*, **3**, 4217   (±)-**Martinellic Acid**
Kamat, V.P., *JCR(S)*, 549   **Maritimun**
Bringmann, G., *T*, **57**, 1269   **Mastigophorene C & D**
Cohen, T., *JACS*, **123**, 30   (±)-**Matatabiether**
Perkins, M.V., *OL*, **3**, 123   (-)-**Membrenone C**
Akaji, K., *T*, **57**, 1749   **MEN 11420**
Konda, Y., *T*, **57**, 4311   **MER-N5075A**
Padwa, A., *JOC*, **66**, 3119   (±)-**Mesembrane**
Taber, D.F., *JOC*, **66**, 143   (-)-**Mesembrine**
Kobayashi, S., *TL*, **42**, 4531   **9-Methoxystrobilurin K**
Kobayashi, S., *BCML*, **11**, 2783   **9-Methoxystrobilurin L**
Garcia, J., *SL*, 120   (-)-**Methylenolactocin**
Bennasar, M.-L., *CC*, 1166   (-)-$N_{(a)}$-**Methylervitsine**
Aoyama, Y., *BMCL*, **11**, 1695   **Methyllinderone**
Munoz, L., *JOC*, **66**, 4206   **Minalemine A**
Singh, G., *SL*, 1983   (-)-**MK 7607**
Jauch, J., *SL*, 87   **Mniopetal E**
Tadano, K., *T*, **57**, 7291   (-)-**Mniopetal F**
Danishefsky, S.J., *JACS*, **123**, 10903   **Monocillin I**
Ogasawara, K., *CC*, 1094   (-)-**Morphine**
Kuehne, M.E., *JOC*, **66**, 1560   **Mossambine**
Kitahara, T., *T*, **57**, 3899   **Mueggelone**
Chida, N., *CC*, 1932   (+)-**Myriocin**

VIII.D.-N, -O

Hirama, M., *JACS*, **123**, 2887, 11294   **N1999-A2**
Omura, S., *OL*, **3**, 2289   **Nafuredin**
Giannis, A., *JACS*, **123**, 11586   **Nakijiquinone C**
Ghosh, S., *T*, **57**, 2011   (±)-β-**Necrodol**
Eustache, J., *TL*, **42**, 7949   **Nectrisine**
Moeller, K.D., *TL*, **42**, 7163   (+)-**Nemorensic Acid**
Tomioka, K., *JOC*, **66**, 8199   (-)-**Neplanocin A**
Ogasawara, K., *H*, **54**, 43   (-)-**Nitraraine**
Bols, M., *ECJ*, **7**, 3744   **Noeuromycin**
Harayama, T., *JCS(P1)*, 523   **Norchelerythrine**
Pattenden, G., *SL*, 1873   **Nostocyclamide**
Wakamiya, T., *BCJ*, **74**, 1743   **NPTX-594**

Panek, J.S., *JOC*, **66**, 2747   **Oligomycin C**

VIII.D.-P

Kilburn, J.D., *JCS(P1)*, 487	6-*epi*-Paeonilactone A
Kilburn, J.D., *JCS(P1)*, 487	Paeonilactone B
Thomas, E.J., *TL*, **42**, 4969	Pamamycin 607
Metz, P., *TL*, **42**, 7801	Pamamycin 607
Lee, E., *JACS*, **123**, 10131	Pamamycin 607
Ogasawara, K., *OL*, **3**, 679	(+)-Pedicularis Lactone
Argade, N.P., *JOC*, **66**, 9038	Pegamine
Gallos, J.K., *TL*, **42**, 5769	Pentenomycin
Shioiri, T., *T*, **57**, 4759	Pentosidine
Garcia, J., *SL*, 120	(-)-Phaseolinic Acid
Smith, A.B., III, *JACS*, **123**, 4834; 10942	(+)-Phorboxazole A
Liebscher, J., *T*, **57**, 4867	Phorbazole C
Ogasawara, K., *TL*, **42**, 1049	(-)-Physostigmine
Ogasawara, K., *TL*, **42**, 1049	(-)-Physovenine
Shiozaki, M., *T*, **57**, 9087	S-lyxo-Phytosphingosine
Molander, G.A., *JOC*, **66**, 4344	(+) & (-)-Pinidinol
Liebeskind, L.S., *OL*, **3**, 3381	Piperemethystine
Keck, G.E., *OL*, **3**, 707	(-)-Pironetin
Mori, K., *TL*, **42**, 2357	Plakoside A
White, J.D., *JACS*, **123**, 8593	Polycavernoside A
Craig, D., *SL*, 1602	(+)-Preussin
Cossy, J., *SL*, 1578	(-)-Prosophylline
Kobayashi, T., *TL*, **42**, 1703	Δ^7-Prostaglandin A$_1$ Methyl Ester
Feringa, B.L., *JACS*, **123**, 5841	Prostaglandin A$_1$ Methyl Ester
Murphy, P.J., *TL*, **42**, 3377	Ptilomycalin A
Spilling, C.D., *JOC*, **66**, 3111	Pukealidin A
Bieber, L.W., *HCA*, **84**, 141	Pumiliotoxin C
Mori, M., *JOC*, **66**, 7873	Pumiliotoxin C
Srikrishna, A., *TL*, **42**, 3929	(-)-2-Pupukeanone
Clive, D.L.J., *JOC*, **66**, 954; *TL*, **42**, 2253	Puraquinonoic Acid
Gin, D.Y., *AG(E)*, **40**, 1128	(+)-Pyrenolide D
Hibino, S., *JOC*, **66**, 8793	(R)-(-)-Pyridindolol
Hibino, S., *JOC*, **66**, 8793	(R)-(-)-Pyridindolol K1 & K2
Baldwin, J.E., *OL*, **3**, 1145	Pyrinodemin A
Snider, B.B., *TL*, **42**, 1639	Pyrinodemin A & B

VIII.D.-Q, -R

Yadav, J.S., *JOC*, **66**, 8370	(+)-*allo* & *talo*-Quercitol
Stork, G., *JACS*, **123**, 3239	Quinine
Danishefsky, S.J., *JACS*, **123**, 10903	Radicirol
Williams, D.R., *OL*, **3**, 1383	(-)-Ratjadone
Kalesse, M., *JOC*, **66**, 1885	(+)-Ratjadone
Harrowven, D.C., *TL*, **42**, 6973	Retrojusticidin B

Rizzacasa, M.A., *JOC*, **66**, 2382   (-)-**Reveromycin B**
Magnus, P., *T*, **57**, 8647   (±)-**Rhazinilam**
Keck, G.E., *AG(E)*, **40**, 231   **Rhizoxin D**
Metz, P., *AG(E)*, **40**, 1058   **Ricciocarpin A & B**
Danishefsky, S.J., *JACS*, **123**, 351   (±)-**Rishirilide B**
Dobler, M.R., *TL*, **42**, 8281   (±)-**Rocaglamide**
Reiser, O., *OL*, **3**, 1315   (-)-**Roccellaric Acid**
Suzuki, T., *TL*, **42**, 1543   (+)-**Rogioloxepane A**
Barluenga, J., *ECJ*, **7**, 3533   (+)-**Rolipram**
Tius, M.A., *JACS*, **123**, 8509   **Roseophilin**
Boger, D.L., *JACS*, **123**, 8515   *ent*-(-)-**Roseophilin**
Goti, A., *OL*, **3**, 1367   (-)-**Rosmarinecine**
Liras, S., *JACS*, **123**, 5918   **Rugulovasine A & B**
Panek, J.S., *JOC*, **66**, 2747   **Rutamycin B**
White, J.D., *JOC*, **66**, 5217   **Rutamycin B**

VIII.D.-S

Galopin, C.C., *TL*, **42**, 5589   (±)-*trans*-**Sabine Hydrate**
Grossman, R.B., *OL*, **3**, 4027   (±)-**Sacarin**
Pinto, B.M., *JOC*, **66**, 2312   **Salacinol**
Furstner, A., *ECJ*, **7**, 5286   (-)-**Salicylihalamide**
Smith, A.B., III, *SL*, 1019   (+)-**Salicylihalamide A**
Snider, B.B., *OL*, **3**, 1817   (-)-**Salicylihalamide A**
Labrecque, D., *TL*, **42**, 2645   **Salicylihalamide A & B**
Martinez, A.G., *TL*, **42**, 7795   **Sarkomycin**
Overman, L.E., *JACS*, **123**, 9033   **Sclerophytin A**
Overman, L.E., *OL*, **3**, 135   **Sclerophytin A & B**
Doskotch, R.W., *JACS*, **123**, 9021   **Sclerophytin A & B**
Tanaka, T., *OL*, **3**, 619; *JOC*, **66**, 4831 (±)-**Scopadulin**
Liras, S., *OL*, **3**, 703   (±)-**Securinine**
Sonnet, P., *TL*, **42**, 1681   **Segetalin A**
Corey, E.J., *OL*, **3**, 3215   **Serratenediol**
Liras, S., *JACS*, **123**, 5918   **Setoclavine**
Overman, L.E., *JACS*, **123**, 4851   **Shahamin K**
Hsung, R.P., *JOC*, **66**, 1049   **Simulenoline**
Calter, M.A., *JOC*, **66**, 7500   **Siphonarienal**
Coelho, F., *T*, **57**, 6901   (±)-**Sitophilate**
Takahashi, T., *SL*, 1935   (±)-**Smenospondiol**
Hagiwara, H., *OL*, **3**, 251   (-)-**Solanapyrone E**
Hertweck, C., *SL*, 1965   D & L-*erythro*-**Sphinganine**
Lin, G., *SL*, 904   **Sphingofungin E**
Shiozaki, M., *TL*, **42**, 2701   **Sphingofungin E**
Shiozaki, M., *T*, **57**, 9087   D-erythro-**Sphingosine**
Kutschy, P., *JOC*, **66**, 3940   (S)-(-)-**Spirobrassinin**
Paterson, I., *AG(E)*, **40**, 4055   **Spongistatin 1**
Harrity, J.P.A., *TL*, **42**, 9055   (±)-**Sporochnol A**
Venkesan, H., *JOC*, **66**, 3653   **SR 121463**

Zanda, M., *JOC*, **66**, 5637   (+) & (-)-Statine
Uang, B.-J., *JOC*, **66**, 5627   *rac*-Staurosporine
Mori, J., *JCS(P1)*, 657   Stellettadine A
Kibayashi, C., *OL*, **3**, 193   (-)-Stellettamide B
Tanaka, T., *JOC*, **66**, 7107   (±)-Stemodinone
Kende, A.S., *OL*, **3**, 2505   (±)-Stemonamide
Williams, D.R., *OL*, **3**, 2721   (-)-Stemospironine
Morimoto, Y., *ECJ*, **7**, 4107   (-)-Stenine
Yamamoto, Y., *SL*, 694   Stevastatin B
Yamada, Y., *TL*, **42**, 9233   Stolonidiol
Martin, S.F., *JACS*, **123**, 8003   (±)-Strychnine
Heathcock, C.H., *OL*, **3**, 4323   Styelsamine B
Mehta, G., *CC*, 1892   Sucatine G

VIII.D.-T

Ikemoto, T., *T*, **57**, 1525   TAK-779
Joullie, M.M., *JACS*, **123**, 4469   Tamandarin A & B
Ciufolini, M.A., *JACS*, **123**, 7534   TAN 1251C
Jokela, R., *TL*, **42**, 6593   (±)-Tangutorine
Ramachandran, P.V., *JOC*, **66**, 2512   Tarchonanthus Lactone
Bates, R.W., *JCS(P1)*, 654   (±)-Tashiromine
Marshall, J.A., *JOC*, **66**, 1373   Tautomycin
Parrain, J.-L., *OL*, **3**, 1713   (±)-Taxifolial A
Boger, D.L., *JACS*, **123**, 1862   Teicoplanin Aglycone
Evans, D.A., *JACS*, **123**, 12411   Teicoplanin Aglycone
Mori, K., *TL*, **42**, 1527   (+)-Testudinariol A
Williams, R.M., *AG(E)*, **40**, 1463   (-)-Tetrazomine
Paquette, L.A., *JACS*, **123**, 4492   Teubrevin G & H
Young, R.N., *JACS*, **123**, 11381   (±)-Thielocin A1β
Boger, D.L., *JACS*, **123**, 561   Thiocoraline
Uyehara, T., *TL*, **42**, 699   2-Thiocyanato
    *neo*Pupukeanane
Srikrishna, A., *JOC*, **66**, 4379   4-Thiocyanato
    *neo*Pupukeanane
Wu, Y.-L., *JOC*, **66**, 853   Tonkinecin
Kuhnert, N., *TL*, **42**, 9261   Tricetanidin
Cook, J.M., *OL*, **3**, 4023   Trinervine
Yang, D., *OL*, **3**, 111   (+)-Triptocallol
Harayama, T., *JCS(P1)*, 523   Trisphaeridine
Pattenden, G., *TL*, **42**, 2573   Trunkamide A
Wee, A.G.H., *JOC*, **66**, 8513   (-)-Turneforcidine
Giralt, E., *JOC*, **66**, 7568   Trunkamide A
Argade, N.P., *JOC*, **66**, 5259   Tyromycin A

VIII.D.-V, -W, -X, -Z

Shing, T.K.M., *JOC*, **66**, 7184   Valienamine
Anderson, R.J., *TL*, **42**, 8697   Variolin B
Argade, N.P., *JOC*, **66**, 9038   (-)-Vascinone
Spilling, C.D., *JOC*, **66**, 3111   Verongamine
Kakinuma, K., *JCS(P1)*, 569   Vicenisamine
Kuehne, M.E., *JOC*, **66**, 1560   Vinblastine
Steglich, W. *SL*, 759   Violacein

Sefkow, M., *TA*, **12**, 987; *JOC*, **66**, 2343 (+)-**Wikstromol**
Yang, D., *OL*, **3**, 1785   (+) & (-)-**Wilforonide**
Overman, L.E., *JACS*, **123**, 9465   *ent*-**WIN 64821**

Jung, M.E., *T*, **57**, 1449   **Xestobergsterol A**

Ino, A., *T*, **57**, 1897   **Yersiniabactin**

Smith, A.B., III, *JACS*, **123**, 12426 (+)-**Zampanolide**
Bracher, F., *M*, **132**, 805   (±)-**Zearalanone**

VIII.E. Reactions in Polar Media

VIII.E.1 Reactions in Aqueous Media

VIII.E.1.-1 Shaughnessy, K.H., Booth, R.S., *OL*, **3**, 2757.
Sterically Demanding, Water-Soluble Alkylphosphines as Ligands for High Activity Suzuki Coupling of Aryl Bromides in Aqueous Solvents

---

VIII.E.1.-2 Huang, T.-S., Li, C.-J., *CC*, 2348.
Conjugate Addition of Arylsilanes to Unsaturated Carbonyl Compounds Catalyzed by Rhodium in Air and Water

---

VIII.E.1.-3 Venkatraman, S., Li. C.-J., *TL*, **42**, 781.
Rhodium Catalyzed Conjugated Addition of Unsaturated Carbonyl Compounds by Triphenylbismuth in Aqueous Media and Under an Air Atmosphere

---

VIII.E.1.-4 Miyaura, N. et al., *CL*, 722.
A Conjugate Addition of Arylboronic Acids to α,β-Unsaturated Carbonyl Compounds Catalyzed by 2β-CD-[Rh(OH)(cod)]$_2$ or [RhCl(cod)]$_2$ in a Singular Aqueous Solvent

---

VIII.E.1.-5 Bong, D.T., Ghadiri, M.R., *OL*, **3**, 2509.
Chemoselective Pd(0)-Catalyzed Peptide Coupling in Water

VIII.E.1.-6 Cai, M.-Z. et al., *SC*, **31**, 3664.
Palladium-Catalyzed Arylation of Allylic Alcohols with Aryl Iodides in Water

VIII.E.1.-7 Villemin, D. et al., *TL*, **42**, 635.
Paladium-Catalyzed Phenylation of Heteroaromatics in Water or Dimethylformamide under Microwave Irradiation

VIII.E.1.-8 Baba, A. et al., *EJOC*, 3207;
A Highly *syn*-Selective Allylation of Aldehydes in Water

VIII.E.1.-9 Ajjou, A.N, *TL*, **42**, 13.
First Example of Water-Soluble Transition-Metal Catalysts for Oppenauer-Type Oxidation of Secondary Alcohols

VIII.E.1.-10 Oshima, K. et al., *BCJ*, **7**, 225.
Radical Reactions by a Combination of Phosphinic Acid and a Base in Aqueous Media

VIII.E.1.-11 Ballini, R. et al., *T*, **57**, 1395.
3-Component Process for the Synthesis of 2-Amino-2-chromenes in Aqueous Media

VIII.E.1.-12 Bigi, F. et al., *TL*, **42**, 5203.
Clean Synthesis in Water. Uncatalyzed Condensation Reaction of Meldrum's Acid and Aldehydes

VIII.E.1.-13 Ribe, S., Wipf, P., *CC*, 299.
Feature Article: Water-Accelerated Organic Transformations

VIII.E.1.-14 Brunel, J.-M., Rodriguez, J. et al., *SL*, 715.
Highly Efficient Phosphazine Base-Catalyzed Michael Addition of β-Ketoesters in Water"

VIII.E.1.-15 Granja, J.R. et al., *OL*, **3**, 2823.
Synthesis of N-(3-Arylpropyl)amino Acid Derivatives by Sonagashira Type of Reaction in Aqueous Media

VIII.E.1.-16 Micskei, K. et al., *TL*, **42**, 7711.
Carbon-Carbon Bond Formation in Neutral Aqueous Medium by Modification of the Naozaki-Hiyama Reaction

VIII.E.1.-17 Yoshida, J. et al., *AG(E)*, **40**, 1074.
2-Pyridyldimethylsilyl as a Removable Hydrophilic Group in Aqueous Diels-Alder Reactions

VIII.E.2 Reactions in Ionic Media

VIII.E.2.-1 Butler, R., *CI(L)*, 532.
Ionic Solution to Chemistry Problems

VIII.E.2.-2 Dyson, P.J. et al., *CJC*, **79**, 705.
A Temperature-Controlled Reversible Ionic Liquid-Water Two Phase-Single Phase Protocol for Hydrogenation Catalysis

---

VIII.E.2.-3 Salunkhe, M.M. et al., *TL*, **42**, 9285.
Coumarin Synthesis *via* Pechmann Condensation in Lewis Acidic Chloroaluminate Ionic Liquid

---

VIII.E.2.-4 Olivier-Bowbigou, H. et al., *CC*, 1360.
Hydroformylation of 1-Hexene with Rhodium in Non-Aqueous Ionic Liquids: How to Design the Solvent and the Ligand to the Reaction

---

VIII.E.2.-5 Parsons, A.F. et al., *CC*, 1350.
Manganes(III) Acetate Mediated Radical Reactions in the Presence of an Ionic Liquid

---

VIII.E.2.-6 Dzyuba, S.V., Bartsch, R.A., *CC*, 1466.
New Room Temperature Ionic Liquids with C_2-Symmetrical Imidazolium Cations

---

VIII.E.2.-7 MacFarlane, D.R. et al., *CC*, 1430.
Low Viscosity Ionic Liquids Based on Organic Salts of the Dicyanamide Anion

---

VIII.E.2.-8 Xiao, J. et al., *OL*, **3**, 295.
Palladium-Catalyzed Regioselective Arylation of an Electron-Rich Olefin by Aryl Halides in Ionic Liquids

---

VIII.E.2.-9 Wasserscheid, P., Gordon, C.M. et al., *CC*, 1186.
Ionic Liquids: Polar, but Weakly Coordinating Solvents for the First Biphasic Oligomerisation af Ethene to Higher α-Olefins with Cationic Ni-Complexes

VIII.E.2.-10 Howarth, J. et al., *SC*, **31**, 2935.
Sodium Borohydride Reduction of Aldehydes and Ketones in the Recyclable Ionic Liquid [BMIM]BF$_6$

VIII.E.2.-11 Branco, L.C., Afonso, C.A.M., *T*, **57**, 4405.
Ionic Liquids as Recyclable Reaction Media for the Tetrahydropyranylation of Alcohols

VIII.E.2.-12 Peng, J., Deng, Y., *TL*, **42**, 403.
Catalytic Beckman Rearrangement of Ketoximes in Ionic Liquids

VIII.E.2.-13 Salunkhe, M.M. et al., *TL*, **42**, 1979.
Fries Rearrangement in Ionic Melts

VIII.E.2.-14 Chiappe, C. et al., *OL*, **3**, 1061.
Stereoselective Halogenations of Alkenes and Alkynes in Ionic Liquids

VIII.E.2.-15 Smietana, M., Miskowski, C., *OL*, **3**, 1037.
Preparartion of Silyl Enol Ethers Using (Bistrimethylsilyl)-acetamide in Ionic Liquids

VIII.E.2.-16 Rosa, J.N., Afonso, C.A.M., Santos, A.G., *T*, **57**, 4189.
Ionic Liquids as a Recyclable Reaction Medium for the Baylis-Hillman Reaction

VIII.E.2.-17 Calo, V. et al., *T*, **57**, 6071.
Heck Reaction of β-Substituted Acrylates in Ionic Liquids Catalyzed by a Pd-Benzothiazide Carbene Complex

VIII.E.2.-18 Kim, M.-J. et al.,. *OL*, **3**, 1507.
Biocatalysis in Ionic Liquids: Markedly Enhanceed Enantioselectivity of Lipase

VIII.E.2.-19 Handy, S.T., Zhang, X., *OL*, **3**, 233.
Organic Synthesis in Ionic Liquids: The Stille Coupling

VIII.E.2.-20 Ren, R.X., Wu, J.X., *OL*, **3**, 3727.
Mild Conversion of Alcohols to Alkyl Halides Using Halide-Based Ionic Liquids at Room Temperature

VIII.E.2.-21 Morrison, D.W., Forbes, D.C. Davis, J.H., Jr., *TL*, **3**, 6053.
Base-Promoted Reactions in Ionic Liquid Solvents. The Knoevenagel and Robinson Annualtion Reactions

VIII.E.2.-22 Buijsman, R.C. et al., *OL*, **3**, 3785.
Ruthenium-Catalyzed Olefin Metathesis in Ionic Liquids

VIII.E.2.-23 Hagiwara, H. et al., *TL*, **42**, 4349.
Heterogeneous Heck Reaction Catalyzed by Pd/C in Ionic Liquid

VIII.E.2.-24 Song, C.E., Choi, J.H. et al., *CC*, 1122.
Ionic Liquids as Powerful Media in Scandium Triflate Catalysed Diels-Alder Reactions: Significant Rate Acceleration, Selectivity Improvement and Easy Recycling of the Catalyst

VIII.E.2.-25 Eckert, C.A. et al., *CC*, 887.
Ionic Liquids as Catalytic Green Solvents for Nucleophilic Displacement Reactions

VIII.F. Combinatorial Chemistry

VIII.F.1 Supports, Linkers and Protecting Groups

VIII.F.1-1 Song, G. et al., *TL*, **42**, 9043.
Microwave-Assisted Preparation of Functionalised Resins for Combinatorial Synthesis.

VIII.F.1-2 Fenniri, H. et al., *JACS*, **123**, 8151.
Barcoded Resins: A New Concept for Polymer-Supported Combinatorial Library Self-Deconvolution.

VIII.F.1-3 Congreve, M.S. et al., *OL*, **3**, 507.
Reporter Resins for Solid-Phase Chemistry.

VIII.F.1-4 Wang, P.G. et al., *JACS*, **123**, 2081; see also: Schmidt, R.R. et al., *OL*, **3**, 747; Takahashi, T. et al., *AG(E)*, **40**, 3230.
Sugar Nucleotide Regeneration Beads (Superbeads): A Versatile Tool for the Practical Synthesis of Oligosaccharides.

VIII.F.1-5 Bradley, M., Kobylecki, R. et al., *AG(E)*, **40**, 938.
Revolutionizing Resin Handlig for Combinatorial Synthesis.

VIII.F.1-6 Sucholeiki, I. et al., *TL*, **42**, 3279.
New Polyoxyalkyleneamine-Grafted Paramagnetic Supports for Solid-Phase Sythesis and Bioapplications.

VIII.F.1-7 Janda, K.D. et al., *TL*, **42**, 2257.
Solid-Phase Synthesis of Oligoesters Using a JandaJel™ Resin.

VIII.F.1-8 Pon, R.T. et al., *JCS(P1)*, 2638; see also: Pon, R.T., Yu, S., *TL*, **42**, 8943 Iyer, R.P. et al., *TL*, **42**, 3669; Azhayev, A.V., Antopolsky, M.L., *T*, **57**, 4977; Paccialli, G. et al., *CC*, 2598 and *SL*, 745; Chu, C.K. et al., *OL*, **3**, 1471; Sekine, M. et al., *TL*, **42**, 8853; Barkley, A., Arya, P., *ECJ*, **7**, 555.
Reusable Solid-Phase Supports for Oligonucleotides and Antisense Therapeutics.

VIII.F.1-9 Kellam, B. et al., *TL*, **42**, 2189; Toth, I. et al., *TL*, **42**, 1159.
A Dde-Based Carboxy Linker for Solid-Phase Synthesis.

VIII.F.1-10 Huang, W. et al., *TL*, **42**, 1973.
2-Polystyrylsulfonyl Ethanol Supports for the Solid-Phase Synthesis of Hydantoins and Ureas.

VIII.F.1-11 Grether, U., Waldmann, H., *ECJ*, 959; **for other examples of "safety-catch" linkers, see also:** Rosenbaum, C., Waldmann, H., *TL*, **42**, 5677; Smythe, M.L. et al., *JOC*, **66**, 7706; Quaderer, R., Hilvert, D., *OL*, **3**, 3181; Main, B.G. et al., *JOC*, **66**, 2240.
An Enzyme-Labile Safety Catch Linker for Synthesis on a Soluble Polymeric Support.

VIII.F.1-12 Zhang, J. et al., *TL*, **42**, 6683.
A Colored Dendrimer as a New Soluble Support in Organic Synthesis: Suzuki Reaction.

VIII.F.1-13 Holmes, C.P. et al., *AG(E)*, **40**, 4488; Pan, Y., Holmes, C.P., *OL*, **3**, 2769.
Use of a Perfluoroalkylsulfonyl (PFS) Linker in a "Tracerless" Synthesis of Biaryls Through Suzuki Cleavage.

VIII.F.1-14 Kim, K., Wang, B., *CC*, 2268; for other examples of
silyl-based linkers, see also: Kim, K. et al., *TL*, **42**, 1463; Lee, Y.,
Silverman, R.B., *T*, **57**, 5339; Lipshutz, B.H., Shin, Y.-J., *TL*, **42**, 5629.
Novel Polymer-Supported 2-(Diphenylmethylsilyl)ethoxymethyl Chloride
(DSEM-Cl) Linker.

VIII.F.1-15 Nicolaou, K.C. et al., *SL*, 900.
A New Photolabile Linker for the Photoactivation of Carboxyl Groups.

VIII.F.1-16 Gibson, S.E. et al., *JCS(P1)* 2526.
Transition Metal Complexes as Linkers for Solid Phase Synthesis:
Chromium Carbonyl Complexes as Linkers for Arenes.

VIII.F.1-17 Gordon, K.H., Balasubramanian, S., *OL*, 3, 53.
Exploring a Benzyloxyaniline Linker Utilizing Ceric Ammonium Nitrate
(CAN) as a Cleavage Reagent: Solid-Phase Synthesis of N-Unsubstituted
β-Lactams and Secondary Amides.

VIII.F.1-18 Harquet, G., Solladie, G. et al., *TL*, **42**, 7563.
Wang *p*-Alkoxyphenylsulfoxide as a New Linker and Pummerer Cleavage
Strategy in Solid-Phase Preparation of 1,2-Diols.

VIII.F.1-19 Melnyk, O. et al., *JOC*, **66**, 4153.
Tartaric Acid-Based Linker for the Solid-Phase Synthesis of C-Terminal
Peptide α-Oxo Aldehydes.

VIII.F.1-20 Najera, C. et al., *TL*, **42**, 7579.
A New Polymer-Supported Reagent for the Fmoc-Protection of Amino
Acids.

VIII.F.1-21 Tamiaki, H. et al., *BCJ*, **74**, 733.
A Novel Protecting Group for Constructing Combinatorial Peptide
Libraries.

VIII.F.1-22 Ley, S.V. et al., *SC*, **31**, 2965.
New Polyethylene Glycol Polymers as Ketal Protecting Groups—A
Polymer Supported Approach to Symmetrically Substituted Spiroketals.

VIII.F.1-23 Somlai, C. et al., *ZN*, **56B**, 526.
Synthesis and Solid-Phase Application of a 9-Xanthenyl
Handle.

VIII.F.1-22 Trivedi, H.S. et al., *SL*, 1932.
A Method for Selective N-Boc Deprotection on Wang Resin.

VIII.F.2 Supported Reagents, Catalysts, Ligands & Scavengers

VIII.F.2-1 Wang, B. et al., *S*, 1611; Barrett, A.G.M. et al., *TL*, **42**, 7899, 8215.
Development and Synthesis of an Alkylboronic Acid-Based Solid-Phase Amidation Catalyst.

VIII.F.2-2 Najera, C. et al., *TL*, **42**, 4487; Lamothe, M. et al., *TL*, **42**, 6703.
Polymer-Bound N-Hydroxysuccinimide as a Solid-Supported Additive for DCC-Mediated Peptide Synthesis.

VIII.F.2-3 Zander, N., Frank, R., *TL*, **42**, 7783.
Polystyrylsulfonyl Chloride Resin: An Efficient Solid-Supported Condensation Reagent for the Solution Phase Synthesis of Esters.

VIII.F.2-4 Ursini, A. et al., *SL*, 388.
Readily Prepared Resin-Bound Thioimidates as Reagents for the Synthesis of Amidines.

VIII.F.2-5 Reek, J.N.H., van Leewen, P.W.N.M. et al., *ECJ*, 7, 1202.
Solid-Phase Synthesis of Homogeneous Ruthenium Catalysts on Silica for the Continuous Asymmetric Transfer Hydrogenation Reaction.

VIII.F.2-6 Fuchikami, T. et al., *SL*, 1623.
Supported Nickel-Catalyzed Hydrogenation of Aromatic Nitriles under Low Pressure Conditions.

VIII.F.2-7 Desai, B., Danks, T.N., *TL*, **42**, 5963.
Thermal and Microwave Assisted Hydrogenation of Electron Deficient Alkenes Using a Polymer Supported Hydrogen Donor.

VIII.F.2-8 Copeland, G.T., Miller, S.J., *JACS*, **123**, 6496; see also:
Reetz, M.T., *AG(E)*, **40**, 285; Senkan, S., *AG(E)*, **40**, 313.
Selection of Enantioselective Acyl Transfer Catalysts from a Pooled Peptide Library Through a Fluorescence-Based Activity Assay: An Approach to Kinetic Resolution of Secondary Alcohols of Broad Structural Scope.

VIII.F.2-9 Lemaire, M. et al., *TA*, **12**, 811.
Enantiopure Poly(glycidyl Methacrylate-Co-Ethylene Glycol Dimethacrylate): A New Material for Supported Catalytic Asymmetric Hydrogen Transfer Reduction.

VIII.F.2-10 Ley, S.V. et al., *SL*, 1555.
Solid-Supported Reagents for the Oxidation of Aldehydes to Carboxylic Acids.

VIII.F.2-11 Kerr, W.J. et al., *SL*, 1257.
Polymer-Supported N-Methylmorpholine N-Oxide as an Efficient and Readily Recyclable Co-Oxidant in the TPAP Oxidation of Alcohols.

VIII.F.2-12 Rademann, J. et al., *AG(E)*, **40**, 1436.
Oxoammonium Resins as Metal-Free Highly Reactive, Versatile Polymeric Oxidation Reagents.

VIII.F.2-13 Giannis, A. et al., *T*, **57**, 4863; Mulbaier, M., Giannis, A., *AG(E)*, **40**, 4393; Rademann, J. et al., *AG(E)*, **40**, 4395; Togo, H. et al., *SL*, 22.
Development of New and Efficient Polymer-Supported Hypervalent Iodine Reagents.

VIII.F.2-14 Gruttadauria, M. et al., *TL*, **42**, 5199
Sol-Gel Entrapped Chromium(VI): A New Selective, Efficient and Recyclable Oxidizing System.

VIII.F.2-15 Alvarez-Builla, J. et al., *S*, 382.
Solid Support Bound 1-Aminoimidazolium Chlorochromate: A Selective, Efficient and Recyclable Oxidant.

VIII.F.2-16 Jackson, R.F.W. et al., *CC*, 2594.
Discovery of New Solid Phase Sulfur Oxidation Catalysts Using Library Screening.

VIII.F.2-17 Giacomelli, G. et al., *OL*, **3**, 855; Deleuze, H. et al., *SC*, **31**, 3207.
A New Supported Reagent for the Photochemical Generation of Radicals in Solution.

VIII.F.2-18 Pears, D., Bradley, M. et al., *JCS(P1)*, 1947.
Ketoester Methylacrylate Resin Secondary Amine Clean-Up in the Presence of Primary Amines.

VIII.F.2-19 Fourkas, J.T., Hoveyda, A.H. et al., *AG(E)*, **40**, 4251; Blechert, S. et al., *SL*, 1547; Brown, J.M. et al., *CJC*, **79**, 1049; Mol, J.C. et al., *TL*, **42**, 7103; Dowden, J., Savovic, J., *CC*, 37.
Immobilization of Olefin Metathesis Catalysts on Monolithic Sol-Gel: Practical, Efficient, and Easily Recyclable Catalysts for Organic and Combinatorial Synthesis.

VIII.F.2-20 Zhang, S.-Y. et al., *TL*, **42**, 5925; Bolm, C., Maischak, A., *SL*, 93.
A Simple and Effective Soluble Polymer-Bound Ligand for the Asymmetric Dihydroxylation of Olefins: DHQD-Phal-OPEG-OMe.

VIII.F.2-21 Salvadori, P. et al., *AG(E)*, **40**, 2519; Gennari, C., Piarulli, U. et al., *ECJ*, **7**, 2628.
An Insoluble Polymer-Bound Bis-Oxazoline Copper(II) Complex: A Highly Efficient Heterogeneous Catalyst for the Enantioselective Mukaiyama Aldol Reaction.

VIII.F.2-22 Burguete, M.I., Luis, S.V. et al., *TL*, **42**, 1673; Calmes, M. et al., *JOC*, **66**, 5859; Mikami, K. et al., *ECJ*, **7**, 730; Uozumi, Y. et al., *TL*, **42**, 407; Li, G.Y. et al., *AG(E)*, **40**, 1106.
A General, Route for the Preparation of Polymer-Supported N-Tosyl Amino Alcohols and their Use as Chiral Auxiliaries.

VIII.F.2-23 Uehlin, L., Wirth, T., *OL*, **3**, 2931; for other selenium based methodologies, see also: Huang, X. et al., *SL*, 1571; Qian, H., Huang,, X., *SL*, 1913; Huang, X., Sheng, S.-R., *TL*, **42**, 9035.
Novel Polymer-Bound Chiral Selenium Electrophiles.

VIII.F.2-24 Marsh, A. et al., *TL*, **42**, 493.
High-Loading Scavenger Resins for Combinatorial Chemistry.

VIII.F.3 Solid-Phase Heterocyclic Synthesis

VIII.F.3-1 Krchnak, V. et al., *CCC*, **66**, 1078 and *TL*, **42**, 1627, 2443.
Solid-Phase Traceless Synthesis of Selected Nitrogen-Containing Heterocyclic Compounds. The Encore Technique for Directed Sorting of Molecular Solid Support.

VIII.F.3-2 Tempest, P., Hulme, C. et al., *TL*, **42**, 4963, 4959.
MCC/S$_N$Ar Methodology. Novel Access to a Range of Heterocyclic Cores.

VIII.F.3-3 Tumelty, D. et al., *OL*, **3**, 83; Lee, J. et al., *TL*, **42**, 2635.
Benzimidazoles

VIII.F.3-4 Houghton, R.A. et al., *TL*, **42**, 5141; Amblard, M. et al.,
TL, **42**, 5389.
1,4-Benzodiazepinediones

VIII.F.3-5 Bradley, M., Chimirri, A. et al., *TL*, **42**, 7683.
2,3-Benzodiazepin-4-ones

VIII.F.3-6 Barlos, K. et al., *TL*, **42**, 2201; Lam, Y. et al., *TL*, **42**, 109.
Benzothiazoles

VIII.F.3-7 Lam, Y. et al., *TL*, **42**, 1167.
1,4-Benzoxazin-3(4H)-ones and 1,4-Benzothiazin-3(4H)-ones

VIII.F.3-8 Meloni, M.M., Taddei, M., *OL*, **3**, 337.
β-Lactams

VIII.F.3-9 Zhang, D., Kiselyov, A.S., *SL*, 1173.
Chromano[4,3-b]quinolines

VIII.F.3-10 Dondoni, A., Massi, A., *TL*, **42**, 7975; Kappe, C.O. et al.,
SL, 741.
Dihydropyrimidinones

VIII.F.3-11 Chen, J.J. et al., *SL*, 1263 and *TL*, **42**, 2269; Houghton,
R.A. et al., *JOC*, **66**, 8673.
Imidazo[1,2-a] Annulated Heterocycles and Imidazo[1,2-a]pyridines

---

VIII.F.3-12 Li, M., Wilson, L.J. *TL*, **42**, 1455; Houghton, R.A. et al.,
TL, **42**, 623.
Imidazolones

VIII.F.3-13 Gray, N.S., Schultz, P.G. et al., *OL*, **3**, 3827; Zhang, H.-C. et al., *TL*, **42**, 4751; Nettekoven, M., *BMCL*, **11**, 2169; Heinelt, U. et al., *BMCL*, **11**, 227.
Indoles

VIII.F.3-14 Sun, Q. et al., *TL*, **66**, 6495.
Isoindolines

VIII.F.3-15 Barrett, A.G.M. et al., *OL*, **3**, 3165; Cereda, E. et al., *TL*, **42**, 4951; Giacomelli, G. et al., *JOC*, **66**, 6823; Kang, K.H., Koh, H.Y., Lee, H.-Y. et al., *TL*, **42**, 1057.
Isoxazoles

VIII.F.3-16 Poulain, R.F. et al., *TL*, **42**, 1495; Rice, K.D., Nuss, J.M., *BMCL*, **11**, 753.
1,2,4-Oxadiazoles

VIII.F.3-17 Kilburn, J.P. et al., *TL*, **42**, 2583; Brain, C.T., Brunton, S.A., *SL*, 382.
1,3,4-Oxadiazoles

VIII.F.3-18 Zwanenburg, B. et al., *EJOC*, 2965.
Oxazolidinones

VIII.F.3-19 Eda, M., Kurth, M.J., *TL*, **42**, 2063; Bursavich, M.G., Rich, D.H., *OL*, **3**, 2625.
Piperidines, Tetrahydropyridines and Pyridium Derivatives

VIII.F.3-20 Gray, N.S., Schultz, P.G. et al., *TL*, **42**, 8751 and *JOC*, **66**, 8273; Brill, W.K.-D., Riva-Toniolo, C., *TL*, **42**, 6515; Brill, W.K.-D. et al., *SL*, 1097; Dorff, P.H., Garigipati, R.S., *TL*, **42**, 2771; Legraverend, M. et al., *TL*, **42**, 8161, 8165, 8169.
Purines

VIII.F.3-21 McCarthy, T.D. et al., *JCS(P1)*, 2817.
Pyrazoles

VIII.F.3-22 Lam, Y. et al., *CL*, 274;
Pyridazins

VIII.F.3-23 Kamal, A. et al., *TL*, **42**, 6969 and *BMCL*, **11**, 387.
Pyrrolo[2,1-c][1,4]benzodiazepine-5,11-diones

VIII.F.3-24 Nettekoven, M. et al., *SL*, 1917.
[1,2,4]-Triazolo-[1,5-a]-Pyridines

VIII.F.3-25 Raveglia, L.F. et al., *EJOC*, 4737.
Pyrimidines

VIII.F.3-26 Molina, P. et al., *JMC*, **44**, 1011.
Pyrido[1,2-c]pyrimidines

VIII.F.3-27 Barrett, A.G.M. et al., *TL*, **42**, 5579; Steger, M. et al.,
BMCL, **11**, 2537.
Pyrrolidines

VIII.F.3-28 Wilson, L.J., *OL*, **3**, 585; Lou, B. et al., *TL*, **42**, 8405;
Sun, Q. et al., *TL*, **42**, 4119; Makino, S. et al., *SL*, 333 and *TL*, **42**, 1749.
**Quinazolines, Quinazolinones, Quinazoline-2,4-diones and
Thioxoquinazolin-4-ones**

VIII.F.3-29 Thomas, A.W. et al., *TL*, **42**, 1645; Huang, X., Liu, Z.,
TL, **42**, 7655.
Quinolones

VIII.F.3-30 Wu, Z., Ede, N.J., *TL*, **42**, 8115.
Quinoxalines

VIII.F.3-31 Jonsson, D. et al., *TL*, **42**, 6953.
Tetrahydropyridones

VIII.F.3-32 Masquelin, T., Obrecht, D., *T*, **57**, 153.
Thiazoles

VIII.F.3-33 Castanedo, G.M., Sutherlin, D.P., *TL*, **42**, 7181.
Thiophenes

VIII.F.3-34 Houghton, R.A. et al., *OL*, **3**, 2797.
1,3,5-Triazepane-2,4-diones

VIII.F.3-35 Larsen, S.D., DiPaolo, B.A., *OL*, **3**, 3341.
1,2,4-Triazoles

VIII.F.3-36 Martinez-Teipel, B. et al., *TL*, **42**, 6455.
1,2,4-Triazin-6-ones

VIII.F.3-37 Hannah, I. et al., *TL*, **42**, 3721.
Tropanes

VIII.F.3-38 Diez, A. et al., *TL*, **42**, 871.
Valerolactams

VIII.F.4 Solid-Supported Organic Reaction Processes

VIII.F.4-1 Heinze, K., *ECJ*, **7**, 2922.
Solid-Phase Organometallic Synthesis.

VIII.F.4-2 Hoveyda, A.H., Snapper, M.L. et al., *JACS*, **123**, 984
Enantioselective Synthesis of Arylamines through Zr-Catalyzed Addition of Dialkylzinc to Imines. Reaction Development by Screening of Parallel Libraries.

VIII.F.4-3 Cossy, J. et al., *SL*, 629.
Solid-Support Synthesis of 1,2-Diols and γ-Lactones through Addition of α-(Benzoyloxy)crotylindium Reagents to Aldehydes.

VIII.F.4-4 Yang, Z. et al., *TL*, **42**, 1815; Conn, C. et al., *BMCL*, **11**, 2565; for other palladium mediated processes, see also: Yun, W. et al., *TL*, **42**, 175; Akaji, K. et al., *T*, **57**, 2293; Begtrup, M. et al., *S*, 909; Delgado, A. et al., *TL*, **42**, 3299.
Opimization Study of Sonogashira Cross-Coupling Reaction on High-Loading Microbeads Using a Silyl Linker.

VIII.F.4-5 Nagashima, T., Davies, H.M.L., *JACS*, **123**, 2695.
Catalytic Asymmetric Solid-Phase Cyclopropanation.

VIII.F.4-6 Raghavan, S. et al., *TL*, **42**, 8383.
Solvomercuration and Demercuration of Allenes on Solid-Phase.

VIII.F.4-7 Sarko, C.R. et al., *TL*, **42**, 8939.
Microwave Assisted Synthesis of a [3+2] Cycloaddition Library.

VIII.F.4-8 Desimone, J., Selva, M., Tundo, P., *JOC*, **66**, 4047.
Nucleophilic Displacement in Supercritical Carbon Dioxide Using Silica-Supported Phase-Transfer Agents.

VIII.F.4-9 Cossy, J. et al., *SL*, 815.
Diastereoselective Synthesis of Homopropargylic Alcohols on Solid Suppport.

VIII.F.4-10 Brase, S. et al., *TL*, **42**, 9179.
Cleavage of Immobilized Disubstituted Triacenes with Electrophiles: Solid Phase Synthesis of Alkyl Halides and Esters.

VIII.F.4-11 Ito, Y. et al., *JACS*, **123**, 4356.
Solid-Phase Synthesis and Asymmetric Reactions of Polymer-Supported Highly Enantioenriched Allylsilanes.

VIII.F.4-12 Cai, J., Wathey, B., *TL*, **42**, 1383; **for other amine syntheses, see also:** Andersson, C.-M. et al., *TL*, **42**, 133; Fisher, M., Brown, R.C.D., *TL*, **42**, 8227; Uriac, P. et al., *TL*, **42**, 6655; Stromgaard, K., Jaroszewski, J.W. et al., *S*, 877; Makara, G.M., Ma, T., *TL*, **42**, 4123; Enders, D. et al., *OL*, **3**, 1241; Laplante, C., Hall, D.G., *OL*, **3**, 1487; Kawatsura, M., Hartwig, J.F., *OM*, **20**, 1960.
A Novel Tracerless Solid Phase Tertiary Amine Synthesis Based on Merrifield Resin.

VIII.F.4-13 Garcia-Ochoa, S. et al., *TL*, **42**, 6675; Sammelson, R.E., Kurth, M.J., *TL*, **42**, 3419.
An Improved Two-Resin Method for the Cleavage of Tertiary Amines from REM Resin.

VIII.F.4-14 Phoon, C.W., Sim, M.M. *SL*, 697; **see also:** Schunk, S., Enders, D., *OL*, **3**, 3177; Huang, K.-T., Sun, C.-M., *BMCL*, **11**, 271; Smith, R.A. et al., *BMCL*, **11**, 2775; Liskamp, R.M.J. et al., *JOC*, **66**, 8454.
Tracerless Synthesis of Urea, Semicarbazide and Carbamate Derivatives Using Bromo-Wang Resin and Bromo-Wang SynPhase™ Lantern.

VIII.F.4-15 Hone, N.D. et al., *TL*, **42**, 1115; Takaya, H., Murahashi, S.-I., *SL*, 991.
Direct Release of Nitriles from Solid Phase.

VIII.F.4-16 Goodman, M. et al., *OL*, **3**, 1133; **for alternate routes to guanidines, see also:** Fan, E. et al., *JOC*, **66**, 2161; Dodd, D.S., Zhao, Y., *TL*, **42**, 1259; Li, M. et al., *TL*, **42**, 2273; Mamai, A., Madalengoitia, J.S., *OL*, **3**, 561; Houghton, R.A. et al., *T*, **57**, 9911.
A Novel Tracerless Resin-Bound Guanidinylating Reagent for Secondary Amines to Prepare N,N-Disubstituted Gaunidines.

VIII.F.4-17 Grimstrup, M., Zaragoza, F., *EJOC*, 3233; Hardcastle, I.R. et al., *TL*, **42**, 1363.
Synthesis of a New Class of Highly Functionalized Benzamides by Threefold Sequential Nucleophilic Substitution at a Resin-Bound Polyelectrophile.

VIII.F.4-18 Scott, W.L., O 'Donnell, M.J. et al., *TL*, **42**, 2073; Herdewijn, P. et al., *CCC*, **66**, 923; Houghton, R.A. et al., *JOC*, **66**, 8268.
Solid-Phase Synthesis of Amino Amides and Peptide Amides with Unnatural Side Chains.

VIII.F.4-19 Attardi, M.A., Tadder, M., *TL*, **42**, 3519.
The Barton Radical Decarboxylation on Solid Phase. A Versatile Synthesis of Peptides Containing Modified Amino Acids.

VIII.F.4-20 Eguchi, M. et al., *TL*, **42**, 1237; Robina, I. et al., *TL*, **42**, 1283; Basso, A., Ernst, B., *TL*, **42**, 6687; Kitagawa, K., Yagami, T. et al., *JOC*, **66**, 1; Nakamura, K. et al., *TL*, **42**, 8337; Volonterio, A., Bravo, P., Zanda, M., *TL*, **42**, 3141; Augustyns, K. et al., *TL*, **42**, 9135; Meldal, M. et al., *ECJ*, 3584; Sanders, J.K.M. et al., *AG(E)*, **40**, 423; Polt, R. et al., *JOC*, **66**, 2327.
Solid-Phase Synthesis and Solution Structure of Bicyclic β-Turn Peptidomimetics: Diversity at the i Position.

VIII.F.4-21 Cox, P.B. et al., *TL*, **42**, 125; for other phoshorous derivatives, see also: Parang, K. et al., *OL*, **3**, 307; Iyer, R.P. et al., *BMCL*, **11**, 2057; Toth, G.K. et al., *OL*, **3**, 1033; Dolle, R.E. et al., *TL*, **42**, 1855; Sasaki, S. et al., *BMCL*, **11**, 2581.
The Solid Phase Synthesis of Unsymmetrical Phosphinic Acids.

VIII.F.4-22 Ellman, J.A. et al., *JACS*, **123**, 10127; for other examples of sulfur derivatives, see also: Xiong, Y. et al., *TL*, **42**, 8423; Solladie, G. et al., *TL*, **42**, 9077; Brase, S. et al., *TL*, **42**, 7833.
Design, Synthesis, and Utility of a Support-Bound *tert*-Butanesulfinamide.

VIII.F.4-23 Dujardin, G. et al., *TL*, **42**, 8049; see also: Graven, A., Meldal, M., *JCS(P1)*, 3198.
Eu(fod)$_3$-Catalyzed Solid-Phase [4+2] Heterocycloadditions: An Efficient Asymmetric Process in Catalyst-Recycling Conditions.

VIII.F.5 Targeted Library Synthesis

VIII.F.5-1 Baldwin, J.E. et al., *JOC*, **66**, 2588.
Parallel Synthesis of Novel Heteroaromatic Acromelic Acid Analogues from Kainic Acid.

VIII.F.5-2 Padmanabhan, S. et al., *BMCL*, **11**, 3151.
Solution-Phase, Parallel Synthesis and Pharmacological Evaluation of Acylguanidine Derivatives as Potential Sodium Channel Blockers.

VIII.F.5-3 Witherington, J. et al., *BMCL*, **11**, 195; Smith, R.A. et al., *BMCL*, **11**, 2951
Solid-Phase Synthesis of Cyclic Alkoxyketones, Inhibitors of the Cysteine Protease Cathepsin K.

VIII.F.5-4 Pisano, C. et al., *OL*, **3**, 1001.
Potent Integrin Antagonists from a Small Library of RGD-Including Cyclic Pseudopeptides.

VIII.F.5-5 Stavenger, R.A., Schreiber, S.L., *AG(E)*, **40**, 3417.
Asymmetric Catalysis in Diversity-Oriented Organic Synthesis: Enantioselective Synthesis of 4320 Encoded and Spacially Segrated Dihydropyrancarboxamides.

VIII.F.5-6 Schreiber, S.L. et al., *JACS*, **123**, 1740.
Split-Pool Synthesis of 1,3-Dioxanes Leading to Arrayed Stock Solutions of Single Compounds Sufficient for Multiple Phenotypic and Protein-Binding Assays.

VIII.F.5-7 Meldal, M. et al., *JACS*, **123**, 2176.
α-Ketocarbonyl Peptides: A General Approach to Reactive Resin-Bound Intermediates in the Synthesis of Peptide Isosteres for Propease Inhibitor Screening on Solid Support.

VIII.F.5-8 Nicolaou, K.C. et al., *ECJ*, **7**, 4280.
Combinatorial Synthesis through Disulfide Exchange: Discovery of Potent Psammaplin A Type Antibacterial Agents Active Against Methicillin-Resistant Staphylococcus Aureus (MRSA).

VIII.F.5-9 Dankwardt, S.M. et al., *BMCL*, **11**, 2085.
Amino Acid Derived Sulfonamide Hydroxamates as Inhibitors of Procollagen C-Proteinase: Solid-Phase Synthesis of Ornithine Analogues.

VIII.F.5-10 Boger, D.L. et al., *JOC*, **66**, 6654.
Parallel Synthesis and Evaluation of 132 (+)-1,2,9,9a-Tetrahydrocyclopropa[c]-benz[e]-indol-4-one (CBI) Analogues of CC-1065 and the Duocarmycins Defining the Contribution of the DNA-Binding Domain.

VIII.F.5-11 Nicolaou, K.C. et al., *ECJ*, **7**, 3824.
Synthesis and Biological Evaluation of Vancomycin Dimers with Potent Activity Against Vancomycin-Resistant Bacteria: Target-Accelerated Combinatorial Synthesis.

VIII.F.5-12 Booth, S. et al., *BMCL*, **11**, 2351.
Parallel Solid-Phase Synthesis of Zatebradine Analogues as Potential I_f Channel Blockers.

VIII.F.5-13 Boger, D.L. et al., *JACS*, **123**, 1280.
Identification of a Novel Class of Small-Molecule Antiangiogenic Agents through the Screening of Combinatorial Libraries which Function by Inhibiting the Binding and Localization of Proteinase MMP2 to Integrin $\alpha_v\beta_3$.

VIII.F.5-14 Willoughby, C.A. et al., *BMCL*, **11**, 3137.
Combinatorial Synthesis of CCR5 Antagonists.

VIII.F.5-15 Verdine, G.L. et al., *JACS*, **123**, 398.
A Synthetc Library of Cell-Permeable Molecules.

VIII.F.5-16 Shair, M.D. et al., *JACS*, **123**, 6740.
Use of Biomimetic Diversity-Oriented Synthesis to Discover Galanthamine-Like Molecules with Biological Properties Beyond Those of the Natural Product.

VIII.F.5-17 Tsantrizos, Y.S. et al., *JOC*, **66**, 4743.
Solid-Phase Synthesis of Peptidomimetic Inhibitors for the Hepatitus C Virus NS3 Protease.

VIII.F.5-18 Juteau, H. et al., *BMCL*, **11**, 747.
Structure-Activity Relationship on the Human EP$_3$ Prostanoid Receptor by Use of Solid-Support Chemistry.

VIII.F.5-19 Kessler, H. et al., *AG(E)*, **40**, 165.
Nonpeptidic α,β_3 Integrin Antagonist Libraries.

VIII.F.5-20 Dervan, P.B. et al., *OL*, **3**, 1201.
Fmoc Solid Phase Synthesis of Polyamides Containing Pyrrole and Imidazole Amino Acids.

VIII.F.5-21 Liu, K.G., Sternbach, D.D. et al., *BMCL*, **11**, 2959.
Identification of a Series of PPARγ/δ Dual Agonists *via* Solid-Phase Parallel Synthesis.

VIII.F.5-22 Burns, C.J. et al., *JOC*, **66**, 3709.
Solution- and Solid-Phase Synthesis of Components for Tethered Bilayer Membranes.

VIII.F.5-23 Orain, D., Bradley, M., *TL*, **42**, 515.
Solid-Phase Synthesis of Trypanothione Reductase Inhibitors—Towards Single Bead Screening.

VIII.F.5-24 Takahashi, T. et al., *JACS*, **123**, 3716.
Parallel Synthesis of a Vitamin D$_3$ Library in the Solid-Phase.

VIII.F.6 Novel Techniques in Combinatorial Chemistry

VIII.F.6-1 Shair, M.D. et al., *JACS*, **123**, 361.
Reaction Microarrays: A Method for Rapidly Determining the Enantiomeric Excess of Thousands of Samples.

VIII.F.6-2 Schreiber, S.L. et al., *AG(E)*, **40**, 3421.
Decoding Products of Diversity Pathways from Stock Solutions Derived from Single Polymeric Macrobeads.

VIII.F.6-3 Kuroda, N. et al., *CPB*, **49**, 1146.
Application of an Automated Synthesis Suite to Parallel Solution-Phase Peptide Synthesis.

VIII.F.6-4 de Mello, A.J. et al., *JCS(P1)*, 514.
Microchip-Based Synthesis and Total Analysis Systems (μSYNTAS): Chemical Microprocessing for Generation and Analysis of Compound Libraries.

VIII.F.6-5 Yoshida, J. et al., *JACS*, **123**, 7941.
"Cation Flow" Method: A New Approach to Conventional and Combinatorial Organic Syntheses Using Electrochemical Microflow Systems.

VIII.F.6-6 Tietze, L.F. et al., *AG(E)*, **40**, 903.
A Novel Concept in Combinatorial Chemistry in Solution with the
Advantages of Solid-Phase Synthesis.

VIII.F.6-7 Raymond, K.N. et al., *AG(E)*, **40**, 733.
Combinatorial Libraries of Metal-Ligand Assemblies with an Encapsulated
Molecule.

VIII.F.6-8 Luis, S.V. et al., *TL*, **42**, 8459.
FT-Raman as a Simple Tool for the Fast Monitering of Reactions on
Silica-Supported Reagents and Catalysts: Application to Silica-Bound
Prolinol and TADDOLS.

VIII.F.6-9 Ellman, J.A. et al., *AG(E)*, **40**, 216.
A Mass Spectrometric Labeling Strategy for High-Throughput Reaction
Evaluation and Optimization: Exploring C-H Activation.

VIII.F.6-10 Davis, B.G., Steel, P.G. et al., *TL*, **42**, 8531.
A Simple Method for the Quantitative Analysis of Resin Bound Thiol
Groups.

VIII.F.7 Reviews

VIII.F.7-1 Arya, P. et al., *AG(E)*, **40**, 339.
Review: "Diversity-Based Organic Synthesis in the Era of
           Genomics and Proteomics."

VIII.F.7-2 Ganesan, A. et al., *PAC*, **73**, 1033
Lecture: "Intrgrating Natural Product Synthesis and
           Combinatorial Chemistry."

VIII.F.7-3 Kirschning, A. et al., *AG(E)*, **40**, 650.
Review: "Functionalized Polymers—Emerging Versatile Tools for
           Solution-Phase Chemistry and Automated Parallel Synthesis."

VIII.F.7-4 Kobayashi, S., Akiyama, R., *PAC*, **73**, 1103.
Lecture: "New Methods for High-Throughput Synthesis."

VIII.F.7-5 Dahmen, S., Brase, S., *S*, 1431.
Review: "Combinatorial Methods for the Discovery and Optimisation of
           Homogenous Catalysts."

VIII.F.7-6 Janda, K.D. et al., *T*, **57**, 4637.
Report: "Polymer-Supported Catalysis in Synthetic Organic Chemistry."

VIII.F.7-7 Cawse, J.N., *ACR*, **34**, 213.
Review: "Experimental Strategies for Combinatorial and High-Throughput Materials Developement.

VIII.F.7-8 Sammelson, R.E., Kurth, M.J., *CRV*, **101**, 137.
Review: "Carbon-Carbon Bond-Forming Solid-Phase Reactions. Part II."

VIII.F.7-9 Yli-Kauhaluoma, J., *T*, **57**, 7053.
Report: "Diels-Alder Reactions on Solid Supports. "

VIII.F.7-10 Souers, A.J., Ellman, J.A., *T*, **57**, 7431.
Reportw: "β-Turn Mimetic Library Synthesis: Scaffolds and Applications."

VIII.F.7-11 Burgess, K. et al., *ACR*, **34**, 826.
Review: "Solid-Phase Synthesis of β-Turn Analogues to Mimic or Disrupt Protein-Protein Interactions."

VIII.F.7-12 Roper, W.R., *AG(E)*, **40**, 2440.
Highlight: "First Metallabenzenes and now a Stable Metallabenzyne."

AUTHOR INDEX

CPSIA information can be obtained
at www.ICGtesting.com
Printed in the USA
BVOW06*1458100117
473079BV00003B/16/P

9 780120 408320